D1382144

S1 to National 4

CHEMISTRY
STUDENT BOOK

Tom Speirs • Bob Wilson

001/11082017

10 9 8 7 6 5 4 3 2 1

ISBN 9780008204501

Published by
Leckie & Leckie Ltd
An imprint of HarperCollinsPublishers
Westerhill Road, Bishopbriggs, Glasgow, G64 2QT
T: 0844 576 8126   F: 0844 576 8131
leckieandleckie@harpercollins.co.uk
www.leckieandleckie.co.uk

Commissioning editor: Clare Souza
Managing editor: Craig Balfour

Special thanks to
Dr Anna Clark (copy edit)
Jayne MacArthur (copy edit)
Jess White (proofread)
Project One Publishing Solutions, Scotland (project management)
Jouve India (layout and illustration)
Ink tank and Ian Wrigley (cover)

Printed and bound by Grafica Veneta S.p.A., Italy

A CIP Catalogue record for this book is available from the British Library.

Acknowledgements

Whilst every effort has been made to trace the copyright holders, in cases where this has been unsuccessful, or if any have inadvertently been overlooked, the Publishers would gladly receive any information enabling them to rectify any error or omission at the first opportunity.

Leckie & Leckie would like to thank the following copyright holders for permission to reproduce their material:

Cover image © Howard Walker / Alamy Stock Photo

Introduction   vi

UNIT 1   CHEMICAL CHANGES AND STRUCTURE   2

1   Rates of reaction 1   4
- *Chemical reactions* • *Speeding up chemical reactions*

2   Rates of reaction 2   13
- *Monitoring the rate of reaction*

3   Chemical structure   22
- *Substances and their states* • *Elements, compounds and mixtures*
- *Elements and the periodic table* • *Chemical formulae of compounds*

4   Atomic structure and bonding related to properties of materials   46
- *Atomic structure* • *Bonding* • *Properties of covalent and ionic substances*
- *Formulae of elements and compounds*

5   Energy changes of chemical reactions   69
- *Exothermic reactions* • *Endothermic reactions*

6   Acids and bases 1   76
- *Acids and alkalis* • *Neutralisation* • *Acidic gases and the environment*
- *Greenhouse gases and global warming*

7   Acids and bases 2   100
- *Bases* • *Carbon dioxide and the cement industry*

Unit 1 practice assessment   110

UNIT 2   NATURE'S CHEMISTRY   **116**

8   Fuels and energy 1: Fuels   118
- *What are fuels?* • *Fossil fuels* • *Hydrocarbons*
- *Fuels that give our bodies energy* • *Alcohol as a fuel*

9   Fuels and energy 2: Controlling fires   130
- *The fire triangle* • *Extinguishing fires safely*

10   Fuels 1: Fossil fuels   138
- *Trapping the Sun's energy* • *Formation of fossil fuels*
- *Fossil fuels and new technologies* • *Burning fuels* • *The law of conservation of mass*

11   Fuels and energy 3: The problems with fossil fuels   154
- *The problem with using fossil fuels* • *Climate change targets*

12   Fuels and energy 4: Meeting energy needs in the future   161
- *Sustainable sources of energy* • *Biomass and biofuels*
- *Alternative technologies in Scotland*
- *Advantages and disadvantages of different sustainable sources of energy*

13   Fuels 2: Solutions to fossil fuel problems   175
- *Combustion of fossil fuels and the carbon cycle* • *Reducing carbon dioxide emissions*
- *Sustainable sources of energy*

14   Fuels 3: Hydrocarbons   194
- *Fractional distillation* • *The alkanes* • *The alkenes* • *Meeting market demand: Cracking*

15 Everyday consumer products 1: Plants for food 208
- Plants for food • Diet and disease • Plants and alcohol
- Improving plant growth with fertilisers

16 Everyday consumer products 2: Cosmetic products 229
- Plants and cosmetics • Using essential oils

17 Everyday consumer products 3: Plants for energy 241
- Plants for food and energy • Carbohydrates – What's in a name? • Digestion
- Alcohol from plants • Effect of temperature and pH on enzyme activity

18 Plants to products 258
- Products from plants • Medicines from plants • Labelling of medicines

Unit 2 practice assessment 269

UNIT 3   CHEMISTRY IN SOCIETY **276**

19 Properties of materials 1 278
- Materials • Fibres and fabrics • Ceramics • Novel materials

20 Properties of materials 2 296
- Polymers and polymerisation • Ceramics • Novel materials

21 Properties of metals 1 321
- Metals • Metals and batteries

22 Properties of metals 2 332
- Reactivity of metals • Alloys • Corrosion
- Chemical cells and the electrochemical series

23 Properties of solutions 352
- Solubility of substances

24 Fertilisers 362
- Growing healthy plants • Natural and synthetic fertilisers
- Environmental impact of fertilisers

25 Nuclear chemistry 373
- Formation of elements • Background radiation

26 Chemical analysis 1 381
- Chemical hazards

27 Chemical analysis 2 389
- Analytical chemists • Simple analytical techniques

Unit 3 practice assessment 400

Appendix: Investigation skills **406**

Answers to Activities and Unit practice assessments at:
www.leckieandleckie.co.uk/page/Resources

Introduction

About this book

This book provides a resource to support you in your study of Chemistry. You may be studying Chemistry as part of your S1–S3 curriculum or as part of your National 3 or National 4 qualifications if you are in S4.

The book covers the Key areas in the Units of National 3 Chemistry and National 4 Chemistry. It also covers all the chemical Experiences and Outcomes (Es and Os) for Curriculum for Excellence (CfE) Science at Third and Fourth Levels.

Companion biology and physics books provide the same support if you are studying biology or physics in any of these courses.

The material has been organised into three units: Chemical changes and structure, Nature's chemistry, and Everyday consumer products. Each unit has been broken down into chapters.

The chapters are packed with information, activities and a range of features to deepen your understanding of Chemistry and to help you prepare for any assessment.

Features

This chapter includes coverage of
Each chapter begins with a brief listing of the topics covered in the chapter, using the N3 and N4 Key area and/or the CfE Es and Os code.

> **This chapter includes coverage of:**
>
> N4 Materials • Materials SCN 4-16a
> • Topical science SCN 4-20a, SCN 4-20b

You should already know
After the list of topics covered in the chapter, sometimes there is a list of the topics **you should already know** from previous study. Some of these topics will have been covered in earlier years at school, while others may depend on other chapters in this book.

> **You should already know:**
>
> • how to use indicators and the pH scale to identify acids, alkalis and neutral solutions
> • that acids can be neutralised by alkalis

Learning intentions
Each section has a set of **Learning intentions**. These are organised according to the N3 and N4 requirements and/or the CfE Levels, so you can see if a section is suitable for the course you are studying. For example, if you are studying for N3 Chemistry, you don't need to read the N4 sections, and if you are studying for N4 Chemistry, you should already know what is in the N3 sections.

To help you navigate your way through the book, different colours are used to indicate National 3/Level 3 content and National 4/Level 4 content.

National 3
Curriculum level 3
Planet Earth: Processes of the planet SCN 3-05b

The problem with using fossil fuels

Learning intention

In this section you will:
- explain the disadvantages of using fossil fuels.

Combustion of fossil fuels and the carbon cycle

National 4
Curriculum level 4
Planet Earth: Energy sources and sustainability SCN 4-04b; Planet Earth: Processes of the planet SCN 4-05b

Learning intentions

In this section you will:
- learn about the carbon cycle
- describe how burning fossil fuels contributes to the carbon cycle
- explain why carbon dioxide levels are increasing in the atmosphere.

This colour-coding is also used in the footer bars and in the Learning checklist at the end of each chapter.

Activities

Each chapter provides many varied and interesting **activities**. There are activities to work on individually (☻) and others which encourage collaboration with a partner (☻☻) or group (☻☻). The activities provide opportunities to:

- consolidate knowledge and understanding

- practise scientific inquiry and investigative skills

- design and carry out experiments

- develop research skills

- develop informed views about topical scientific issues.

> **GO! Activity 27.4**
>
> ☻ **1. (a)** State why an environmental scientist would use a portable battery-operated pH meter.
>
>   **(b)** A farmer wants to grow a crop which grows best over a very narrow pH range. Suggest why the farmer should use a pH meter rather than universal indicator to test the soil.
>
> ☻☻ **2.** The pH of garden soil has an effect on how well certain vegetables will grow.
>   *You will need: pH paper, pH meter (optional).*
>   Collect samples of soil from two different areas in the school grounds and test their pH. You should follow the instructions from the garden centre on page 396.
>   Ask your teacher if you can use a pH meter.

The 👤✓ graphic indicates that an activity may need teacher input.

The ⚠ graphic is also used to highlight a general safety alert or to indicate a stage in an experiment when particular care needs to be taken.

Chemistry in action

Chemistry in action boxes show real-life examples of chemistry and areas of chemistry which have applications in real-life.

☀ Chemistry in action: Gas welding torches

Gas welding torches use a mixture of oxygen and a hydrocarbon called acetylene. Acetylene is the only hydrocarbon gas that gives a flame hot enough to cut through steel or to melt the edges of metal parts, welding them together.

Figure 10.13 *The temperature of an oxy-acetylene flame is hot enough to cut through steel.*

Word bank

Word banks give short, simple explanations and definitions of important words and phrases.

📖 Word bank

• **Catalyst**

A catalyst speeds up a chemical reaction without being used up itself.

Hint

Hints are usually designed to help you with a particular activity or task.

🔍 Hint

You may wish to use the information in Table 23.4 to help you.

Did you know ...?

Did you know boxes give you extra points of interest about chemistry and chemists.

❓ Did you know ...?

It has been estimated that the energy used to produce, deliver and dispose of junk mail produces more greenhouse gases than 2.8 million cars do.

Make the link

Lots of areas of chemistry are related to other areas of chemistry and to other subjects as well. **Make the link** boxes show how different aspects of chemistry link to each other and to other subjects.

⚗ Make the link

For more about cracking long-chain hydrocarbons see Chapter 14.

Keep up to date!

Chemistry is a subject which advances all the time. **Keep up to date** boxes show topical points for discussion which you can explore with your own research.

⟳ Keep up to date!

Scientists are aiming to have edible cling film made from milk on the market by 2019. You can check how they are progressing by searching for 'edible plastics' online.

You need to know

This feature is used to give background or helpful information.

★ You need to know

• **State symbols**

Symbols, instead of words, can be used to show the state in which substances exist.

solid: (s)

liquid: (ℓ)

gas: (g)

solution: (aq)

Examples

Examples are used to highlight a method in working out a problem.

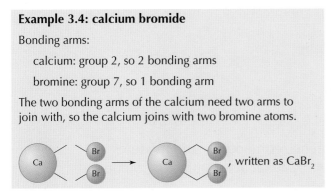

Example 3.4: calcium bromide

Bonding arms:

calcium: group 2, so 2 bonding arms

bromine: group 7, so 1 bonding arm

The two bonding arms of the calcium need two arms to join with, so the calcium joins with two bromine atoms.

, written as $CaBr_2$

Learning checklist

Each chapter closes with a summary of learning statements showing what you should be able to do when you complete the chapter. Each statement is colour coded to show which course it relates to. Each statement is also cross-referenced to at least one activity, so you can use the **Learning checklist** to check you have a good understanding of the topics covered in the chapter. Traffic light icons can be used for your own self-assessment.

Learning checklist

After reading this chapter and completing the activities, I can:

N3 | L3 | **N4** | **L4**

- state that the higher a metal is in the reactivity series the faster it will react with oxygen. **Activity 22.1 Q1(c)**   ◯ ◯ ◯

- state that very reactive metals like potassium, sodium and calcium react quickly with water. **Activity 22.1 Q1(a)**   ◯ ◯ ◯

- state that unreactive metals like copper, silver, gold and platinum do not react with acid. **Activity 22.1 Q1(b)**   ◯ ◯ ◯

- use the results of experiments involving metals and oxygen, water and acid to arrange metals in order of reactivity. **Activity 22.1 Q4(a)**   ◯ ◯ ◯

- use the reactivity series to predict how a metal will react with oxygen, water and acid. **Activity 22.1 Q4(b)**   ◯ ◯ ◯

Practice assessments

Each unit has a **practice assessment** which is designed to assess a sample of all the knowledge and skills within that unit. These assessments can be self-marked using the answers provided online. These questions allow you to check your progress.

Appendix: Investigation skills

There is also an **Appendix** on pages 406–412 which gives guidance on some of the essential scientific skills you will need as you work through this book. You might find it useful to refer to this when carrying out activities.

Answers

Answers to activity questions and assessments are available online at:
www.leckieandleckie.co.uk/page/Resources

1 Rates of reaction 1

Chemical reactions • Speeding up chemical reactions

2 Rates of reaction 2

Monitoring the rate of reaction

3 Chemical structure

Substances and their states • Elements, compounds and mixtures • Elements and the periodic table • Chemical formulae of compounds

4 Atomic structure and bonding related to properties of materials

Atomic structure • Bonding • Properties of covalent and ionic substances • Formulae of elements and compounds

5 Energy changes of chemical reactions

Exothermic reactions • Endothermic reactions

6 Acids and bases 1

Acids and alkalis • Neutralisation • Acidic gases and the environment • Greenhouse gases and global warming

7 Acids and bases 2

Bases • Carbon dioxide and the cement industry

Unit 1 practice assessment

UNIT 1
Chemical changes and structure

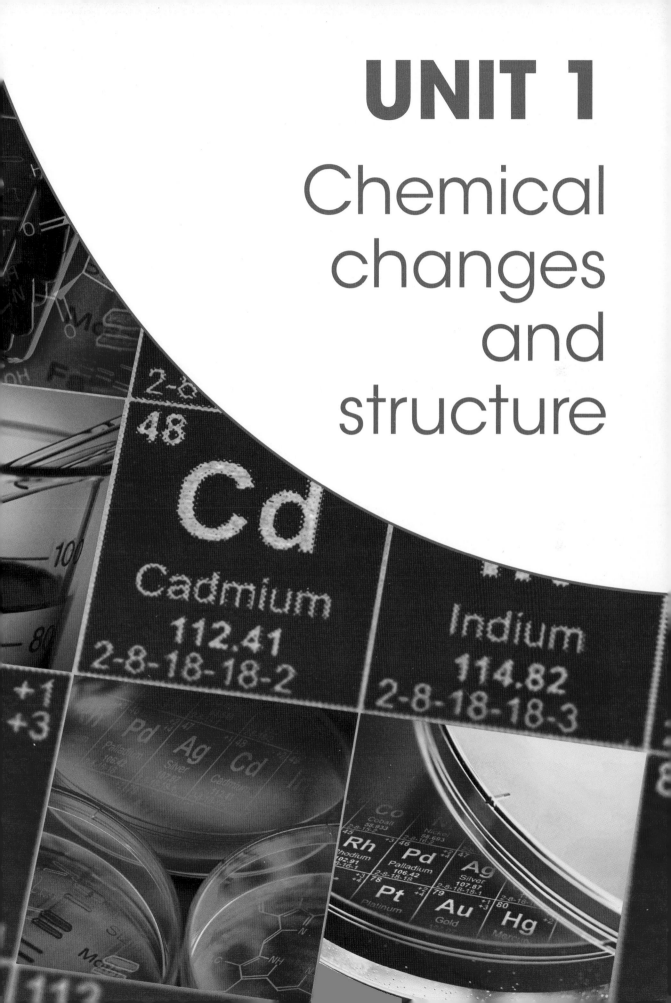

1 Rates of reaction 1

You should already know:

- that a chemical reaction takes place when different materials are made.

National 3

Curriculum level 3

Materials: Chemical changes SCN 3-19a

Chemical reactions

Learning intention

In this section you will:

- identify indicators of chemical reactions having happened.

A chemical reaction is a chemical change which results in one or more new substances being formed. Changes which can indicate that a chemical reaction has taken place include:

- a **colour change**
- a **solid being produced**
- a **temperature change**
- a **gas being produced**.

Three examples of chemical reactions are magnesium burning in air, pieces of zinc being added to hydrochloric acid, and adding potassium iodide to lead nitrate.

bright white light

white solid formed

Figure 1.1 *Magnesium burning in air produces a bright white flame and a white powder, indicating a chemical reaction is taking place.*

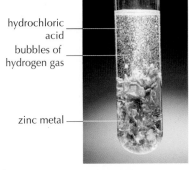

hydrochloric acid

bubbles of hydrogen gas

zinc metal

Figure 1.2 *Bubbles of gas coming from the zinc metal indicate a reaction is taking place.*

Figure 1.3 *Mixing lead nitrate and potassium iodide produces a yellow solid which indicates a reaction is taking place.*

? Did you know ...?

Cooking an egg is an example of a chemical reaction. The colourless jelly part of the egg (albumen) changes into a white solid when it's heated.

Figure 1.4 *A chemical reaction takes place when an egg is fried.*

⚠ Beware!

Not all changes indicate a reaction has taken place. For example when nail varnish dries it looks different on the nails to how it looked in the bottle but no new substance has been produced so no chemical reaction has taken place. The same when hot tea cools; the temperature changes but no new substance is produced.

GO! Activity 1.1

Answers to all activity questions and assessments in this Unit are available online at:
www.leckie andleckie.co.uk/page/Resources

☺ **1.** State which of the following are chemical reactions and say why.

  (A) baking a cake

  (B) boiling water

  (C) paint drying

  (D) food digesting

2. A student wrote a report about an experiment her teacher showed the class.

  Mrs Thompson dropped a piece of very thin metal into a jar which had chlorine in it. The metal shrivelled up. Then it went on fire. When it stopped burning there was a green solid in the jar which wasn't there at the start.

  (a) From the student's report give **two** pieces of evidence which indicate that a chemical reaction has taken place.

  (b) The experiment had to be carried out in a fume cupboard. What does this suggest about chlorine?

National 3

Curriculum level 3

Materials: Chemical changes SCN 3-19a

Speeding up chemical reactions

The conditions under which a reaction takes place can be changed in order to increase the rate (speed it up). The reaction of magnesium with hydrochloric acid can be used to investigate the factors which can affect the rate of a reaction.

The experimental conditions which can be changed, such as temperature, are known as **variables**. When one variable is changed all the others have to stay the same. This ensures that any difference observed in the rate of a reaction is due only to the variable which has been changed.

Concentration

Variable changed: concentration of acid.

Variables unchanged: temperature and volume of the acid; mass (weight) and particle size of the magnesium.

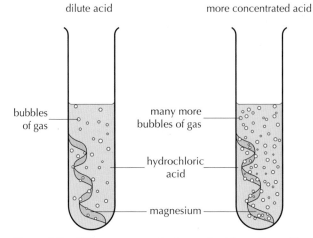

Figure 1.5 *Different concentrations of acid reacting with magnesium.*

Conclusion: Increasing the concentration increases the rate of a reaction.

★ You need to know

- **Concentration**

Concentration is a measure of how much of a substance is dissolved in a liquid.

For example, a concentrated acid solution has more acid dissolved in water than a dilute acid.

Temperature

Variable changed: temperature of acid.

Variables unchanged: concentration and volume of the acid; mass (weight) and particle size of the magnesium.

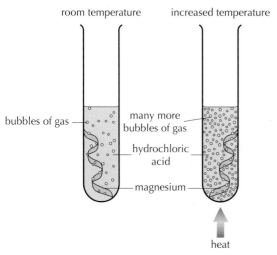

Figure 1.6 *Magnesium and hydrochloric acid reacting at different temperatures.*

Conclusion: Increasing the temperature increases the rate of a reaction.

Particle size (surface area)

Variable changed: particle size of magnesium.

Variables unchanged: concentration, volume and temperature of the acid; mass (weight) of the magnesium.

Small pieces of magnesium have greater surface area than a single piece of magnesium of the same mass (weight).

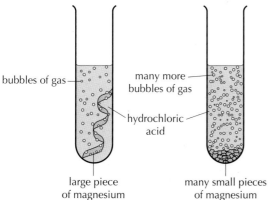

Figure 1.7 *Different particle size reacting with acid.*

Conclusion: Decreasing the particle size increases the surface area of the particles and increases the rate of a reaction.

? Did you know ...?

In flour mills there is a danger of explosion because the flour dust particles are so small that they can act like a gas. Even the tiniest spark can ignite the particles causing an explosively fast reaction.

Figure 1.8 *This giant fireball was produced by burning flour dust in air as part of an outdoors demonstration on the dangers of dust build-up in factories.*

⟳ Keep up to date!

Use the internet to find out the cause of the explosion at the Bosley Wood Flour Mill in July 2015. The incident, in which four people were killed, was investigated by the police and the Health and Safety Executive (HSE).

Word bank

• Catalyst

A catalyst speeds up a chemical reaction without being used up itself.

Adding a catalyst

Many chemical reactions can be speeded up by adding a **catalyst**. For example, hydrogen peroxide slowly produces oxygen gas when left in the bottle. However, when a catalyst is added, the oxygen is produced much more quickly.

The concentration, volume and temperature of the hydrogen peroxide are kept the same.

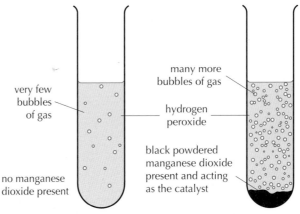

very few bubbles of gas

no manganese dioxide present

many more bubbles of gas

hydrogen peroxide

black powdered manganese dioxide present and acting as the catalyst

Figure 1.9 *The effect of adding a catalyst to hydrogen peroxide.*

Conclusion: Adding a catalyst increases the rate of a reaction. The catalyst does not get used up.

Figure 1.10 *Dramatic effect of adding a catalyst to hydrogen peroxide in the laboratory.*

? Did you know ...?

An unusual idea to help cut down air pollution is to coat clothing with a catalyst to speed up the breakdown of polluting gases in the air.

Find out more about **catalytic clothing** by searching the internet.

Figure 1.11 *The dye used in denim jeans is good at absorbing the catalyst which can break down atmospheric pollutants.*

★ You need to know

• Catalysts in industry

Catalysts are widely used in industry. They speed up chemical processes and reduce energy costs.

N3 | L3 | N4 | L4

GO! Activity 1.2

☻ **1.** Pieces of zinc were added to nitric acid. Bubbles of gas were given off. Suggest four things which could be done to speed up the rate at which the bubbles were produced.

☻☻ **2.** The reaction of marble pieces with hydrochloric acid can be used to show how changing variables in a reaction can affect the rate of a reaction. Bubbles of carbon dioxide gas are produced during the reaction.

The plan below details how you could carry out an experiment to show that changing the size of the marble pieces affects the rate of reaction with hydrochloric acid.

Read the plan carefully then follow the instructions, making sure you follow all of the safety rules as you would normally be expected to do when carrying out practical work.

You can work in a small group but you should record your own observations and complete the answers yourself.

Experiment

 Remember! You must follow the normal safety procedures when carrying out practical work.

Method

(a) Collect the apparatus shown in the diagram.

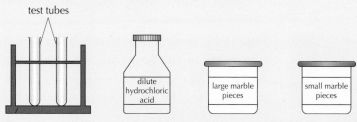

(b) Add one spatula full of large marble pieces to one test tube and one spatula full of small marble pieces to the other.

(c) Half fill each test tube with hydrochloric acid.

Results and conclusions

(d) Write down what you observe (see) happening in a table.
Under 'Observations' describe how fast any gas was produced.

Particle size	Observations

(e) From your observations, write down your conclusion about what effect changing the particle size has on the rate of the reaction.

(f) Suggest one thing you could do to improve the experiment in order to get more accurate results.
(Hint: Is there anything you could have measured more accurately?)

☼ Chemistry in action: Reactions all around

Chemical reactions are happening at different rates all around us in our everyday life. Here are some examples.

Fireworks displays involve powdered metals like magnesium reacting explosively fast. The different colours are due to different metals being used.

Figure 1.12 *Fireworks.*

Rusting is a very slow chemical reaction which happens when unprotected iron and steel are left outside and exposed to the weather.

The paper these words are written on is slowly reacting with some of the chemicals used to make the paper and will over time lose its colour and crumble. Some historically important documents have to be treated with chemicals to slow down the chemical reactions which are damaging the paper.

Burning the gas in a gas cooker to produce heat to cook food is a very fast chemical reaction.

Figure 1.13 *Padlock rusting.*

Figure 1.14 *Milk in fridge.*

Milk and other foods are kept in a fridge at around 4°C, which slows down the chemical reactions which cause food to go off.

Industrial chemical reactions have to be fast to make them economical. Ammonia is an important chemical used to make fertilisers which are needed to grow healthy crops. The gases used to make ammonia are passed over a heated catalyst. The catalyst is made into small pieces so it has a big surface area for the reaction to take place on.

All modern cars are fitted with air bags which inflate during a collision to protect the driver and passengers. The gas used to inflate the bags is produced by a number of explosively fast chemical reactions.

All cars with petrol engines have to be fitted with catalytic converters which remove harmful exhaust gases like carbon monoxide and oxides of nitrogen. The honeycomb structure of the catalyst in a catalytic converter provides a large surface area for reactions to take place on.

Figure 1.15 *An air bag inflating.*

GO! Activity 1.3

1. Ask your teacher to pick an application of chemistry involving the rate of a reaction and which has an effect on society or the environment for you to investigate. You can select your own if you wish.

   Your task is to investigate the application and write a short report on how it affects our lives or the environment.

   You can work in a small group to collect information but your report should be your own work.

   You can get some information from the Chemistry in action section 'Reactions all around' on page 10.

   You could also look at a website for extra information.

   You should note your sources of information.

   (Hint: Type a phrase such as 'Examples of everyday chemical reactions' into a search engine.)

2. Lots of different colours are seen in a fireworks display. Look at flame colours in the table below and suggest which metals could produce the following colours: red, green, yellow and lilac.

Metal	Flame colour
barium	green
calcium	orange-red
copper	blue-green
lithium	red
potassium	lilac
sodium	yellow
strontium	red

3. (a) Why do you think the gas used to inflate an airbag has to be produced explosively fast?

   (b) Why do you think it is important for catalysts to have a large surface area?

Learning checklist

After reading this chapter and completing the activities, I can:

N3 L3 N4 L4

- state that a new substance is formed during a chemical reaction. **Activity 1.1 Q1** ○ ○ ○

- recognise that a chemical reaction may have taken place because one or more of the following may be seen: a colour change; bubbles of gas; a solid formed; an energy change. **Activity 1.1 Q2(a)** ○ ○ ○

- state that the rate of a chemical reaction can be increased by increasing any one of the following: the concentration of the reactants; the temperature of reaction mixture; the surface area of reactants; and by adding a catalyst. **Activity 1.2 Q1** ○ ○ ○

- give examples of everyday reactions which take place at different rates. **Activity 1.3 Q1** ○ ○ ○

- *think analytically.* **Activity 1.1 Q2(b), Activity 1.3 Q3(a)** ○ ○ ○

- *work with others to plan and carry out a practical activity.* **Activity 1.2 Q2** ○ ○ ○

- *select information from a table of data.* **Activity 1.3 Q2** ○ ○ ○

- *apply knowledge in an unfamiliar situation.* **Activity 1.3 Q3** ○ ○ ○

2 Rates of reaction 2

This chapter includes coverage of:

N4 Rates of reaction

You should already know:

- that during a chemical reaction one or more of the following may be seen: a colour change; bubbles of gas; a solid formed; an energy change
- that the rate of a chemical reaction can be increased by increasing any one of the following: the concentration of the reactants; the temperature of reaction mixture; the surface area of reactants; and by adding a catalyst.

Monitoring the rate of reaction

National 4

Learning intentions

In this section you will:
- look at ways to monitor the rate of a chemical reaction
- interpret rate-of-reaction graphs.

The rate of chemical reactions can be monitored in a number of ways. A simple way is to measure the **volume of gas** produced in a reaction over time, then draw a graph of the results. A suitable reaction is marble chips (pieces of calcium carbonate) reacting with hydrochloric acid. Carbon dioxide gas is produced.

Both of the arrangements shown in Figure 2.1 and Figure 2.2 can be used – bubbling through water is the most common method used in the laboratory but using a syringe gives more accurate measurements and can be used for any gas.

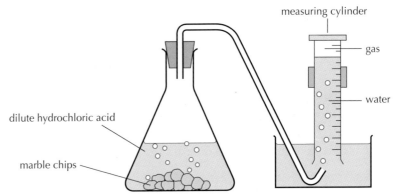

Figure 2.1 *The volume of gas produced during a chemical reaction can be measured by collecting the gas over water.*

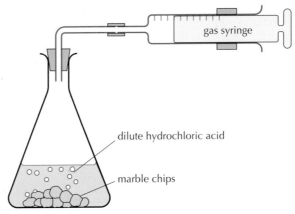

Figure 2.2 *The volume of gas produced during a chemical reaction can be measured by collecting the gas in a syringe.*

The **loss in mass** over time can also be used to monitor the progress of a reaction. The arrangement shown in Figure 2.3 could be used. A computer connected to an electronic balance is used to collect data from the experiment.

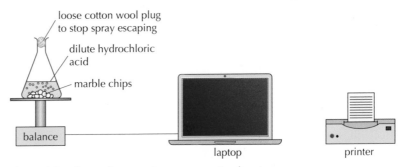

Figure 2.3 *Set-up for experiment measuring loss in mass.*

As the carbon dioxide is produced, it escapes into the air and so the mass of the reaction mixture decreases.

Both experiments do the same thing – they measure the amount of carbon dioxide produced over time.

The rate of reaction was monitored in a number of experiments using marble and hydrochloric acid. The variables concentration, temperature and particle size were changed but the volume of acid used and the mass of marble used were the same in each experiment and all of the marble reacted. This means the total volume of gas produced in each experiment is the same. For each variable there were two experiments, so that the effect of changing the variable could be measured.

Changing concentration

Experiment 1

The **volume of gas** collected every minute was measured using the arrangement shown in Figure 2.1. Table 2.1 shows the results. A graph of volume of gas against time was drawn.

Look at sections A, B and C of Figure 2.4.

- At A, near the start of the reaction: the graph is almost a straight line with a fairly steep slope. The gas was being produced quickly. The reaction was **fast**.

- At B, towards the end of the reaction: the graph starts to level off. The gas was not being produced as quickly. The reaction was **slowing down**.

- At C, at the end of the reaction: the graph has completely levelled off. All of the marble has reacted. No more gas was being produced. The reaction has **stopped**.

Experiment 2

More concentrated hydrochloric acid was reacted with the marble chips. The graph shown in Figure 2.5 was obtained – the original graph shown in Figure 2.4 is included to make it easier to compare the results.

- At A, the graph is again a straight line but with a much steeper slope than with the lower concentration of acid. The reaction was much **faster**.

- At B, the graph starts to level off more quickly than in the first experiment. The gas was not being produced as quickly.

- At C, the graph has completely levelled off at the same volume as the first experiment. This is because, although the concentration of the acid has changed, the mass of marble reacting is the same. The gas may have been produced more quickly but the same final volume of gas was produced in both experiments.

Table 2.1 *Volume of gas collected over time.*

Time (min)	Gas volume (cm³)
0	0
1	20
2	35
3	45
4	50
5	52
6	53
7	53
8	53

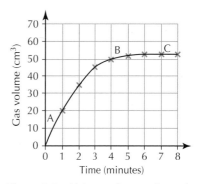

Figure 2.4 *Volume of gas collected over time.*

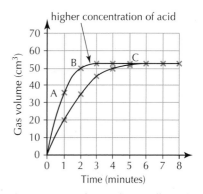

Figure 2.5 *Volume of gas collected over time.*

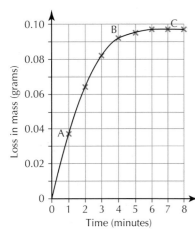

Figure 2.6 *Loss in mass of marble over time.*

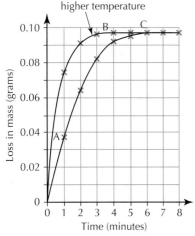

Figure 2.7 *Loss in mass of marble over time.*

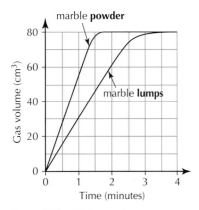

Figure 2.8 *Volume of gas collected over time.*

Changing temperature

Experiment 3

Experiment 1 was repeated but instead of collecting the gas, the **loss in mass** was measured over time using the arrangement shown in Figure 2.3. Table 2.2 shows the results.

The graph is similar to the one obtained in experiment 1 and the shape of the graph at points A, B and C can be explained in the same way (Figure 2.6).

Experiment 4

When the experiment was carried out with the acid at a **higher temperature** the graph shown in Figure 2.7 was obtained. The graph from Figure 2.6 is included to make a comparison easier.

- At A, the graph is again a straight line but with a much steeper slope than with the acid at room temperature. The reaction was much **faster**.

- At B, the graph starts to level off more quickly than in the first experiment. Mass was not being lost as quickly.

- At C, the graph has completely levelled off at the same mass as the experiment with the acid at room temperature. This is because, although the temperature of the acid has changed, the mass of marble reacting was the same. Mass was lost more quickly at the higher temperature but the same amount of mass was lost overall in both experiments.

Table 2.2 *Loss in mass of marble over time.*

Time (min)	Loss in mass (g)
0	0.000
1	0.037
2	0.064
3	0.082
4	0.092
5	0.095
6	0.097
7	0.097
8	0.097

Changing the particle size

Experiment 1 was repeated using **lumps** of marble. Experiment 2 used **powdered** marble. All other variables were kept the same. The results were monitored and the graphs are shown in Figure 2.8.

The shape of each graph can be explained as in the previous experiments. The slope of the graph with the powdered marble is much steeper indicating that the reaction was faster when the marble is powdered as it has a larger surface area for the acid to react with.

Endpoint and quantities produced and reacting

The endpoint of a reaction is the time at which the graph levels out (goes flat). This means no more gas is being produced – the reaction has stopped.

Look at Figure 2.8, which shows the volume of gas produced when powdered marble and lumps of marble react with hydrochloric acid.

In each reaction, the volume of gas produced at the endpoint is $80\,cm^3$. The volumes are the same because the same mass of marble is reacting in each experiment.

However, the time taken to reach the endpoint in each reaction is different:

- marble powder: approximately 1.6 min
- marble lumps: approximately 3.5 min.

The marble powder reaches the endpoint more quickly (it takes less time), so it is the faster reaction.

The volume of gas produced at other times can also be obtained from the graph.

- After 2 min, $60\,cm^3$ of gas is produced with marble lumps.
- The time taken to produce $30\,cm^3$ of gas using marble powder is 0.5 min.

Look at Figure 2.6, which shows loss in mass when marble reacts with hydrochloric acid. The endpoint of the reaction occurs after 6 min and the loss in mass of reactants is just below $0.10\,g$ ($0.097\,g$).

Sometimes the change in the total mass of the reaction flask and its contents over time is measured and a graph plotted. Figure 2.9 shows the type of curve obtained.

Although the shape of the graph is different to the other graphs in this section it can be interpreted in the same way.

- At A, the graph is a steep straight line, showing that the reaction is fast. The reactants are being used up quickly.
- At B, the graph is less steep, showing that the reaction is slowing down because the reactants are being used up.
- At C, the graph has levelled off. At least one of the reactants is completely used up and the reaction has stopped. This is the endpoint of the reaction.

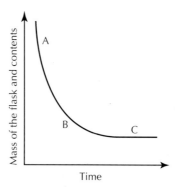

Figure 2.9 *Change in total mass over time.*

GO! Activity 2.1

☺ **1.** Powdered zinc metal can be reacted completely with hydrochloric acid. One of the products is hydrogen gas. The volume of hydrogen produced can be measured and the results used to see how the rate of reaction changes over time.

(a) Describe how you could carry out an experiment to measure the volume of gas produced over time. Include a labelled diagram.

(b) The rate-of-reaction graph produced from the results of an experiment like the one you described in part (a) is shown.

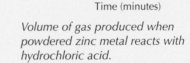

(i) Is the reaction faster at point A or point B?

(ii) What is it about the line on the graph that helped you to answer part **(b) (i)**?

(iii) What has happened to the reaction at point C on the graph?

Volume of gas produced when powdered zinc metal reacts with hydrochloric acid.

(c) What volume of hydrogen gas was produced after 3 minutes?

(d) (i) Sketch the graph above and add another line to show what happens when the reaction is carried out at a higher temperature.

(You do not need to use graph paper or include the scales on the graph.)

(ii) Explain the shape of this graph compared to the original.

(Hint: Experiment 4 on page 16 will help you answer this question.)

(e) The table below shows results for a similar reaction.

Time (min)	Volume of hydrogen (cm³)
0	0
1	15
2	25
3	31
4	34
5	34
6	

Predict the volume of gas collected after 6 minutes.

2. The rate of a chemical reaction can be monitored by measuring the loss in mass over time.

   (a) Describe how you could carry out an experiment to measure the loss in mass over time. Include a labelled diagram.

   (b) The graphs below show loss in mass over time when the same mass of powdered chalk and lumps of chalk react with hydrochloric acid. All the chalk reacts in each reaction.

   (i) Which graph, X or Y, shows the reaction with powdered chalk?

   (ii) Explain your answer to part (b) (i).

   (c) How many minutes does it take for the reaction represented by graph X to finish?

   (d) From the graphs, work out the total loss in mass in both experiments.

   (e) Suggest why the total loss in mass is the same in each experiment.

   (f) If a more concentrated acid had been used than the one used to produce graph X, how would the shape of the graph near the start of the reaction compare to graph X?

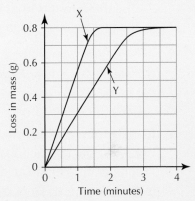

Loss in mass when powdered chalk and lumps of chalk react with hydrochloric acid.

Activity 2.2

Magnesium reacts with hydrochloric acid to produce hydrogen gas as one of the products. By collecting the gas and measuring the volume given off at set time intervals the rate of the reaction can be monitored.

You can work in a small group but you should record your own observations and complete the answers yourself.

Experiment

Plan

(a) Make up a **plan** outlining how you could carry out an experiment to monitor the change in the rate over the course of the reaction between magnesium ribbon and hydrochloric acid.

In your plan include a labelled diagram of the experimental arrangement you would use and show clearly how you would collect the gas. State what measurements you would need to take and how you would record and present your results.

(b) State the variables which should be kept unchanged in the experiment.

(Hint: Look at Figure 2.1 and 'Changing concentration' on page 15 if you need help with your planning.)

(c) Ask your teacher to check your plan and ask if you can carry out the experiment.

(continued)

Experiment and results

⚠️ Remember! You must follow the normal safety procedures when carrying out practical work.

(d) Record your results in a table. Use them to draw a graph of volume of hydrogen collected against time.

(e) Repeat the experiment to show the effect of using a smaller particle size or use the results from a group who have used smaller particles.

Use these results to draw a graph of volume of hydrogen collected against time on the graph you drew in part **(d)**.

Conclusions

(f) State any conclusions you can make from the results of both experiments.

(g) Discuss in the group anything you could do to improve the experiment to get more accurate results.

(h) The group could also make up a plan outlining how you could monitor the effect on the rate of reaction of changing the concentration of the acid and changing the temperature of the acid.

(i) Discuss your results in the group and compare the graphs obtained from each set of experiments.

Learning checklist

After reading this chapter and completing the activities, I can:

N3 L3 **N4** L4

* describe how the rate of a chemical reaction can be monitored by measuring the volume of gas produced over time. **Activity 2.1 Q1(a)** ◯ ◯ ◯

* describe how the rate of a chemical reaction can be monitored by measuring the loss in mass of reactants over time. **Activity 2.1 Q2(a)** ◯ ◯ ◯

* state that the slope of a rate-of-reaction graph is steep near the start of a reaction which indicates that the reaction is fast at this point. **Activity 2.1 Q1(b)(i), (ii)** ◯ ◯ ◯

* state that the slope of a rate-of-reaction graph becomes less steep when a reaction is slowing down. **Activity 2.1 Q1(b)(i), (ii)** ◯ ◯ ◯

* state that the slope of a rate-of-reaction graph eventually levels off because the reaction has stopped. **Activity 2.1 Q1(b)(iii)** ◯ ◯ ◯

N3 | L3 | **N4** | **L4**

- work out the quantity of product produced during a reaction from a rate-of-reaction graph. **Activity 2.1 Q1(c)** ⃝ ⃝ ⃝

- work out the loss in mass during a reaction from a rate-of-reaction graph. **Activity 2.1 Q2(d)** ⃝ ⃝ ⃝

- state that when the temperature of a reaction mixture is increased the slope of a rate-of-reaction graph will be steeper at the start of the reaction compared to the lower temperature. **Activity 2.1 Q1(d)(i), (ii)** ⃝ ⃝ ⃝

- state that when the concentration of reactants is increased the slope of a rate-of-reaction graph will be steeper at the start of the reaction compared to the lower concentration. **Activity 2.1 Q2(f)** ⃝ ⃝ ⃝

- state that when the particle size of a solid reactant is decreased (larger surface area), the slope of a rate-of-reaction graph will be steeper at the start of the reaction compared to the larger particle size (smaller surface area). **Activity 2.1 Q2(b)(i), (ii)** ⃝ ⃝ ⃝

- sketch a graph to show how increasing the temperature affects the rate of a reaction compared to the lower temperature. **Activity 2.1 Q1(d)(i)** ⃝ ⃝ ⃝

- *work with others to plan and carry out a practical activity.* **Activity 2.2** ⃝ ⃝ ⃝

- *interpret information presented in a graph.* **Activity 2.1 Q2(b), (c)** ⃝ ⃝ ⃝

- *draw conclusions.* **Activity 2.1 Q2(e)** ⃝ ⃝ ⃝

- *make predictions and generalisations.* **Activity 2.1 Q1(e)** ⃝ ⃝ ⃝

3 Chemical structure

You should already know:

- that dissolving can be speeded up by using powdered solid and stirring
- that the amount of substance that can be dissolved in water can often be increased by warming the water.

National 3

Curriculum level 3

Materials: Properties and uses of substances SCN 3-16b

★ You need to know

Substances can exist in three states: solid, liquid and gas.

📖 Word bank

- **Soluble**

A substance that dissolves in water is soluble.

- **Solute**

A solute is a substance that dissolves.

- **Solvent**

A solvent is the liquid in which a substance dissolves.

- **Solution**

A solution is formed when a substance dissolves.

Substances and their states

Learning intentions

In this section you will:
- identify the states of matter
- use the terms solute, solvent and solution when describing a substance dissolving
- name different solvents and give examples of how they can be used in everyday life.

Everything in the world could be described as being a substance. Many substances can dissolve in water – they are said to be **soluble**. The substance dissolving is known as the **solute**. The liquid doing the dissolving is called the **solvent**. A **solution** is formed when a substance dissolves. Water is our most important solvent.

coffee granules (solute)

stir to dissolve

dissolved coffee (solution)

water (solvent)

Figure 3.1 *Making a cup of instant coffee is an everyday example of making a solution.*

In the laboratory, copper sulfate can be dissolved in water to form a solution, as shown in Figure 3.2.

There is a limit to how much of a solute can dissolve in a solvent. When no more solute can dissolve the solution is said to be **saturated**.

The Dead Sea in Israel has six times the amount of salts dissolved in it than other seawater. So much salt is dissolved in it that you can easily float in the sea. The water evaporates naturally in shallow pools, leaving behind a mixture of salts which can be collected, separated and changed into more useful chemicals.

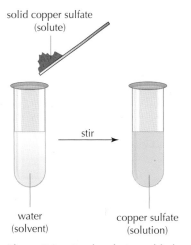

solid copper sulfate
(solute)

stir

water
(solvent)

copper sulfate
(solution)

Figure 3.2 *A solute being added to a solvent to make a solution.*

Figure 3.3 *The Dead Sea has such a high concentration of dissolved salts that you can easily float in the water.*

> ### 📖 Word bank
>
> • **Saturated**
> A solution is saturated when no more solute can be dissolved in it.

GO! Activity 3.1

☺ **1.** Give the states in which matter can exist.

2. (a) Vinegar is made by dissolving ethanoic acid in water. State which is:

    (i) the solvent   **(ii)** the solute   **(iii)** the solution.

  (b) State what is meant by a saturated solution.

Elements, compounds and mixtures

Learning intentions

In this section you will:

• describe the difference between elements, compounds and mixtures

• describe how to break down certain compounds into their elements

• name compounds made from two elements

• write word equations showing reactants forming products

• select methods for separating mixtures

• give an example of separating mixtures in everyday life.

National 3

Curriculum level 3

Materials: Properties and uses of substances
SCN 3-15b, SCN 3-16a, SCN 3-16b

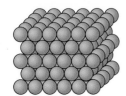

Figure 3.4 *Atoms in the element gold are tightly packed in an organised way.*

📖 Word bank

- **Atom**

What everything is made from.

- **Element**

A substance made up of the same kind of atom.

- **Mixture**

Made from different substances which are not chemically joined.

- **Compound**

Substance made up of different atoms chemically joined.

Everything is made of **atoms**. If a substance is made of the same kind of atom then it is known as an **element**. The elements are listed in the periodic table (pages 34–35). Gold is an example of an element that exists as a solid.

Mixtures of elements can undergo chemical reactions to form compounds. In a **mixture** the atoms of the different elements are not joined together but in a **compound** they are.

A mixture of iron and sulfur can easily be separated by passing a magnet over the mixture. The grey iron filings stick to a magnet but the yellow sulfur doesn't. When they are heated the atoms of the elements join to form a dark compound. The iron filings can no longer be separated out using a magnet. It is difficult to separate the atoms in a compound.

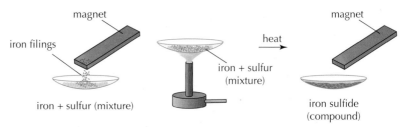

Figure 3.5 *Iron can easily be separated from a mixture of iron and sulfur (left-hand side) but not when the elements react to form a compound (right-hand side).*

Figure 3.6 shows how atom models can be used to show the elements hydrogen and chlorine, which are gases, being mixed then reacting to form a compound. Notice that in the model of the compound the hydrogen and chlorine atoms are joined together.

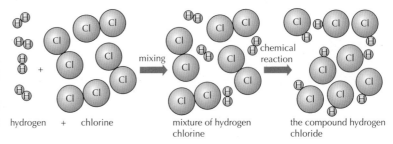

Figure 3.6 *A mixture of hydrogen and chlorine reacts to form the compound hydrogen chloride.*

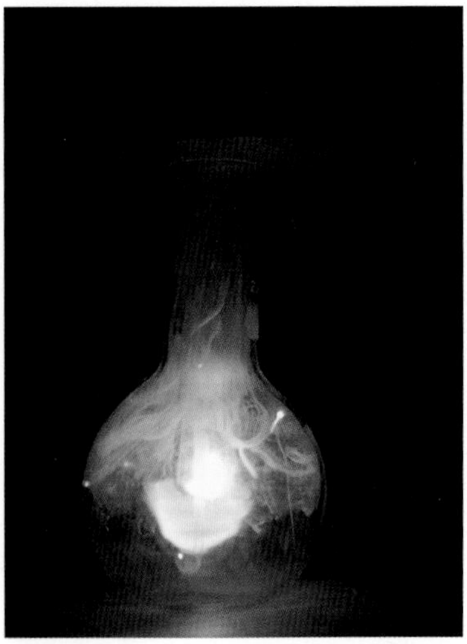

The properties of compounds are very different to the properties of the elements they are made from. Sodium chloride (common table salt) is a good example. Sodium is a soft, very reactive metal, and chlorine is a green, very reactive gas. They react violently to form the white compound sodium chloride, which is a very unreactive and useful compound that we use to flavour foods.

Figure 3.7 *The elements sodium and chlorine react violently to form the compound sodium chloride.*

| N3 | L3 | N4 | L4 |

One way of separating the elements in some compounds is to use electricity. When copper chloride is dissolved in water and electricity is passed through it (see Figure 3.8) solid copper metal is formed at one of the carbon rods and chlorine gas at the other.

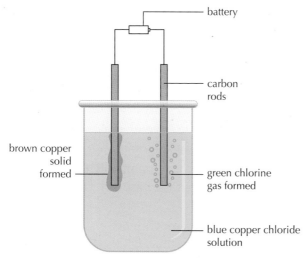

Figure 3.8 *Copper chloride can be broken down into its elements – copper and chlorine – by passing electricity through a solution of copper chloride.*

Naming compounds

The names of compounds come from the names of the elements. The name ending is **-ide** when atoms of two different elements join. In the examples in Figures 3.5, 3.6 and 3.8:

- iron reacts with sulfur to produce iron sulf**ide**

- hydrogen reacts with chlorine to produce hydrogen chlor**ide**

- copper chlor**ide** breaks down to form copper and chlorine.

Table 3.1 gives some other examples.

Table 3.1 *Examples of compounds and the elements in them.*

Name of compound	Elements in compound
sodium fluor**ide**	sodium and fluorine
zinc ox**ide**	zinc and oxygen
lead brom**ide**	lead and bromine
magnesium nitr**ide**	magnesium and nitrogen

Word equations

Chemical reactions can be summarised using word equations.

Reactants → **Products**

When iron reacts with sulfur, iron sulfide is produced.

Word equation: iron + sulfur → iron sulfide

When hydrogen reacts with chlorine, hydrogen chloride is produced.

Word equation: hydrogen + chlorine → hydrogen chloride

When copper chloride solution is broken down, the elements copper and chlorine are formed.

Word equation: copper chloride → copper + chlorine

GO! Activity 3.2

☻ 1. Copy the summary paragraph and use the following words to help you complete it.

    joined   easily   atom   electricity   separated   mixture   elements

    A substance can be described as an element if it contains only one kind of **(a)** ____.
    A compound contains atoms of two or more different **(b)** _____ chemically **(c)** ____.
    In a **(d)** _____ there is more than one type of substance but they are not joined.
    Mixtures can be easily **(e)** _____.

    Compounds are generally not **(f)**_____ broken down into their elements.
    Some compounds can be broken down by passing **(g)** _____ through them.

2. Diagrams **A–D** show atom models.

    A   B   C   D

    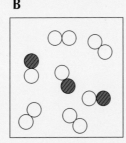 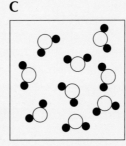 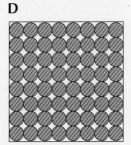

    (a) Which two models represent an element?

    (b) Which model represents a compound?

    (c) Which model represents a mixture?

3. Give the names of the compounds formed when the following elements react.

    (a) Magnesium and fluorine

    (b) Zinc and chlorine

    (c) Sodium and sulfur

    (d) Lead and oxygen

4. Look at Table 3.1. Write word equations for each of the reactions taking place.

5. Write word equations for the reactions taking place in Question 3.

Separating mixtures

Many mixtures can be easily separated. A mixture of sweets can be separated by picking out those that look the same.

mixture   separated

Figure 3.9 *Mixtures of sweets can be easily separated because of their colour.*

Metal can be separated from plastics during recycling. The mixture is placed on a conveyor belt and as it moves through the separator magnets pick up steel cans and use sensors ('magic eyes') to pick out aluminium cans, which are then blown to one side by jets of air for collection.

In the laboratory a mixture of solid and liquid can be separated by **filtering**. The solid stays in the filter paper and the liquid drips through.

> **📖 Word bank**
>
> • **Filtering**
> A technique for separating a mixture of solid and liquid.

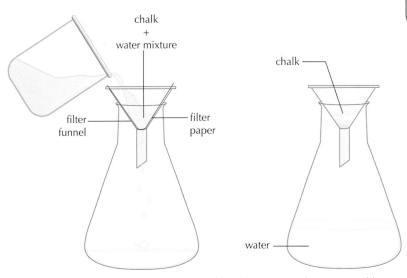

chalk + water mixture

filter funnel

filter paper

chalk

water

Figure 3.10 *A mixture of solid chalk and liquid water can be separated by filtering.*

A puddle of rainwater seems to quickly disappear on a warm day. The liquid water actually changes into gas and mixes with the air. This is called **evaporation**. This idea can be used to separate out a dissolved substance. For example, if a solution of copper sulfate is left for a few days the water evaporates and crystals of copper sulfate are left behind (see 'Problem solving in the laboratory' on page 29).

The idea of evaporation has been used for thousands of years to obtain salt from seawater. The seawater is collected in small ponds (pans) and the water evaporates, leaving the salt behind.

> **📖 Word bank**
>
> • **Evaporation**
> The process which happens when a liquid changes into a gas.

Figure 3.11 *Salt pans on the island of Gozo, Malta.*

? Did you know ...?

The town of Prestonpans in East Lothian got part of its name from the fact that in the twelfth century, monks started to produce salt from seawater in the area. They used large pans to collect the seawater and left the water to evaporate, leaving the sea salt behind in the pan. At one time in the fifteenth century, over 11 tonnes of salt a week was being produced. The first part of the town's name comes from the word 'priest'.

📖 Word bank

• **Distillation**

A technique used to separate mixtures of liquids.

Mixtures of liquids can be separated by **distillation**. Figure 3.12 shows a laboratory distillation apparatus.

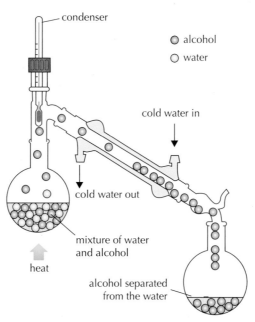

Figure 3.12 *A particle picture showing a mixture of water and alcohol being separated by distillation.*

★ You need to know

Boiling point is the temperature at which a liquid becomes a gas or a gas becomes a liquid.

The liquids have different **boiling points**. The alcohol has the lower boiling point so it changes to a gas (evaporates) first. The condenser acts as a cold water jacket and changes the alcohol from gas to liquid (condensation), which is then collected in a separate container.

Problem solving in the laboratory

The problem

A technician accidentally mixed powdered copper carbonate and copper sulfate. He must separate the mixture so that he gets pure samples of solid copper carbonate and copper sulfate.

Solving the problem

The technician looked at the solubility table (Table 3.2) to check the solubility of each of the compounds.

Table 3.2 *The solubility of different compounds.*

	carbonate	chloride	hydroxide	nitrate	sulfate
copper	insoluble	very soluble	insoluble	very soluble	very soluble
iron	insoluble	very soluble	insoluble	very soluble	very soluble
potassium	very soluble	very soluble	very soluble	very soluble	very soluble
silver	insoluble	insoluble	insoluble	very soluble	soluble
sodium	very soluble	very soluble	very soluble	very soluble	very soluble

Result: copper carbonate: insoluble; copper sulfate: very soluble.

Conclusion

The technician concluded that:

1. If the mixture was added to water, and stirred, the copper sulfate would dissolve and form a solution.

2. The insoluble copper carbonate could be separated from the solution by filtering.

3. The copper sulfate solution could then be heated to boil off most of the water then left at room temperature until the rest of the water evaporated and crystals formed.

4. The solid copper carbonate could be washed with water and allowed to dry.

This is shown in Figure 3.13.

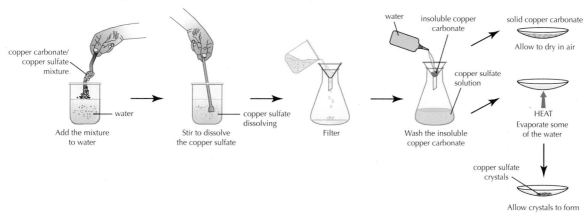

Figure 3.13 *Separating a mixture of soluble and insoluble compounds.*

Activity 3.3

☺ **1. (a)** Choose the correct word in bold to complete each sentence.

    (i) Powdered chalk can be separated from water by **filtration / evaporation**.

    (ii) Water can be separated from a copper sulfate solution by **filtration / evaporation**.

  (b) **Distillation** is based on the fact that liquids have different **boiling points**. Give the meaning of the words in bold.

2. (a) Use the information in Table 3.2 to find the solubility of iron hydroxide and silver chloride.

  (b) Using your answer to Question 2 part **(a)** state whether the method of separating a mixture shown in Figure 3.13 would work for a mixture of iron hydroxide and silver chloride.

  (c) Explain your answer to Question 2 part **(b)**.

3. Seawater contains a number of salts dissolved in the water. Discuss with others how you could separate a mixture of seawater and sand and end up with dry sea salt.

Write out your plan and ask your teacher to check it.

Ask if you can carry out the experiment.

> 🔍 **Hint**
>
> Figure 3.13 will help you.

☀ Chemistry in action: Separating mixtures in everyday life

Figure 3.14 *Making filter coffee – the insoluble part of the coffee stays in the filter paper and the dissolved coffee drips through.*

Making coffee

Making filter coffee at home or in a coffee shop involves filtering. A lot of the ground coffee bean is insoluble, so when hot water is added, the soluble part of the coffee bean dissolves in the water and filters through while the solid insoluble part is left behind.

Part of the process by which instant coffee is made from coffee beans involves dissolving the soluble part of the bean, filtering off the solution and then evaporating off the water, leaving the solid soluble coffee behind. Instead of evaporating off the water the coffee solution is sometimes frozen and the ice separated from the coffee. This is known as freeze-dried coffee.

| N3 | L3 | N4 | L4 |

Cleaning drinking water

Filtering is an important part of the purification of drinking water. The water is stored in reservoirs that have giant filter beds made up of layers of stones, gravel and sand. When water passes through the filter bed, solid particles are trapped and the water passes through.

Figure 3.15 *Water treatment plants use filter beds to filter and purify water in reservoirs.*

Figure 3.16 *Traditional copper whisky stills in a distillery on the Isle of Islay.*

Making whisky

Scotland is famous the world over for making whisky. The production of whisky involves distillation. In its early stages whisky is a mixture of water and alcohol. The alcohol is separated by distilling the mixture in copper stills. The alcohol has a lower boiling point than water so will turn into a gas at a lower temperature than the water. This means that as the mixture is heated the alcohol comes off as a gas before the water. It is cooled (condensed), turned into a liquid and is collected. The alcohol is then usually stored in sherry or oak casks for a number of years to allow it to develop its golden colour and flavour from the cask.

Making fuels from oil

We rely on distillation for our petrol and diesel for cars and lorries. We get many of our fuels from crude oil, which is a complicated mixture of compounds. When it comes out of the ground in its raw (crude) state, the oil is not useful. The oil mixture can be separated by distilling it into smaller, more useful mixtures.

Making fresh water from seawater

In 2022 the FIFA World Cup will be held in the Gulf State of Qatar. Qatar, like some other Gulf states, relies on a type of distillation called desalination to provide water for drinking and growing crops. Qatar is on the Gulf of Iran, so has an unlimited supply of seawater but is one of the driest places on Earth, so has little natural drinking water. Freshwater is produced by heating seawater. The pure water evaporates, condenses and is collected. The salts are left behind. It takes a lot of energy to desalinate water and work is ongoing to produce freshwater using solar power.

? Did you know ...?

Over 300 million people worldwide rely on the desalination of seawater for their daily water needs, even though it is an energy-intensive and expensive process. But what if you could use desalination to make energy instead of use it? A research team from Singapore has done just this. They have designed a solar-powered reactor that converts seawater into clean drinking water and hydrogen fuel.

GO! Activity 3.4

☺ 1. Read the Chemistry in action section 'Separating mixtures in everyday life'.

   Copy out the table and complete it.

Separation technique	Everyday examples
filtration	
distillation	

2. Desalinated water is drinkable water produced from seawater. The graph shows the percentage of the world's desalinated water produced by different countries.

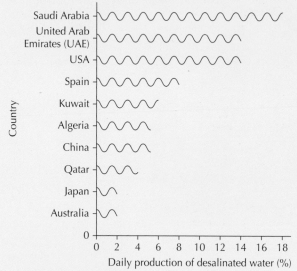

(a) Complete the table using information from the graph.

Country	World daily production of desalinated water (%)

(b) Look at a map of the world and find Qatar and Saudi Arabia. From its position on the map, suggest why Saudi Arabia needs to produce so much drinking water by desalination.

(c) Why do you think we don't need to desalinate water in Scotland?

Elements and the periodic table

National 3

Curriculum level 3

Materials: Properties and uses of substances
SCN 3-15a

Learning intentions

In this section you will:

- gain an understanding of the way the periodic table is set out
- relate the properties of elements to their position in the periodic table
- relate the properties of elements to their uses.

Elements are substances made from the same atoms. The periodic table shows elements with similar chemical properties arranged in families. Elements in the same vertical **group** react in a similar way – they have similar chemical properties. Each element has its own symbol and a number used to identify it called the atomic number (see page 47).

A periodic table is on pages 34 and 35; it highlights some of the elements and their uses.

📖 Word bank

- **Group**
A column of elements in the periodic table.

❓ Did you know ...?

Throughout the nineteenth century scientists tried to arrange elements in some sort of order. A number of ideas were put forward but it wasn't until 1869 that the Russian chemist Dmitri Mendeleev seemed to have the answer. He put forward his ideas as a periodic table that put elements with similar chemical properties together. What Mendeleev realised was that where gaps appeared in his table it was because the missing elements hadn't yet been discovered. He correctly predicted the properties of many of the missing elements. Our modern periodic table is based on Mendeleev's work.

Figure 3.17 *A statue of Dmitri Mendeleev, often referred to as 'the father of the modern periodic table', with his periodic table in the background.*

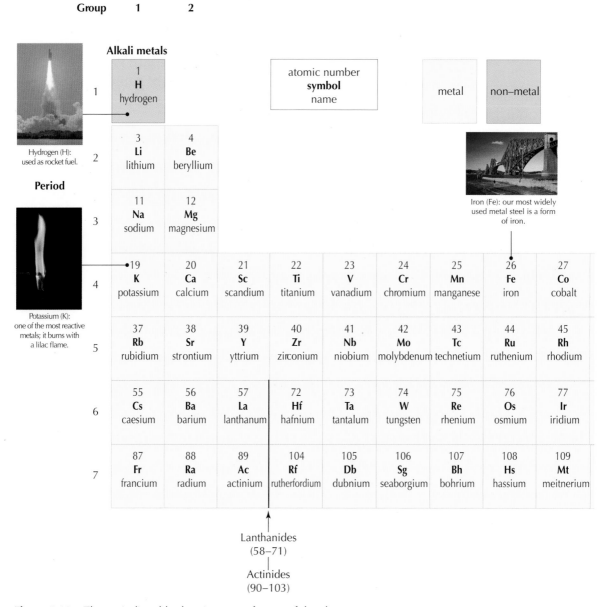

Figure 3.18 *The periodic table showing uses of some of the elements.*

? Did you know ...?

- Only elements 1–94 in the periodic table occur naturally on Earth.
- Elements with atomic numbers 95–118 have all been made by scientists.
- Scientists predict that as the technology improves, elements 119 and 120 will be made.
- Element 119 has been given the temporary name ununennium (Uue) and 120 unbinilium (Ubn).

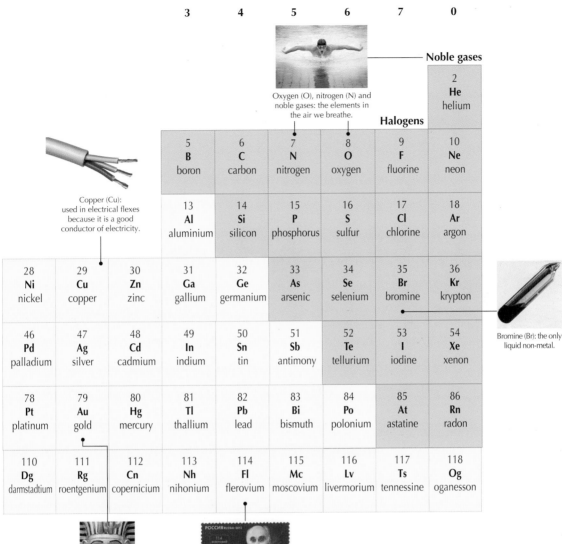

		3	4	5	6	7	0

Oxygen (O), nitrogen (N) and noble gases: the elements in the air we breathe.

Noble gases

| 2 **He** helium |

Halogens

5 **B** boron	6 **C** carbon	7 **N** nitrogen	8 **O** oxygen	9 **F** fluorine	10 **Ne** neon
13 **Al** aluminium	14 **Si** silicon	15 **P** phosphorus	16 **S** sulfur	17 **Cl** chlorine	18 **Ar** argon

Copper (Cu): used in electrical flexes because it is a good conductor of electricity.

28 **Ni** nickel	29 **Cu** copper	30 **Zn** zinc	31 **Ga** gallium	32 **Ge** germanium	33 **As** arsenic	34 **Se** selenium	35 **Br** bromine	36 **Kr** krypton
46 **Pd** palladium	47 **Ag** silver	48 **Cd** cadmium	49 **In** indium	50 **Sn** tin	51 **Sb** antimony	52 **Te** tellurium	53 **I** iodine	54 **Xe** xenon
78 **Pt** platinum	79 **Au** gold	80 **Hg** mercury	81 **Tl** thallium	82 **Pb** lead	83 **Bi** bismuth	84 **Po** polonium	85 **At** astatine	86 **Rn** radon
110 **Dg** darmstadtium	111 **Rg** roentgenium	112 **Cn** copernicium	113 **Nh** nihonium	114 **Fl** flerovium	115 **Mc** moscovium	116 **Lv** livermorium	117 **Ts** tennessine	118 **Og** oganesson

Bromine (Br): the only liquid non-metal.

Gold (Au): a precious metal known for thousands of years.

Flerovium (Fl): one of the most recently named man-made elements, named after the Russian scientist Georgy Flyorov in 2012.

↻ Keep up to date!

It takes years after the discovery of an element before it is named. Names are agreed by the International Union of Pure and Applied Chemistry (IUPAC). Check their website to see if any new elements have been made and still have to be given official names.

Figure 3.19 *In 2006, element 111 received its official name, roentgenium (Rg), 12 years after first being made.*

Metals and non-metals

Some periodic tables have a zigzag line that separates the metal elements from the non-metal elements.

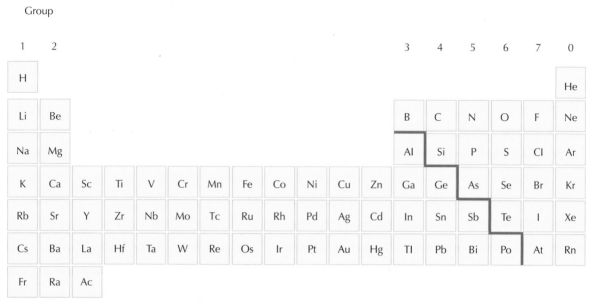

Figure 3.20 *In the periodic table the **metals** are to the left of the zigzag line and the **non-metals** to the right.*

Table 3.3 compares some of the properties of metals and non-metals.

Table 3.3 *Properties of metals and non-metals.*

Metals	Non-metals
all but one are solid, e.g. gold (Au), iron (Fe), aluminium (Al) are all solid, while mercury (Hg) is the only liquid	can be solid, liquid or gas, e.g. sulfur (S) is solid, bromine (Br) is liquid (the only one), helium (He) is gas
good conductors of electricity, e.g. copper wires	the only good conductor of electricity is carbon in the form of graphite
good conductors of heat, e.g. aluminium pots	most are poor conductors of heat
shiny when polished	mostly dull
many are strong and can be bent or beaten into new shapes, e.g. car bodies made of iron (steel)	solids tend to be hard and break easily

Some important groups of metals and non-metals are highlighted in Figure 3.21.

Group

1	2											3	4	5	6	7	0
H																	He
Li	Be											B	C	N	O	F	Ne
Na	Mg											Al	Si	P	S	Cl	Ar
K	Ca	Sc	Ti	V	Cr	Mn	Fe	Co	Ni	Cu	Zn	Ga	Ge	As	Se	Br	Kr
Rb	Sr	Y	Zr	Nb	Mo	Tc	Ru	Rh	Pd	Ag	Cd	In	Sn	Sb	Te	I	Xe
Cs	Ba	La	Hf	Ta	W	Re	Os	Ir	Pt	Au	Hg	Tl	Pb	Bi	Po	At	Rn
Fr	Ra	Ac															

Group 1
Alkali metals
A group of very reactive metals

Group 7
Halogens
A group of very reactive non-metals

Group 0
Noble gases
A group of very unreactive non-metals

Figure 3.21 *Periodic table highlighting the alkali metals (group 1), the halogens (group 7) and the noble gases (group 0).*

Group 1: the alkali metals

All the elements in group 1:

- are soft metals which are shiny when cut but quickly go dull in air
- react quickly with cold water and burn with coloured flames
- are so reactive they have to be stored under oil.

Figure 3.22 *Sodium metal is so soft it can be cut with a knife.*

Figure 3.23 *Iodine reacts violently with powdered aluminium.*

Group 7: the halogens

All the elements in group 7:

- are non-metals
- react quickly with metals.

? Did you know ...?

Halogen bulbs contain very small amounts of bromine and iodine gas. These bulbs appear brighter than ordinary bulbs because the halogens allow higher temperatures to be used without shortening the lifespan of the tungsten filament inside. They're also more efficient than ordinary bulbs, producing more light from the same amount of power.

Figure 3.24 *Halogen bulbs are so bright they are used in floodlights.*

Figure 3.25 *Helium-filled balloons rise because helium is less dense (lighter) than air.*

Group 0: the noble gases

All the elements in group 0:

- are non-metal gases
- are very unreactive.

? Did you know ...?

Around 30 000 tonnes of helium per year are used in a number of ways, from MRI body scanners to finding leaks in gas pipelines. However, the world's supply of helium is running out. Helium is a natural resource and there is not a lot of it. Once used it can't be recycled because it is so light that when it is released into the atmosphere it escapes into space.

In 2016 scientists based in the UK found what has been described as the world's biggest helium gas deposit, in the East Africa's Great Rift Valley in Tanzania. It has been found in gas bubbling to the surface in natural hot water (geothermal) pools. The gas is mainly nitrogen, but it contains around 10% of helium. It has been estimated there is enough helium there to meet world demand for 7 years.

Figure 3.26 *The helium discovered in the East Africa's Great Rift Valley could meet global demand for 7 years.*

GO! Activity 3.5

☺ **1.** The elements are listed in an ordered way in a special chart.

   (a) What name is given to this chart?

   (b) In each sentence select the correct word highlighted in bold:

   (i) Metals appear on the **left / right** of the chart and non-metals on the **left / right**.

   (ii) Elements with similar chemical properties are arranged in the same **group / row** in the chart.

2. The diagram shows a set-up used to test if a substance is an electrical conductor.

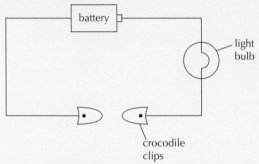

   (a) Describe what you would do to test a substance to see if it is an electrical conductor.

   (b) How would you know if a substance is an electrical conductor?

   (c) Which of the following elements would be electrical conductors?

   > 🔍 **Hint**
   >
   > Look at the periodic table on pages 34 and 35.

   (i) Silver

   (ii) Sulfur

   (iii) Tin

   (iv) Calcium

   (v) Silicon

   (vi) Zinc

   (vii) Nickel

   (d) What do the elements you listed in part **(c)** have in common?

   (e) Give the symbols for the elements listed in part **(c)**.

   (f) Graphite is a form of carbon that conducts electricity. Why would you not expect graphite to conduct electricity?

(continued)

3. The table shows the properties of elements X, Y and Z.

Element	Properties		
	Conductor?	State?	Reactive?
X	non-conductor of electricity	liquid	very reactive with metals
Y	conductor of electricity	solid	very reactive with water
Z	non-conductor of electricity	gas	very unreactive

(a) (i) Which of the elements is a metal?

(ii) Explain your answer to part (a) (i).

(b) Which element belongs to:

(i) Group 1 (ii) Group 7 (iii) Group 0?

(c) Give another name for the elements in:

(i) Group 1 (ii) Group 7 (iii) Group 0.

(d) Name element X and give its symbol.

4. Helium is used in balloons because it is less dense (lighter) than air. Suggest another reason for using helium.

📖 Word bank

- **Byproduct**

When the process of making one thing results in a second product as well, the second product is called a byproduct.

☀ Chemistry in action: Metals in danger of running out

Figure 3.27 *Some rare metals are essential for use in smartphones.*

You might not realise it, but almost everywhere around you are rare metals, many of which are in short supply. In your phone, computer or any other LCD (liquid crystal display) screen you will find **indium**. **Gallium**, which can give out light from a shock of electricity, is used in LEDs (light-emitting diodes), lasers and the solar industry. **Rhenium**, one of the rarest elements in Earth's crust, is most commonly needed in jet engines.

In our daily lives, we rely on many metals that are uncommon. Often they are found only in places such as China, Bolivia or the Democratic Republic of Congo. This can mean that we can't always rely on being able to get them. Some elements, such as **arsenic** and **selenium**, can't even be mined alone; they are usually the **byproduct** of other mining processes.

Supply restrictions could affect the availability of metals such as **chromium** and **niobium**, which go into forming important types of steel, and **tungsten** and **molybdenum**, which are used for high-temperature alloys.

Some scientists think there is a need for greater electronics recycling programs. The more these metals are reused, the lower will be the demand for fresh mining. Product designers need to spend more time thinking about what happens after their products are no longer being used.

GO! Activity 3.6

☺ 1. Read the section 'Metals in danger of running out'. List the metals shown in **bold** in a table, alongside their symbol and atomic number. The first one has been done for you. Use the periodic table on pages 34 and 35 to help you.

Element	Symbol	Atomic number
indium	In	49

2. **(a)** Give **two** reasons why we could possibly run out of some important metals.

   (b) State **one** way in which we could help ensure a supply of these metals.

Chemical formulae of compounds

National 3
Curriculum level 3

Materials: Properties and uses of substances
SCN 3-15b

Learning intentions

In this section you will:

- work out chemical formulae of two-element compounds
- write chemical formulae from names of compounds where prefixes are present.

The symbols of elements can be used to make a chemical formula for a compound. A chemical formula uses chemical symbols and numbers as a shorthand way of showing which elements the compound is made of.

The atoms of each element can be thought of as having 'bonding arms', which can be used to work out how many atoms of each element combine. The number of bonding arms depends on which group in the periodic table the element is in. This is shown in Table 3.4.

Table 3.4 *The number of bonding arms an atom has depends on which group the element is in.*

Group	1	2	3	4	5	6	7	0
Number of bonding arms	1	2	3	4	3	2	1	0

★ You need to know

The number 1 is not used in formulae because the fact that the symbol is written means there must be at least one atom of that element in the formula.

Example 3.1: hydrogen chloride

Bonding arms:

hydrogen: group 1, so 1 bonding arm

chlorine: group 7, so 1 bonding arm

The bonding arms join with each other, as shown in the diagram below.

, written as HCl

Example 3.2: magnesium oxide

Bonding arms:

magnesium: group 2, so 2 bonding arms

oxygen: group 6, so 2 bonding arms

The two arms from the magnesium join with the two arms of the oxygen.

, written as MgO

Example 3.3: sodium sulfide

Bonding arms:

sodium: group 1, so 1 bonding arm

sulfur: group 6, so 2 bonding arms

The two bonding arms of the sulfur need two arms to join with, so the sulfur joins with two sodium atoms.

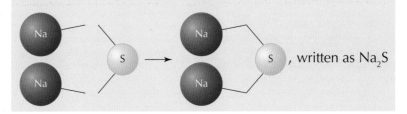

, written as Na_2S

| N3 | L3 | N4 | L4 |

Example 3.4: calcium bromide

Bonding arms:

calcium: group 2, so 2 bonding arms

bromine: group 7, so 1 bonding arm

The two bonding arms of the calcium need two arms to join with, so the calcium joins with two bromine atoms.

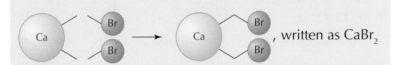, written as $CaBr_2$

The names of some compounds tell us the number of atoms of each element present. They have a prefix at the beginning of the name which tells you how many atoms there are of a particular element. Table 3.5 shows some common prefixes and the number each represents.

Table 3.5 *Common prefixes and the numbers they represent.*

Prefix	Number
mono-	1
di-	2
tri-	3
tetra-	4
penta-	5

Example 3.5
(a) Carbon **mono**xide

mono = **1**, so one carbon is joined to one oxygen; the formula is **CO**

(b) Carbon **di**oxide

di = **2**, so one carbon is joined to two oxygens; the formula is CO_2

(c) Phosphorus **tri**chloride

tri = **3**, so one phosphorus is joined to three chlorines; the formula is PCl_3

GO! Activity 3.7

☺ **1.** Use the periodic table on pages 34 and 35 and information in Table 3.4 to draw bonding arm diagrams to work out the formulae of the following compounds.

  (a) Potassium iodide   **(b)** Barium sulfide

  (c) Magnesium chloride   **(d)** Aluminium bromide

2. Write formulae for the following compounds.

  (a) Nitrogen monoxide   **(b)** Sulfur dioxide

  (c) Sulfur trioxide   **(d)** Carbon tetrachloride

  (e) Phosphorus pentachloride

Learning checklist

After reading this chapter and completing the activities, I can:

N3 L3 N4 L4

- state that the three states of matter are solid, liquid and gas. **Activity 3.1 Q1** ○ ○ ○

- state that when a solute dissolves in a solvent a solution is formed. **Activity 3.1 Q2(a)** ○ ○ ○

- state that a solution in which no more solute can dissolve is described as a saturated solution. **Activity 3.1 Q2(b)** ○ ○ ○

- identify elements because they are made up of the same kind of atoms. **Activity 3.2 Q1** ○ ○ ○

- identify compounds because they are formed when the atoms of different elements join. **Activity 3.2 Q1** ○ ○ ○

- state that mixtures are not chemically joined and can easily be separated. **Activity 3.2 Q1** ○ ○ ○

- identify mixtures, elements and compounds from atom diagrams. **Activity 3.2 Q2** ○ ○ ○

- state that most compounds are not easily broken down into their elements. **Activity 3.2 Q1** ○ ○ ○

- state that when they are dissolved in water some compounds can be broken down into their elements by passing electricity through the solution. **Activity 3.2 Q1** ○ ○ ○

- state that the names of compounds made from two elements come from the names of the elements and end in -ide. **Activity 3.2 Q3** ○ ○ ○

- write word equations to show reactants forming products. **Activity 3.2 Q4, Q5** ○ ○ ○

- state that solids can be separated from liquids or solutions by filtering. **Activity 3.3 Q1(a)(i)** ○ ○ ○

- state that water can be separated from a solution by evaporation. **Activity 3.3 Q1(a)(ii)** ○ ○ ○

- state that mixtures of liquids can be separated by distillation. **Activity 3.3 Q1(b)** ○ ○ ○

- state that the boiling point is the temperature at which a liquid becomes a gas or a gas becomes a liquid. **Activity 3.3 Q1(b)** ○ ○ ○

N3 L3 N4 L4

- give examples of everyday methods of separating mixtures such as making filter coffee, filter beds in reservoirs, whisky distillation, distillation of crude oil and desalination of seawater. **Activity 3.4 Q1**

- state that elements are arranged in an ordered way in the periodic table and they each have their own symbol. **Activity 3.5 Q1(a), Q2(e), Q3(d)**

- state that metal elements appear on the left of the periodic table and non-metals on the right. **Activity 3.5 Q1(b)(i)**

- state that elements with similar chemical properties are in the same group in the periodic table. **Activity 3.5 Q1(b)(ii)**

- state that group 1 elements are known as the alkali metals and are very reactive. **Activity 3.5 Q3(b)(i), (c)(i)**

- state that group 7 elements are known as the halogens and are very reactive with metals. **Activity 3.5 Q3(b)(ii), (c)(ii)**

- state that group 0 elements are known as the noble gases and are very unreactive. **Activity 3.5 Q3(b)(iii), (c)(iii)**

- state that metals are good conductors of electricity. **Activity 3.5 Q2(c), (d), Q3(a)**

- state that carbon, in the form of graphite, is the only non-metal conductor of electricity. **Activity 3.5 Q2(f)**

- identify metals as conductors of electricity because the bulb in an electrical circuit lights up when they are tested. **Activity 3.5 Q2(a), (b)**

- work out the formulae of two-element compounds by using the 'bonding arms' method. **Activity 3.7 Q1**

- work out chemical formulae for compounds with prefixes in their name, such as carbon monoxide, CO. **Activity 3.7 Q2**

- *present information in the form of a table.* **Activity 3.4 Q2(a), Activity 3.6 Q1**

- *interpret information presented in a graph.* **Activity 3.4 Q2(a)**

- *extract scientific information from a report.* **Activity 3.6 Q2**

- *apply my scientific knowledge in an unfamiliar situation.* **Activity 3.3 Q2(b), (c), Activity 3.5 Q4**

- *obtain information from tables of data.* **Activity 3.3 Q2(a)**

- *work with others to plan and carry out a practical activity.* **Activity 3.3 Q3**

4 Atomic structure and bonding related to properties of materials

This chapter includes coverage of:

N4 Atomic structure and bonding related to properties of materials • Materials SCN 4-15a

You should already know:

- what an element is
- about the periodic table and how it is set out
- that elements in the same group have similar chemical properties
- that metals are on the left-hand side of the periodic table and non-metals on the right
- how to workout chemical formulae of two-element compounds using 'bonding arms'
- how to write chemical formulae from names of compounds where prefixes are present.

National 4

Curriculum level 4

Materials: Properties and uses of substances SCN 4-15a

Atomic structure

Learning intentions

In this section you will:

- name the three particles that make up an atom
- state the mass and charge of each particle in an atom
- draw a diagram showing the position of the particles in an atom
- explain why an atom has a neutral charge
- work out atomic number and mass number of an atom
- draw a diagram to show how electrons are arranged in an atom.

📖 Word bank

- **Atom**

An atom is the smallest, most basic unit of an element, containing particles called protons, neutrons and electrons.

All substances are made up of atoms. **Atoms** contain particles called protons (p), neutrons (n) and electrons (e⁻). Figure 4.1 shows how these particles are arranged in atoms. The left-hand diagram gives the idea that atoms are not flat and electrons are moving. The right-hand diagram is called a 'target diagram' and shows the electrons arranged in shells, like layers in an onion.

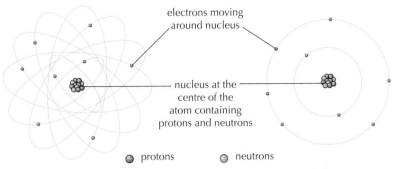

Figure 4.1 *The arrangement of protons (p), neutrons (n) and electrons (e⁻) in an atom.*

Each particle has a charge and mass, as shown in Table 4.1.

Table 4.1 *Charge and mass of particles.*

Particle	Charge	Mass
proton (p)	one positive (+)	1
electron (e⁻)	one negative (–)	almost zero
neutron (n)	no charge (0)	1

The number of protons in an atom identifies which element the atom belongs to. The number of protons is known as the **atomic number**. The elements are arranged in order of increasing atomic number in the periodic table (see the periodic table on pages 34 and 35).

In an atom the number of protons is the same as the number of electrons. The positive charge of the protons is balanced by the negative charge of the electrons so, overall, an atom is neutral (neutrons have no charge).

If the atomic number is known, the element can be identified and the number of protons and electrons in each atom of an element can be worked out.

> ### 📖 Word bank
>
> • **Atomic number**
> The atomic number of an element is given by the number of protons in an atom of that element. Atomic number is unique to each element.

Example 4.1

An element has an atomic number of 11. Using the periodic table, the element can be identified as sodium. Each sodium atom must have 11 protons and 11 electrons.

If the name of the element is known, then the atomic number can be found in the periodic table, and this gives the number of protons and electrons.

Example 4.2

Fluorine has an atomic number of 9, so a fluorine atom has nine protons and nine electrons.

If the number of protons in an atom is known, the atomic number of the element will be the same, and the name of the element can be found from the periodic table.

Example 4.3

An atom has 18 protons. Its atomic number must be 18, and so the element is argon.

The mass of an atom is due to the mass of the protons and neutrons. The mass of the protons added to the mass of the neutrons is known as the **mass number**. The electrons are so light that they don't affect the mass.

mass number = number of protons + number of neutrons

The number of neutrons can be calculated by subtracting the atomic number from the mass number.

number of neutrons = mass number – atomic number

> 📖 **Word bank**
>
> • **Mass number**
> The mass number of an atom is the number of protons and neutrons added together.

Example 4.4

An atom of lithium has 4 neutrons. Calculate its mass number.

From the periodic table, the atomic number of lithium is 3. The atom must have 3 protons.

mass number = number of protons + number of neutrons

$$= \quad 3 \quad + \quad 4$$
$$= \quad 7$$

GO! Activity 4.1

☺ **1. (a)** Copy and complete the following paragraph. You may wish to use the information in Table 4.1 to help you.

The atom is made up of **(a)**_____ (p), neutrons **(b)**_____ and **(c)**_____ (e⁻).
The protons and **(d)**_____ are found in the **(e)**_____ at the centre of the atom.
The **(f)**_____ move around the nucleus. Protons have a **(g)**_____ charge and a
mass of **(h)**_____. Neutrons have **(i)**_____ charge and a mass of 1. Electrons
have a **(j)**_____ charge and a mass of almost **(k)**_____.

(b) Draw a 'target diagram' to show the arrangement of the particles that make up a hydrogen atom with 1 proton, 2 neutrons and 1 electron. Label your diagram clearly.

2. (a) The table below gives information about the atoms of three elements. Use the periodic table on pages 34 and 35 to help you complete entries **(a)–(o)**.

Element	Symbol	Atomic number	Protons	Electrons	Neutrons	Mass number
lithium	(a)	(b)	(c)	(d)	4	(e)
(f)	(g)	(h)	17	(i)	(j)	35
(k)	(l)	10	(m)	(n)	11	(o)

(b) Explain why atoms have a neutral charge, even although they are made up of positive and negative particles.

| N3 | L3 | **N4** | L4 |

The story of the atom

Three British scientists and a New Zealander working in Britain played key roles in developing the idea of atoms and discovering the particles that make up the atom.

1809: John Dalton first used the word **atom** to describe the particle which everything is made of.

1897: Joseph John Thomson discovered a very light negatively charged particle, which was called the **electron**.

John Dalton

JJ Thomson

1909: Ernest Rutherford, born in New Zealand but with Scottish parents, put forward the idea of the existence of positive particles, which are much heavier than electrons. These particles were called **protons**.

1911: Ernest Rutherford explained how protons and electrons are arranged in an atom. Protons are in the centre of the atom, called the **nucleus**. The electrons move around the nucleus. At this time the neutron had not been discovered, but the idea of its existence had been put forward.

1932: James Chadwick discovered the **neutron** in the nucleus of the atom.

Ernest Rutherford

James Chadwick

National 4

Curriculum level 4

Materials: Properties and uses of substances SCN 4-15a

Bonding

Learning intentions

In this section you will:

- find out what a molecule is
- give examples of elements and compounds which exist as molecules
- make models of molecules and draw structures based on the models
- describe how a covalent bond is formed between non-metal atoms
- draw diagrams to show how covalent bonds are formed
- describe how ions are formed
- describe how an ionic bond is formed.

📖 Word bank

- **Molecule**
A molecule is a group of two or more non-metal atoms joined together.

- **Diatomic molecule**
A diatomic molecule has only two non-metal atoms joined together.

Molecules

The noble gases are the only elements to exist as individual atoms. All the other elements exist as structures in which the atoms are joined in some way. The non-metal elements (except the noble gases) exist as different-sized groupings of atoms. The smaller groups of atoms are called **molecules**. The smallest molecules have just two atoms joined together. They are called **diatomic molecules**. Table 4.2 shows the diatomic elements and their states at room temperature.

Table 4.2 *The state of some diatomic molecules at room temperature.*

Diatomic molecule	State
hydrogen	gas
nitrogen	gas
oxygen	gas
fluorine ⎤ chlorine ⎬ halogens bromine ⎮ iodine ⎦	gas gas liquid solid

📖 Word bank

- **Compound**
A compound is formed when the atoms of two or more elements join together.

Other non-metal elements form larger molecules. Phosphorous has four atoms in each molecule and sulfur eight. Both are solids at room temperature.

When the atoms of two or more elements join together, a **compound** is formed.

When the elements are non-metals, molecules are again formed.

You can make 'ball and stick' models of molecular elements and compounds, and draw diagrams of the structures. Table 4.3 shows diagrams of 'ball and stick' models of some molecular elements and compounds. The symbols of the elements in the molecule are often used instead of a coloured circle.

Table 4.3 *Ball and stick models of some elements and compounds.*

Element	chlorine	phosphorus	sulfur
Ball and stick model	O—O	(model)	(model)
Structural diagram	Cl — Cl	(P diagram)	(S diagram)

Compound	hydrogen chloride	water	nitrogen hydride
Ball and stick model	O—○	(model)	(model)
Structural diagram	H — Cl	(H O H)	(N H H H)

Activity 4.2

☺ **1.** Use the words in the following list to help you complete the summary paragraph.

diatomic   non-metal   water

Molecules are formed when atoms of **(a)** _____ elements join. Group 7 (halogens), hydrogen, nitrogen and oxygen are examples of elements which exist as **(b)** _____ molecules. **(c)**_____ is an example of a compound which exists as molecules.

☺☺**2.** *You will need: a molecular model kit or computer program.*

Use a molecular model kit or computer program to make models of:

(a) the molecules shown in Table 4.3

(b) (i) fluorine (two fluorine atoms)

  (ii) phosphorus trichloride (one phosphorous and three chlorines)

(c) Draw diagrams of the models you made in part **(b)**, in the same style as in Table 4.3.

Covalent bonding

The atoms in molecules are held together by forces of attraction called **covalent bonds**. In a covalent bond two atoms share a pair of electrons. They do this to obtain the same electron arrangement as the closest noble gas. Noble gases are extremely stable (unreactive) because they have a full outer shell of electrons.

📖 Word bank

• **Covalent bond**

A covalent bond is formed when non-metal atoms share a pair of electrons.

N3	L3	N4	L4

The sharing of a pair of electrons in a molecule of fluorine and a molecule of hydrogen chloride is shown in Figures 4.2 and 4.3, respectively. These are known as 'dot and cross' diagrams – the dots and crosses represent electrons. Only the electrons in the outer shell are shown. This is the same as the group number in the periodic table. For example, fluorine is in group 7 so has seven electrons in its outer energy shell. Hydrogen is in group 1 so has one electron in its outer shell.

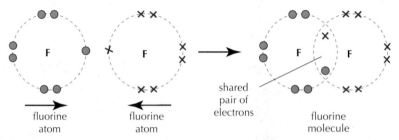

Figure 4.2 *Two fluorine atoms share a pair of electrons to form a covalent bond.*

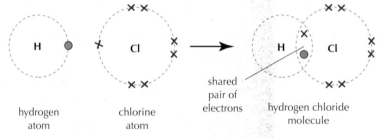

Figure 4.3 *A hydrogen and a chlorine atom share a pair of electrons to form a covalent bond.*

The covalent bond is often shown as a line between the symbols of the elements in the molecule. This is shown in Table 4.3.

★ You need to know

Atoms of non-metal elements react to form covalent bonds.

Covalent bonds exist between non-metal atoms in elements and compounds.

Covalent substances exist as molecules.

GO! Activity 4.3

☺ 1. (a) Draw dot and cross diagrams to show how the electrons are shared when:

    (i) two chlorine atoms join to form a chlorine molecule

    (ii) a hydrogen and a fluorine atom join to form a hydrogen fluoride molecule.

  (b) Draw a chlorine molecule and a hydrogen chloride molecule using a line between the symbol for each element to represent the bond.

  (c) Name the type of bonding formed between non-metal atoms.

🔍 Hint

Remember! The number of electrons in the outer shell of an atom of an element is the same as the group number of the element in the periodic table.

Ions and ionic bonding

When metal atoms react with non-metal atoms, the metal atom transfers electrons to the non-metal atom.

When a metal atom loses an electron it becomes positively charged. The non-metal atom that gains the electron becomes negatively charged. Charged atoms are called **ions**.

> 📖 **Word bank**
>
> • **Ion**
>
> An ion is a charged atom. It can be positive or negative.

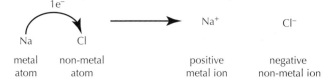

Na	Cl		Na$^+$	Cl$^-$
metal atom	non-metal atom		positive metal ion	negative non-metal ion

The attraction between the positive and negative ions results in the formation of an **ionic bond**. The ions attract each other in all directions, and this results in the formation of an ionic crystal structure. This is shown in Figure 4.4.

> 📖 **Word bank**
>
> • **Ionic bond**
>
> An ionic bond is formed when positive and negative ions attract each other.

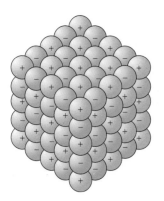

Figure 4.4 *A model showing how the ions in sodium chloride are arranged.*

Figure 4.5 *Sodium chloride crystals are cubic shaped.*

Table 4.4 gives more examples of ion formation and shows the change in the number of electrons in the ion compared with the atom.

Table 4.4 *Ion formation of the atoms of some elements.*

Element	Atom symbol	Number of electrons in atom	Electrons lost or gained	Ion symbol	Number of electrons in ion
calcium	Ca	20	loses 2e$^-$	Ca^{2+}	18
sulfur	S	16	gains 2e$^-$	S^{2-}	18
aluminium	Al	13	loses 3e$^-$	Al^{3+}	10

> ⭐ **You need to know**
>
> Metal atoms lose electrons and form positive ions.
>
> Non-metal atoms gain electrons and form negative ions.
>
> Positive metal ions attract negative non-metal ions to form ionic bonds.
>
> Ionic compounds exist as ionic crystal structures.
>
> There are **no** ionic molecules.

😊 Activity 4.4

1. The table below gives information about the atoms and ions of two elements. Use the information in Table 4.4 to help you complete entries **(a)–(h)**.

Element	Atom symbol	Number of electrons in atom	Electrons lost or gained	Ion symbol	Number of electrons in ion
potassium	**(a)**	**(b)**	**(c)**	K⁺	**(d)**
(e)	O	**(f)**	gains 2e⁻	**(g)**	**(h)**

2. When lithium reacts with chlorine, lithium chloride is formed.

(a) What type of bond is formed between lithium and chlorine?

(b) With the aid of a diagram explain how the bond between lithium and chlorine is formed.

3. What type of bonding is found in the following compounds?

(a) Potassium bromide (KBr)

(b) Calcium chloride ($CaCl_2$)

(c) Nitrogen iodide (NI_3)

(d) Water (H_2O)

National 4
Curriculum level 4
Materials: Properties and uses of substances SCN 4-15a

Properties of covalent and ionic substances

Learning intentions

In this section you will:

- identify a substance as being covalent molecular or ionic, from its melting point and boiling point
- use melting point and boiling point to predict the state of a substance at room temperature
- describe how to test the electrical conductivity of a substance
- work out the type of bonding in a compound from the results of conductivity testing.

★ You need to know

Melting point is the temperature at which a substance changes from a solid to a liquid or from a liquid to a solid.

Boiling point is the temperature at which a substance changes from liquid to gas or from gas to liquid.

Melting points and boiling points

Elements and compounds made up from covalently bonded molecules have low melting points and boiling points. They can exist as gases, liquids and solids at room temperature (20 °C). Table 4.5 gives some examples.

Table 4.5 *The melting point and boiling point of some substances.*

Substance	Melting point (°C)	Boiling point (°C)	State at room temperature
chlorine	–101	–34	gas
bromine	–7	59	liquid
sulfur	115	445	solid
ammonia	–78	–33	gas
water	0	100	liquid
salol	42	173	solid

Figure 4.6 *The halogens exist as covalent molecules. Chlorine is a green gas, bromine is a brown liquid and iodine is a purple solid.*

A substance will be solid at temperatures lower than the melting point, liquid at temperatures between the melting and boiling points, and gas at temperatures higher than the boiling point. Figure 4.7 shows the different states of water at different temperatures.

Water is a liquid between 0°C and 100°C.

Temperature (°C)

0   100

Ice is the solid form of water below 0°C.

Water exists as a gas above 100°C.

Figure 4.7 *Water exists as a solid, a liquid or a gas, depending on the temperature.*

Ions are only found in compounds formed between metals and non-metals. Ionic compounds have very high melting and boiling points so they exist only as solids at room temperature (20 °C). Table 4.6 gives some examples.

Table 4.6 *The melting point and boiling point of some compounds.*

Name of compound	Melting point (°C)	Boiling point (°C)
barium chloride	961	1560
calcium oxide	2614	2850
lithium bromide	550	1265
magnesium chloride	714	1412
potassium iodide	681	1323
sodium chloride	801	1465

Many ionic compounds, like copper sulfate, are highly coloured.

Figure 4.8   *Copper sulfate is a highly coloured ionic compound.*

Conductivity

Substances that allow electricity to pass through them are called **conductors**. Those that do not let electricity pass through them are called **insulators**. A simple conduction tester circuit is shown in Figure 4.9.

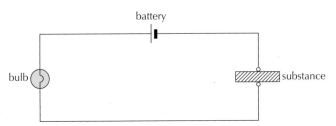

Figure 4.9   *A simple circuit to test conduction.*

If the bulb lights when a substance is connected into the circuit, then the substance is a conductor. If the bulb doesn't light, then the substance is an insulator.

Table 4.7 shows the results of testing a number of substances in different states.

Table 4.7   *Conductivity of some substances.*

Substances that don't conduct in any state	Substances that only conduct when liquid or in solution
bromine	sodium chloride
sulfur	lithium bromide
ammonia	potassium iodide
salol	magnesium chloride

The substances in Table 4.7 that don't conduct in any state are all covalent. The substances that only conduct when dissolved in water or when liquid are ionic.

Bonding and properties summary

Table 4.8 summarises the properties of ionic and covalent substances.

Table 4.8 *Properties of ionic and covalent substances.*

	Type of bond	
	Ionic bond	**Covalent bond**
Found in	• compounds of metals and non-metals	• elements and compounds made up of non-metals
How formed	• metals lose electrons and non-metals gain electrons to form ions	• pairs of electrons in outer shells are shared between atoms
Structure	• all have an ionic crystalline structure • oppositely charged ions attract each other	• most have a covalent molecular structure
Properties	• conduct electricity when liquid and in solution • non-conductors when solid • high melting point and boiling point; all solids at room temperature	• non-conductors of electricity in any state • covalent molecules have low melting point and boiling point
Examples	• sodium chloride, calcium oxide	• fluorine, hydrogen chloride

GO! Activity 4.5

☺ **1.** Copy the table and complete the last column, identifying **(a)**–**(f)**.

Substance	Melting point (°C)	Boiling point (°C)	State at room temperature (20°C)
chlorine	−101	−34	**(a)**
bromine	−7	59	**(b)**
sulfur	115	445	**(c)**
ammonia	−78	−33	**(d)**
calcium oxide	2614	2850	**(e)**
salol	42	173	**(f)**

(continued)

☺☺**2.** The table shows the properties of two substances, A and B.

Substance	Melting point	Boiling point
A	low	low
B	high	high

(a) Use the information in the table to state what kind of bonding is likely to be present in each of A and B.

(b) A group of students decided to test the conductivity of substances A and B to confirm the type of bonding in each. They decided to test the substances as solids, liquids and in solution.

Suggest why they tested the substances in different states.

3. Solid calcium chloride was found to be an electrical insulator when tested.

A student concluded that calcium chloride must be covalent.

Comment on her conclusion.

National 4

Curriculum level 4

Materials: Properties and uses of substances SCN 4-15a

★ You need to know

You need to know how to write formulae from the names of compounds where prefixes are present (page 43).

★ Memory aid

A memory aid to help you remember the names and formulae of the diatomic elements is:

I	I_2
Brought	Br_2
Clay	Cl_2
For	F_2
Our	O_2
New	N_2
Home	H_2

Formulae of elements and compounds

Learning intentions

In this section you will:

- work out the formula of an element
- use valency rules to work out the formula of a compound
- work out the formula of a compound from models and structures
- find the relative atomic mass (RAM) of an element
- use RAM to calculate the formula mass of a substance from its formula
- write word and formula equations
- write formula equations with state symbols.

Elements

Most elements have their symbol as their formula, e.g. sodium: Na; copper: Cu; sulfur: S. The exceptions are the diatomic molecules (see page 50): H_2, N_2, O_2, F_2, Cl_2, Br_2 and I_2.

Compounds

You should already know that the names of two-element compounds end in **-ide**. (See 'Naming compounds' in Chapter 3, page 25.)

| N3 | L3 | **N4** | **L4** |

Compounds with names ending in **-ate** and **-ite** contain more than two elements, one of which is oxygen. Table 4.9 gives some examples.

Table 4.9 *The names of some compounds made from atoms of more than two elements.*

Name of compound	Name of elements
sodium sulfite	sodium, sulfur and oxygen
calcium nitrite	calcium, nitrogen and oxygen
magnesium carbonate	magnesium, carbon and oxygen
copper sulfate	copper, sulfur and oxygen
potassium nitrate	potassium, nitrogen and oxygen

GO! Activity 4.6

☺ **1.** Write down the chemical formulae for these elements:

(a) magnesium

(b) phosphorus

(c) hydrogen

(d) oxygen

2. Write down the elements present in these compounds:

(a) sodium nitrite

(b) aluminium sulfate

(c) calcium carbonate

(d) potassium nitrate

(e) magnesium sulfite

(f) lithium hydrogencarbonate

Working out formulae of compounds

You should already know that the idea of 'bonding arms' can be used to work out chemical formulae for compounds (see pages 41–43), and also that chemical formulae can sometimes be worked out from the name of the compound (see page 43).

Using valency

A chemical formula is a shorthand way of showing which elements are present in a compound and the ratio in which they exist. Element symbols and numbers are used.

Valency rules can be used to work out the formula of many compounds. Valency is another word for 'combining power' and is the same as the number of bonding arms – it tells us how many bonds atoms of one element can form with atoms of another element. The valency of an element depends on its group number in the periodic table and is summarised in Table 4.10.

📖 Word bank

• **Valency**

The valency of an atom of an element indicates how many bonds it can form with atoms of other elements.

Table 4.10 *The valency of an element depends on which group it is in in the periodic table.*

Group 1	Group 2	Group 3	Group 4	Group 5	Group 6	Group 7	Group 0
H							He
Li	Be	B	S	N	O	F	Ne
Na	Mg	Al	Si	P	S	Cl	Ar
K	Ca	Ga	Ge	As	Se	Br	Kr
Valency 1	**Valency 2**	**Valency 3**	**Valency 4**	**Valency 3**	**Valency 2**	**Valency 1**	**Valency 0**

Once the valency of an atom is identified, it is simply a case of swapping the numbers to get the number of atoms of each element in the compound. Table 4.11 shows how the chemical formulae for some two-element compounds are worked out using valency rules. The term 'formula ratio' used in the table indicates the proportion of one element compared with the other in the compound.

Table 4.11 *Examples of how valency rules are used to work out the chemical formulae of compounds.*

Elements	Mg	Br	H	Cl	B	H
Group	2	7	1	7	3	1
Valency	2	1	1	1	3	1
Formula ratio	1	2	1	1	1	3
Formula	$MgBr_2$		HCl		BH_3	
Name	magnesium bromide		hydrogen chloride		boron hydride	

If the valency of each element is the same then the numbers cancel down. Table 4.12 shows this when working out the formula for calcium oxide.

Table 4.12 *Cancelling down when elements have the same valency.*

Elements	Ca	O
Group	2	6
Valency	2	2
Formula ratio	1 2	2 1
Formula	CaO	
Name	calcium oxide	

If the valency of each element is not the same but can be divided by the same number then the numbers cancel down. Table 4.13 shows this when working out the formula for magnesium carbide.

Table 4.13 *Cancelling down when the valencies are different but can be divided by the same number.*

Elements	Mg	C
Group	2	4
Valency	2	4
Formula ratio	2 4	2 1
Formula	Mg_2C	
Name	magnesium carbide	

Using models and structural formulae

Chemical formulae can be worked out from molecular models. Simply identify the elements in the model and how many atoms of each element there are. Figure 4.10 gives some examples.

(a)

H_2O

(b)

NH_3

(c)

CO_2

(d)

CCl_4

Figure 4.10 *Chemical formulae of compounds can be worked out from their molecular models.*

Structural formulae show how atoms are bonded to each other. Chemical formulae can be worked out in the same way as for molecular models. Some examples are shown in Figure 4.11.

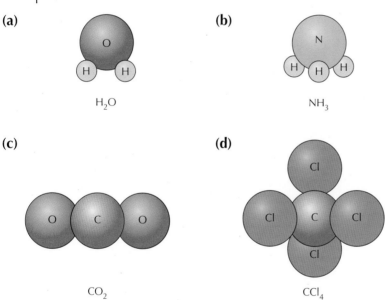

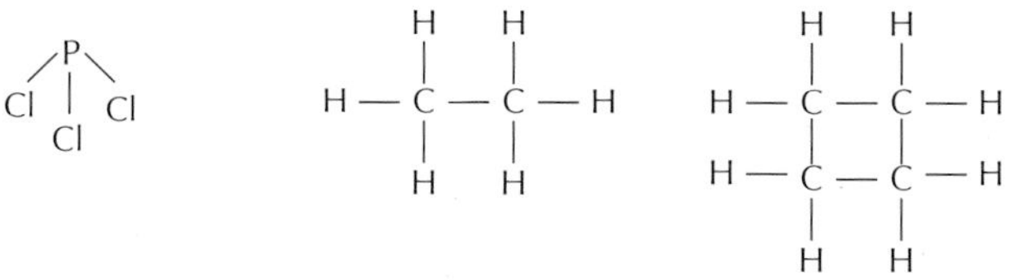

PCl_3 C_2H_6 C_4H_8

Figure 4.11 *Chemical formulae can be worked out from structural formulae.*

GO! Activity 4.7

☺ **1.** Use the valency rules to work out the formula of each compound.

    (a) Sodium nitride

    (b) Aluminium oxide

    (c) Calcium sulfide

    (d) Boron phosphide

2. Work out the chemical formulae from the molecular models and structures shown.

    (a)

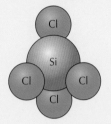

    (b)

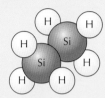

    (c)

$$H-\overset{\underset{\displaystyle H}{\displaystyle |}}{\overset{\displaystyle |}{\underset{\displaystyle H}{C}}}-\overset{\underset{\displaystyle H}{\displaystyle |}}{\overset{\displaystyle |}{\underset{\displaystyle H}{C}}}-\overset{\underset{\displaystyle H}{\displaystyle |}}{\overset{\displaystyle |}{\underset{\displaystyle H}{C}}}-\overset{\underset{\displaystyle H}{\displaystyle |}}{\overset{\displaystyle |}{\underset{\displaystyle H}{C}}}-H$$

    (d)

H, H on top, C, then H—C—C—H, with H below

    (e)

H on top, C, with H, H, H

☺☺**3.** *You will need: a molecular model kit or computer program.*

    (a) Use a molecular model kit or computer program to make models of compounds with the following chemical formulae. Ask your teacher to check your models. Keep your models for part **(b)**.

      (i) SiH_4

      (ii) H_2S

      (iii) NCl_3

      (iv) C_3H_8

    (b) From the models you made in part **(a)**, draw the structure of each compound, using a short line (–) to represent a covalent bond.

Relative atomic mass and formula mass

The **relative atomic mass** (RAM) is the average mass of the element. The mass of an atom is due to the mass of the protons and neutrons – the electrons are so light they do not contribute to the total mass. However, most elements have atoms with different numbers of neutrons, so an average has to be taken.

For example, 75% of chlorine atoms have a mass of 35 and 25% have a mass of 37. This averages out at 35.5. This means the RAM of chlorine is 35.5.

📖 Word bank

• **Relative atomic mass**

Relative atomic mass (RAM) is the average mass of an element.

N3	L3	N4	L4

RAM values have no units and are seldom whole numbers. They are often rounded to the nearest 0.5, as shown in Table 4.14.

Table 4.14 *Relative atomic mass of some elements.*

Element	Symbol	Relative atomic mass
aluminium	Al	27
argon	Ar	40
bromine	Br	80
calcium	Ca	40
carbon	C	12
chlorine	Cl	35.5
copper	Cu	63.5
fluorine	F	19
gold	Au	197
helium	He	4
hydrogen	H	1
iodine	I	127
iron	Fe	56
lead	Pb	207
lithium	Li	7

Element	Symbol	Relative atomic mass
magnesium	Mg	24.5
mercury	Hg	200.5
neon	Ne	20
nickel	Ni	58.5
nitrogen	N	14
oxygen	O	16
phosphorus	P	31
platinum	Pt	195
potassium	K	39
silicon	Si	28
silver	Ag	108
sodium	Na	23
sulfur	S	32
tin	Sn	118.5
zinc	Zn	65.5

The formula mass of a substance can be calculated using the RAM of the elements in the chemical formula of the substance.

Example 4.5

(a) Fluorine

Formula: F_2; formula mass = 19 + 19 = 38

(b) Sodium chloride

Formula: NaCl; formula mass = 23 + 35.5 = 58.5

(c) Carbon dioxide

Formula: CO_2; formula mass = 12 + (2 × 16)

$$= 12 + 32 = 44$$

(d) Aluminium sulfide

Formula: Al_2S_3; formula mass = (2 × 27) + (3 × 32)

$$= 54 + 96 = 150$$

GO! Activity 4.8

☺ 1. Calculate the formula mass of the following substances, using the RAM values in Table 4.14.

(a) Bromine: Br_2

(b) Nitrogen dioxide: NO_2

(c) Magnesium chloride: $MgCl_2$

(d) Aluminium oxide: Al_2O_3

2. The table shows the relative atomic masses and melting points of some group 1 elements.

Element symbol	Relative atomic mass (RAM)	Melting point (°C)
Li	7	180
Na	23	98
K	39	64
Rb	85.5	39

(a) Describe the trend in melting point as the relative atomic mass increases.

(b) Predict the approximate melting point of caesium (RAM 133).

3. Work out the formula mass for each substance from the given formula.

(a) N_2

(b) CaO

(c) HCl

(d) PH_3

(e) MgI_2

Writing chemical equations using formulae

You should already know that a chemical reaction can be summarised in a word equation (see page 25).

A chemical reaction can also be summarised by a formula equation. When doing this, it is important to use the rules for working out formulae.

Example 4.6

Word equation: magnesium + oxygen → magnesium oxide

(element) (diatomic (compound)
element)

Working out the formula for each substance:

Magnesium: most elements have their symbol as their formula (Mg)

Oxygen: one of the seven elements which exist as diatomic molecules (O_2) (see page 50)

Magnesium oxide: a compound, so the valency rules are used:

Elements	Mg O
Group	2 6
Valency	2 ⤫ 2
Formula ratio	1 2 ⤫ 2 1
Formula	MgO
Name	magnesium oxide

Formula equation: Mg   + O_2 →   MgO

State symbols are sometimes added to formula equations:

$$Mg(s) \; + \; O_2(g) \; \rightarrow \; MgO(s)$$

> ★ **You need to know**
>
> • **State symbols**
> Symbols, instead of words, can be used to show the state in which substances exist.
> solid: (s)
> liquid: (ℓ)
> gas: (g)
> solution: (aq)

Example 4.7

Word equation: hydrogen   +   chlorine   →   hydrogen
                (diatomic   (diatomic   chloride
                element)   element)   (compound)

Working out the formula for each substance:

Hydrogen: diatomic molecule (H_2)

Chlorine: diatomic molecule (Cl_2)

Hydrogen chloride: a compound so the valency method is used (see Table 4.11 on page 60).

Formula equation, with state symbols added:

$$H_2(g) \; + \; Cl_2(g) \; \rightarrow \; HCl(g)$$

Example 4.8

Word equation: carbon   +   oxygen   →   carbon
                (element)   (diatomic   monoxide
                              element)   (compound)

Working out the formula for each substance:

Carbon: element (C)

Oxygen: diatomic molecule (O_2)

Carbon monoxide: the valency rules are not needed to work out the formula for carbon monoxide as mono- indicates 'one', i.e. carbon with one oxygen (CO)

Formula equation, with state symbols added:

$$C(s) \; + \; O_2(g) \; \rightarrow \; CO(g)$$

GO! Activity 4.9

☺ 1. Write a formula equation for each of these:

   (a) sodium   +   oxygen   →   sodium oxide
   (b) aluminium   +   chlorine →   aluminium chloride
   (c) nitrogen   +   hydrogen →   nitrogen hydride
   (d) lithium   +   bromine   →   lithium bromide

2. Mercury is the only liquid metal at room temperature. It is a very toxic element. If mercury is spilled in a laboratory, powdered sulfur can be spread over it. The mercury and sulfur react to form solid mercury sulfide.

   (a) Write a word equation for the reaction of mercury with sulfur.

   (b) Write a formula equation for the reaction of mercury with sulfur. (Take the valency of mercury as 2.)

   (c) Add state symbols to the formula equation in part (b).

Learning checklist

After reading this chapter and completing the activities, I can:

N3 | L3 | **N4** | **L4**

| | • state that atoms are made of protons (p), neutrons (n) and electrons (e⁻). **Activity 4.1 Q1(a)** | ○ ○ ○ |

| | • state that protons have a mass of 1 and a positive charge. **Activity 4.1 Q1(a)** | ○ ○ ○ |

| | • state that electrons have practically no mass and a negative charge. **Activity 4.1 Q1(a)** | ○ ○ ○ |

| | • state that neutrons have a mass of 1 and no charge. **Activity 4.1 Q1(a)** | ○ ○ ○ |

| | • state that protons and neutrons are found in the nucleus and the electrons move around the nucleus. **Activity 4.1 Q1(a)** | ○ ○ ○ |

| | • draw a 'target diagram' to show how protons, neutrons and electrons are arranged in an atom. **Activity 4.1 Q1(b)** | ○ ○ ○ |

| | • work out the atomic number of an element using: atomic number = number of protons. **Activity 4.1 Q2(a)** | ○ ○ ○ |

| | • work out the mass number of an atom using: mass number = number of protons + number of neutrons. **Activity 4.1 Q2(a)** | ○ ○ ○ |

N3 | L3 | **N4** | **L4**

- state that an atom has a neutral charge because there are the same number of protons and electrons in an atom and their charges balance each other. **Activity 4.1 Q2(b)**   ○ ○ ○

- state that a molecule is a small grouping of non-metal atoms bonded together. **Activity 4.2 Q1**   ○ ○ ○

- state that the halogens (group 7), hydrogen, nitrogen and oxygen are elements which exist as diatomic molecules. **Activity 4.2 Q1**   ○ ○ ○

- state that water is an example of a compound which exists as molecules. **Activity 4.2 Q1**   ○ ○ ○

- draw the structure of a molecule from its molecular model. **Activity 4.2 Q2**   ○ ○ ○

- state that covalent bonds are formed when non-metal atoms join. **Activity 4.3 Q1(c), Activity 4.4 Q3(c), (d)**   ○ ○ ○

- draw a diagram to show how non-metal atoms share a pair of electrons to form a covalent bond. **Activity 4.3 Q1(a)**   ○ ○ ○

- state that an ionic bond is formed between metals and non-metals. **Activity 4.4 Q2(a), Q3(a), (b)**   ○ ○ ○

- state that positive ions are formed when metal atoms lose electrons and negative ions are formed when non-metal atoms gain electrons. **Activity 4.4 Q1**   ○ ○ ○

- state that positive ions and negative ions attract each other to form an ionic bond. **Activity 4.4 Q2(a)**   ○ ○ ○

- draw a diagram to show how a metal atom transfers electrons to a non-metal atom to form an ionic bond. **Activity 4.4 Q2(b)**   ○ ○ ○

- state that covalent molecular substances have low melting and boiling points. **Activity 4.5 Q2(a)**   ○ ○ ○

- state that ionic compounds have high melting and boiling points. **Activity 4.5 Q2(a)**   ○ ○ ○

N3 L3 **N4 L4**

- use melting and boiling points to predict the state of a substance at room temperature. **Activity 4.5 Q1**
  ○ ○ ○

- state that covalent substances are non-conductors of electricity (insulators) in all states. **Activity 4.5 Q3**
  ○ ○ ○

- state that ionic compounds conduct electricity in solution or when liquid but not in the solid state. **Activity 4.5 Q3**
  ○ ○ ○

- state that elements have their symbol as their formula except for the diatomic molecules – hydrogen (H_2), oxygen (O_2), nitrogen (N_2) and the group 7 elements. **Activity 4.6 Q1**
  ○ ○ ○

- state that a compound with a name ending in -ite or -ate will contain oxygen in addition to other elements. **Activity 4.6 Q2**
  ○ ○ ○

- use valency rules to work out formulae of compounds. **Activity 4.7 Q1**
  ○ ○ ○

- work out chemical formulae from molecular models and drawn structures. **Activity 4.7 Q2, Q3**
  ○ ○ ○

- work out the formula mass of a substance from its formula, using the relative atomic mass (RAM) of the elements. **Activity 4.8 Q1, Q3**
  ○ ○ ○

- write formula equations for chemical reactions. **Activity 4.9 Q1, Q2(b)**
  ○ ○ ○

- include state symbols in formula equations: solid (s); liquid (ℓ); gas (g); solution (aq). **Activity 4.9 Q2(c)**
  ○ ○ ○

- *apply my scientific knowledge to solve a problem.* **Activity 4.5 Q2(b)**
  ○ ○ ○

- *obtain information from tables of data.* **Activity 4.8 Q1, Q3**
  ○ ○ ○

- *make predictions and generalisations.* **Activity 4.8 Q2**
  ○ ○ ○

5 Energy changes of chemical reactions

This chapter includes coverage of:

N4 Energy changes of chemical reactions • Materials SCN 4-19a

You should already know:

- that a change in temperature could indicate that a chemical reaction has taken place.

Exothermic reactions

Learning intentions

In this section you will:

- identify when a chemical reaction results in heat energy being given out
- give examples of reactions which give out energy.

An **exothermic** reaction is one in which energy is given out to the surroundings. We can tell when an exothermic reaction is taking place because there is a temperature rise during the reaction.

Some portable hand-warmers rely on chemical reactions to produce heat. One type involves a concentrated solution of a chemical. In the hand-warmer sachet, there is a metal disc. When this is pressed, it causes the solution to begin to change back into the solid (crystal) form of the chemical. At the same time, heat energy is given out over a few hours. The hand-warmer can be used again by putting it in hot water to dissolve the crystals, then letting the solution cool to room temperature.

Although not very common, you can get cans of food or drink that are self heating. Some of these have cans with two chambers: one chamber contains the food or drink, and the other contains a chemical mixture that reacts and produces heat when the can is activated. The chemicals do not mix with the food.

📖 Word bank

- **Exothermic**
In an exothermic reaction, energy is given out to the surroundings.

Figure 5.1 *Some portable hand-warmers work because heat is given out during an exothermic chemical reaction.*

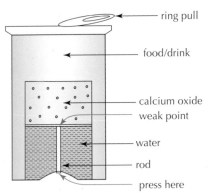

ring pull

food/drink

calcium oxide
weak point

water

rod

press here

Figure 5.2 *Some self-warming drinks work because an exothermic chemical reaction is activated and warms the liquid.*

Make the link

See pages 83–84 for more about neutralisation reactions.

An example of an exothermic reaction that can be carried out in the laboratory is the reaction of hydrochloric acid and sodium hydroxide (an alkali). This is a neutralisation reaction.

Word equation:

hydrochloric + sodium → sodium + water
acid   hydroxide   chloride

Equal volumes of the acid and alkali with the same concentration are used. The temperature of each is measured before they are mixed. The highest temperature reached after the two solutions are mixed is measured. Plastic beakers are used to reduce heat loss to the surroundings.

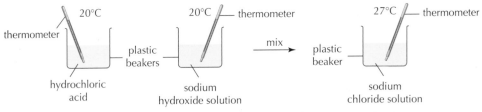

20°C

thermometer

plastic beakers

hydrochloric acid

20°C — thermometer

sodium hydroxide solution

mix

plastic beaker

27°C — thermometer

sodium chloride solution

Figure 5.3 *An increase in temperature during a neutralisation reaction indicates the reaction is exothermic.*

Word bank

• **Combustion**

Combustion is the reaction between oxygen and a fuel such as methane, in which energy is given out.

A flame is another indicator of an exothermic chemical reaction taking place. Everyday examples are natural gas burning in a cooker or a Bunsen burner and petrol burning in a car engine.

Natural gas is mainly methane. When it burns it reacts with oxygen to form carbon dioxide and water. This is known as **combustion**.

Make the link

See Chapter 10 for more about combustion as an example of an exothermic reaction.

Figure 5.4 *The heat needed to fill a hot air balloon comes from burning gas, an exothermic reaction.*

☀ Chemistry in action: Explosions in China

Deadly chemical disaster in Tianjin, China

13 August 2015

Map of China showing the port city of Tianjin.

More than 140 people were killed and hundreds more injured when a number of explosions occurred in the city of Tianjin in northern China. The explosions were so huge that they could be seen from outer space by a Japanese weather satellite. According to Chinese state media, the blasts occurred when a shipment of chemicals blew up in the port city. It is not clear how the first explosion happened but the explosions which followed may have been caused

because the firefighters used water to put out the flames.

One of the first firefighters at the scene said 'When we arrived we immediately started spraying water onto the flames. However, this seemed to make the situation worse.'

Eye-witnesses said it felt like an earthquake was happening. Others thought an atomic bomb had been dropped. Buildings up to a mile away felt the effects of the explosions.

The explosion at the port of Tianjin wrecked buildings and vehicles over a wide area.

The news report above describes a number of explosions (very exothermic reactions).

The explosions described in the report could have been caused by one of the chemicals, calcium carbide, reacting with water to produce calcium hydroxide and a flammable gas called ethyne. The ethyne may then have exploded and also caused another chemical, ammonium nitrate, to explode.

In China, calcium carbide is used to make ethyne, which, in turn, is used to make polyvinyl chloride (PVC), a very useful plastic (see Chapter 20). China has lots of coal, a source of calcium carbide, so it is cheaper for China to produce ethyne from calcium carbide than from imported oil. The pie chart in Figure 5.5 shows the huge amount of ethyne used by China compared with other countries.

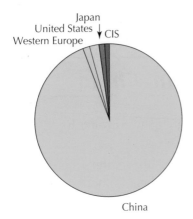

Japan
United States ↓ CIS
Western Europe

China

Figure 5.5 *The pie chart shows the huge amount of ethyne used by China compared to other countries.*

Production of calcium carbide in China has been increasing because the country's steady economic growth has led to strong demand for ethyne to make PVC, mainly for use in the construction industry.

In the USA and Europe consumption of calcium carbide is generally decreasing because oil is used to obtain ethyne. This is because it is easier and more economical in Europe to obtain ethyne from oil.

? Did you know ...?

Most industrial chemical reactions are exothermic. If the heat produced is not controlled there is the risk of a runaway exothermic reaction, also known as 'thermal runaway'. This is when so much heat is produced that there is the risk of explosion. Thermal runaway is thought to be one of the factors that caused one of the world's worst industrial disasters. It happened at a pesticide plant in India. A toxic gas known as MIC was released into the air and blown towards the city of Bhopal. It resulted in thousands of deaths, and severe and permanent disablement among the population.

Figure 5.6 *The memorial in Bhopal for those killed and disabled in the 1984 disaster.*

Endothermic reactions

National 4

Curriculum level 4
Materials: Chemical changes SCN 4-19a

Learning intentions

In this section you will:
* identify when a chemical reaction results in heat energy being taken in from the surroundings
* give examples of reactions which take in energy.

📖 Word bank

* **Endothermic**
In an endothermic reaction, energy is taken in from the surroundings.

An **endothermic** reaction is one in which energy is taken in from the surroundings. There is a drop in temperature during the reaction.

The hand-warmer, described in the section on exothermic reactions (see page 69), can be used again by putting it in hot water to dissolve the crystals. This is an example of an endothermic process.

Some sports injury cold-packs have a small plastic bag containing water; inside this bag there is another bag containing a mixture of solid ammonium nitrate and ammonium chloride. When the bag is squeezed, the inner bag bursts and the chemicals dissolve in the water. Heat energy is taken from the water during the process so its temperature drops. This is an endothermic process.

Photosynthesis in plants is the most important endothermic reaction for life.

Figure 5.7 *The temperature drop in chemical cold-packs is due to an endothermic process.*

Word equation:

$$\text{carbon dioxide} + \text{water} \xrightarrow[\text{chlorophyll}]{\text{light}} \text{glucose} + \text{oxygen}$$

More energy from the sun is taken in by plants than is given out when glucose and oxygen are made. It has been estimated that six times more energy is taken in by plants than is used by the human race in a year.

Endothermic reactions can be carried out in the laboratory. When solid barium hydroxide and ammonium nitrate are mixed in a beaker, the chemical reaction that occurs causes the temperature to fall dramatically.

Figure 5.8 *Photosynthesis is the most important endothermic reaction for life.*

GO! Activity 5.1

☺ **1. (a)** Choose the correct word(s) in bold in each sentence.

   (i) Heat energy is given out during an **exothermic / endothermic** reaction.

   (ii) Heat energy is **taken in from / given out to** the surroundings during an endothermic reaction.

   (b)(i) Give **two** examples of exothermic reactions.

   (ii) Give an example of an endothermic reaction.

2. Three experiments involving solutions were carried out and the results are shown in the table.

Experiment	Temperature of reactants before mixing (°C)	Temperature after mixing (°C)
1	17	21
2	18	11
3	16	18

   (a)(i) State whether each reaction is exothermic or endothermic.

   (ii) Justify each of your answers in part **(a)(i)**.

   (b)(i) Which reaction is the most exothermic?

   (ii) Justify your answer to part **(b)(i)**.

(continued)

Make the link

See Chapter 10 for more about photosynthesis.

3. Experiment

Chemical reactions involve energy changes taking place. Your task is to plan and carry out an experiment to find out whether the reaction between potassium hydroxide solution and nitric acid is exothermic or endothermic.

You can work in a small group, but you must present your findings on your own.

Plan

Your plan should make clear:

- what equipment and chemicals you need (solutions are likely to have been made up for you) and how you are going to carry out the experiment – you may wish to include a diagram
- observations and measurements you will make
- how you will present your results
- how you will be able to conclude whether the reaction is exothermic or endothermic.

You must ask your teacher to check your plan before carrying out any practical work.

You must follow the normal safety procedures when carrying out an experiment.

Report

Your report should include:

- your results and observations
- any conclusions you can make from your results
- statements describing any changes you could make to improve the accuracy of your results.

Make sure you discuss your final report with your teacher.

4. Read the section that includes the newspaper report which describes how an explosion happened in Tianjin, China, in 2015 (pages 71–72).

(a) Write a short report about how the explosions are thought to have occurred.

(b) Explain why China uses so much calcium carbide.

(c) (i) Write a word equation for the reaction of calcium carbide with water.

   (ii) Name the elements present in calcium carbide.

(d) Europe does not use much calcium carbide as a source of ethyne.

   State the main source of ethyne in Europe.

N3   L3   **N4**   **L4**

Learning checklist

After reading this chapter and completing the activities, I can:

N3 L3 **N4 L4**

- state that heat energy is given out to the surroundings during an exothermic reaction. **Activity 5.1 Q1(a)(i)** ○ ○ ○

- identify that an exothermic reaction has taken place because of a rise in temperature. **Activity 5.1 Q2(a), (b)** ○ ○ ○

- state that burning (combustion) and the reaction between an acid and an alkali (neutralisation) are examples of exothermic reactions. **Activity 5.1 Q1(b)(i)** ○ ○ ○

- state that heat energy is taken in from the surroundings during an endothermic reaction. **Activity 5.1 Q1(a)(ii)** ○ ○ ○

- identify that an endothermic reaction has taken place because of a drop in temperature. **Activity 5.1 Q2(a)** ○ ○ ○

- state that photosynthesis is an example of an endothermic reaction. **Activity 5.1 Q1(b)(ii)** ○ ○ ○

- *extract scientific information from a report.* **Activity 5.1 Q4** ○ ○ ○

- *work with others to plan and carry out a practical activity.* **Activity 5.1 Q3** ○ ○ ○

6 Acids and bases 1

This chapter includes coverage of:

N3 Acids and bases • Planet Earth SCN 3-05b • Materials SCN 3-18a

You should already know:

- concentration is a measure of the amount of solute dissolved in a solvent.

National 3

Curriculum level 3

Materials: Chemical changes SCN 3-18a

Acids and alkalis

Learning intentions

In this section you will:

- learn the names of everyday acids and alkalis
- learn how indicators can be used to find out if a solution is acid, alkali or neutral
- learn how to use the pH scale to work out if a solution is acid, alkali or neutral
- learn about the effect of adding water to an acid and an alkali
- learn the importance of pH in food manufacture.

You have probably tasted foods that have a sour taste. This is most likely due to chemicals called **acids**. Vinegar is a solution of ethanoic acid. Vinegar gives the sharp taste to ketchup and brown sauce, and is often added to meals like fish and chips, for added flavour. Fizzy drinks, citrus fruits and even tea contain acids.

Figure 6.1 *Many foods and drinks contain acids.*

Figure 6.2 *Some common laboratory acids.*

Your stomach contains hydrochloric acid, which helps break down food.

Alkalis are different types of chemical substances than acids. When an alkali reacts with an acid, a salt and water are formed.

Acid indigestion can be caused by eating certain foods. The discomfort of indigestion can be relieved by taking remedies containing calcium bicarbonate, which reacts with the acid. A solution of calcium bicarbonate is an example of an alkali.

Alkalis are also used in cleaning materials such as oven cleaner because they are good at removing grease. Caustic soda (sodium hydroxide) is used to clear blocked drains because it dissolves grease deposits and breaks down other waste material like hair. It is also used as an industrial metal cleaner and paint stripper.

Liquids or solutions that are neither acid nor alkali are said to be **neutral**. Pure water and alcohol are neutral liquids.

Indicators and the pH scale

Chemicals called **indicators** can be used to tell if a solution is an acid, an alkali or is neutral. An indicator changes colour depending on the solution to which it is added.

Litmus is a chemical that can be used as an indicator to tell if a solution is an acid or an alkali. It can be used in the form of a solution or strips of paper.

Figure 6.3 *Many indigestion tablets are alkaline and react with stomach acid.*

Figure 6.4 *Caustic soda is an alkali used in some household cleaners.*

> ## 📖 Word bank
>
> • **Indicator**
> A chemical indicator can be used to tell if a solution is acidic, alkaline or neutral.

Figure 6.5 *Litmus solution turns red/pink in acid and blue/purple in alkali.*

A number of plants can be used to make indicators. Table 6.1 gives some examples.

Table 6.1 *Some plant sources of chemical indicators.*

Plant used to make indicator	Colour of indicator	Colour in acid	Colour in alkali
red cabbage	purple	red	blue
beetroot	red	red	purple
cherries	red	red	blue

📖 Word bank

- **Acid**

Acids have a pH below 7.

- **Alkali**

Alkalis have a pH above 7.

An alkali is a water-soluble base. You can read more about bases on page 101.

- **Neutral**

Neutral solutions have a pH of 7.

❓ Did you know ...?

The colour of the flowers of the hydrangea plant can be blue or pink, depending on the pH of the soil.

Figure 6.7 *Blue hydrangeas can be made to go pink by changing the pH of the soil.*

Figure 6.8 *The production of acids in milk makes it go sour.*

Figure 6.9 *Testing the pH of urine can provide information about your health.*

The chemical indicators in Table 6.1 only tell us if a solution is an acid or an alkali. Universal indicator turns different colours depending on how acidic or alkaline a solution is. Each colour matches to a number known as the **pH**. The pH scale runs from less than zero to above 14.

On the pH scale:

- **acid** solutions have a pH below 7

- **alkali** solutions have a pH above 7

- **neutral** solutions have a pH of 7.

Most solutions found at home or in the laboratory are in the pH range 1–14, which is the range shown for universal indicator in Figure 6.6.

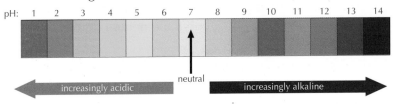

Figure 6.6 *Colours shown by universal indicator at different pH.*

Table 6.2 gives the pH of some common household and laboratory solutions.

Table 6.2 *pH of solutions often found in the home and the laboratory.*

Solution	pH
hydrochloric acid, battery acid	1
lemon juice, vinegar	3
pure water, alcohol	7
milk of magnesia, bicarbonate of soda	11
caustic soda, oven cleaner	14

Fresh cow's milk has a pH just below 7, which means it is slightly acidic. Over time, bacteria cause chemical reactions to take place in the milk, which causes the pH to drop – the milk becomes more acidic. This is what makes the milk taste sour. Milk kept in a fridge takes longer to go sour because the reactions taking place in the milk are much slower at a lower temperature.

The pH of your urine ranges from 5 to 8, with 7.5 being the average. The pH of urine varies during the day. It is usually lower first thing in the morning than at night. Certain conditions can cause your urine pH to be unusually high or low. High urine pH can be caused by kidney failure or urinary tract infection. Low urine pH can be caused by untreated diabetes.

The pH of your stomach also varies during the day. In the morning your stomach has a pH of around 5. When you eat your stomach releases hydrochloric acid, along with other chemicals, to help digest the food. This causes the pH of your stomach to drop to between 1 and 2, that is, it becomes more acidic. As the food is digested the pH rises again, that is, it becomes less acidic.

Diluting acids and alkalis

A concentrated acid or alkali is a solution with a high proportion of chemical dissolved in the water. Adding more water to a concentrated acid or alkali dilutes the solution, that is, it lowers the concentration.

This makes an acid less acidic so the pH rises towards 7. Alkaline solutions become less alkaline, so the pH falls towards 7.

If you look at bottles of acid and alkali in the laboratory you are likely to see the concentration beside the name written as, for example, 1.0 mol/l. 'mol/l' stands for 'moles per litre'. The important thing to know is the higher the number, the higher the concentration. An acid with a concentration of 1.0 mol/l is more concentrated than a 0.5 mol/l acid and is more acidic, so has a lower pH.

Acid spillages are diluted with water to increase the pH, that is, to make it less acidic, and then treated with alkali (see page 86).

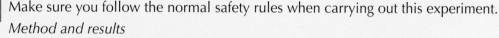

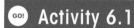 Activity 6.1

 **1. Experiment**

In this experiment you will make an indicator from red cabbage.

You will need: chopped red cabbage, boiling water, plastic jug, stirrer, sieve, everyday acids and alkalis, test tube, dropper.

You can work with others on this activity but must write your report yourself.

⚠️ Make sure you follow the normal safety rules when carrying out this experiment.

Method and results

(a) Weigh out 30 g of chopped red cabbage and add it to a plastic jug.

(b) Add enough boiling water (from a kettle) to just cover the cabbage.

(c) Stir the mixture until the water turns purple.

(d) Let the mixture cool, then filter it through a small sieve to separate the cabbage from the purple solution.

Keep the purple solution to use as an indicator to test the pH of some acids and alkalis found in the home (e.g. vinegar, lemon juice, bicarbonate of soda and milk of magnesia).

(continued)

(e) Collect a solution to be tested and add about 1cm depth to a test tube.

(f) Use a dropper to add drops of your indicator to the solution in the test tube.

(g) Record your observations (what you see) and conclusions in a table.

Solution tested	Colour change	Acid or alkali?
vinegar	purple →	

(h) Suggest one thing you could do to improve the experiment in order to get more accurate results.

2. The colour chart shows the range of colours that can be obtained when universal indicator is added to solutions of different pH.

A   B   C   D

(a) Copy and complete the table below by matching the name of each chemical with the letters **A–D** and estimate the pH of each solution. Table 6.2 on page 78 will help you.

Solution	A,B,C or D	pH
caustic soda		
citric acid		
milk of magnesia		
hydrochloric acid		

(b) Which of the solutions is the most acidic?

(c) Which of the solutions is the most alkaline?

(d) (i) State the pH of pure water.

(ii) State the colour of universal indicator in pure water.

3. The graph shows how the pH of milk changes when it is left out for a few days.

(a) After how many days did the pH of the milk start to fall?

(b) After how many days did the pH show a sharp drop?

(c) (i) State whether the milk is becoming more or less acidic as time goes by.

(ii) Explain how you came to your conclusion.

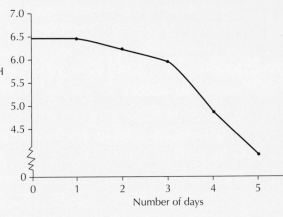

(d) Calculate the change in the pH of the milk from day 1 to day 4.

N3   L3   N4   L4

4. A swimming pool technician was going to test the pH of the water in the pool. She realised that the water level needed to be topped up so waited until water was added before testing the water.

    (a) Suggest why the pH is tested after more water is added.

    (b) The technician found that the water had a pH of 8.

        (i) What does this pH indicate about the water?

        (ii) Describe how the technician would have tested the pH of the water.

5. Select the correct word from those highlighted in bold in the sentences below.

    (a) Adding water to an acid or alkali makes it more **dilute / concentrated**.

    (b) When water is added to an acid the pH moves **up / down** towards 7.

    (c) When water is added to an alkali the pH moves **up / down** towards 7.

The importance of acids and pH in foodstuffs

Many foods and drinks are acidic and so have a pH of less than 7. Acids occur naturally in many fruit and vegetables. The tangy flavour of certain foods is due to acids. Many foods would have little flavour without acids.

Figure 6.10 shows a tomato ketchup label, showing the acids that have been added.

Table 6.3 gives some naturally occurring acids found in foods.

Figure 6.10   *A label from a tomato ketchup bottle. Vinegar and citric acid are ingredients.*

Table 6.3   *Examples of some acids that are found in foods.*

Acid	Food
citric	lemon, orange
malic	apple
tartaric	grapes, potatoes
oxalic	tea, cocoa

Acids are also deliberately added to many commercially produced (packaged) foods to make flavours sharper and also to act as **preservatives**. In Europe food additives are given E numbers, which are used to identify them.

Table 6.4 gives some acids used as preservatives with their E numbers.

> 📖 **Word bank**
>
> • **Preservative**
> A preservative stops moulds forming in products like bread.

Table 6.4   *Acids used as preservatives.*

Acid	E number	Food
sorbic	E 200	cheese, cakes, salad dressing
benzoic	E 210	soft drinks, ketchup
propionic	E 280	bread, pizza
ascorbic (vitamin C)	E 300	jam, sausages

Figure 6.11 *Many cola drinks contain carbonated water, phosphoric acid and citric acid.*

The body is able to process acids in our food and pushes them out as waste materials. However, eating foods with a lot of acids like oxalic and benzoic acids over a long time could harm the kidneys. Some acids are also thought to be the possible cause of irritation and inflammation in the body.

Fizzy drinks contain carbonated water, which is made by dissolving the acidic gas carbon dioxide in the water to get the 'fizz'. Colas often have phosphoric acid added to give them a tangy flavour. As a result, colas are more acidic than lemon juice or vinegar, but so much sugar is added that it hides the acidity. The acid attacks teeth and weakens the enamel, which leads to tooth decay.

Food manufacturers are interested in measuring the pH of foods and in keeping pH at certain levels to control mould growth and prevent the product going off.

pH plays an important part in the manufacture and preservation of foodstuffs. The pH of the milk used to make cheese plays a part in deciding whether the cheese will be soft or hard. With soft cheese the formation of moulds can be really slowed down if the pH is kept between 4 and 5. Table 6.5 shows the pH ranges used in the processing of various foods and drinks.

Word bank

- **Aroma**
A particular smell, usually pleasant.

- **Shelf-life**
The time that a food can be stored for and still be eaten.

Table 6.5 *Range of pH in the processing of some foods and drinks.*

Food	Dairy products: cheese, yoghurt, butter	Meat	Bread and pasta	Cold salads Fresh fruit and vegetables	Drinks: beer, wine
pH	4–5	5–7	4–6	3–6	6 (beer) 3–4 (wine)
Effect of pH	Acts as preservative and gives flavour	Indicates freshness. Too high a pH causes loss of **aroma** and darkens meat	Gives longer **shelf-life**	Small amounts of lemon juice/vinegar added to slow down growth of **microorganisms** and give longer shelf-life	Affects fermentation processes, taste and increases shelf-life

Word bank

- **Microorganism**
A living thing that can only be seen with a microscope, e.g. bacteria.

N3 L3 N4 L4

(GO!) Activity 6.2

☻ 1. Write a report about acids found in food and drink. You should include the name of one acid:

   (a) which is found naturally in foods/drinks and in which food/drink it is found

   (b) which our body has difficulty in processing

   (c) which is added to some foods/drinks and why it is added.

   Your report should not be more than 50 words long.

   Discuss your report with a partner then your teacher.

   You may wish to present your report to others in the class – think about using PowerPoint to help you deliver your presentation.

2. Explain how carbon dioxide is used in the drinks industry and what effect it can have on our health.

Neutralisation

Learning intentions

In this section you will:

- identify when a neutralisation reaction has taken place
- describe how a pH indicator can be used to follow the course of a neutralisation reaction
- give examples of neutralisation used in everyday life
- name the products of neutralisation
- describe the effect adding water has on the pH of an acid and alkali.

National 3

Curriculum level 3

Materials: Chemical changes SCN 3-18a

The pain of indigestion caused by too much acid in the stomach can be relieved by taking antacid tablets. They contain an alkali, which reacts with the acid. This reaction is called **neutralisation**.

In a neutralisation reaction the pH of the acid rises towards 7 (neutral) and the pH of the alkali drops towards 7. Neutralisation can be followed by adding universal indicator to the acid and alkali and noting the colour change when they are mixed. This is shown in Figure 6.12.

📖 Word bank

- **Neutralisation**
  A reaction between an acid and an alkali, producing a salt and water.

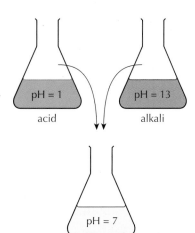

Figure 6.12 *The change in pH during a neutralisation reaction can be followed using universal indicator.*

Naming salts

Water and a salt are always formed when an acid is neutralised by an alkali. The chemical name of the salt used in cooking is sodium chloride. There are many other salts, and their names depend on the elements in the acid and alkali used in the reaction. The first part of the name of a salt comes from the alkali used. The second part comes from the acid, as shown in Table 6.6.

Table 6.6 *The second part of the name of a salt comes from the acid.*

Acid	Salt name ending
hydrochloric	chloride
sulfuric	sulfate
nitric	nitrate

Word equations can be written for neutralisation reactions.

Word equation: alkali + acid → water + salt

Example 6.1

sodium hydroxide + hydrochloric acid → water + sodium chloride

lithium hydroxide + sulfuric acid → water + lithium sulfate

potassium hydroxide + nitric acid → water + potassium nitrate

---

Uses of salts

Figure 6.13 *A mixture of sodium chloride and magnesium chloride is spread on roads to de-ice them.*

Sodium chloride is not only used to give food extra flavour, it is also an important industrial chemical used to make chlorine and caustic soda, and is also widely used to treat icy roads.

Table 6.7 gives examples of some common salts and their uses.

Table 6.7 *The uses of some common salts.*

Salt	Use
calcium sulfate	plaster for setting broken bones
magnesium chloride	low-temperature de-icing of roads
barium sulfate	'barium meal' for showing up stomach X-rays
sodium nitrate	food preservative

N3 | L3 | N4 | L4

🟢 Activity 6.3

🙂 1. (a) State what happens to the pH of an acid as an alkali is gradually added to it.

(b) Describe how you could follow the change in pH of an acid as alkali is added.

(c) State the name of the reaction taking place.

(d) State what happens to the pH of an alkali as an acid is added to it.

2. Tom and Sarah added a piece of magnesium to hydrochloric acid. Hydrogen and magnesium chloride were produced.

Word equation:

magnesium + hydrochloric acid → hydrogen + magnesium chloride

In his report, Tom said it was an example of neutralisation because a salt was formed. Sarah said it wasn't neutralisation, even though a salt was produced.

State who was correct and explain why.

3. Copy and complete the following word equations.

(a) calcium hydroxide + sulfuric acid → water + _____ _____

(b) magnesium hydroxide + hydrochloric acid → water + _____ _____

(c) sodium hydroxide + nitric acid → water + _____ _____

💥 Chemistry in action: Using neutralisation

Our stomach contains hydrochloric acid, which can have a pH as low as 1. The stomach has a lining that stops the acid damaging it. Sometimes, however, the acid can get into the digestive tract, which is not as well protected as the stomach, and a burning sensation can be felt. This is known as indigestion or heartburn.

Indigestion remedies can be taken to treat the condition. These are often called antacids because they contain mild alkalis, such as magnesium hydroxide, which neutralise the acid.

Our saliva is alkaline and it neutralises some of the acid made by bacteria in our mouth. This acid attacks the enamel of our teeth, which results in tooth decay. We clean our teeth with toothpaste to get rid of plaque and trapped food, which cause the production of acid. Toothpastes contain alkalis such as sodium hydrogencarbonate and sodium hydroxide, which also neutralise acid in the mouth.

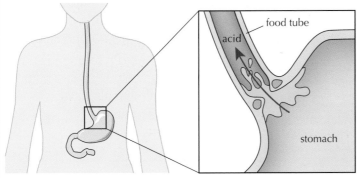

Figure 6.14 *Excess acid in the stomach causes indigestion and heartburn.*

Figure 6.15 *Toothpastes contain alkalis, which neutralise acid in the mouth.*

(continued)

| N3 | L3 | N4 | L4 |

Acid spillages are cleared up by diluting the acid with water then adding alkali such as bicarbonate of soda. The spillage is diluted with water first. This increases the pH of the acid towards 7, that is, makes it less acidic. The alkali then neutralises the acid. The pH of an alkali decreases towards 7 when water is added.

Acid spill hits road and rail

Firefighters diluting spilt acid with water before neutralising it with alkali.

Restrictions imposed after a chemical spillage at a storage yard in Cumbernauld have been lifted.

The town's railway station was closed after the leak on Sunday night, along with a number of roads in the area.

Fire officers said the 'massive spillage' happened at the Ellis and Everard storage yard in Lenziemill Road in the south of the town.

Dozens of firefighters were called to the yard after a 30 000 litre tank containing hydrochloric acid ruptured. The firefighters used water jets to reduce the risk of vapours getting into the air and were helped by rain on Sunday night and Monday.

An exclusion zone was established round the yard, with drivers and rail passengers advised to stay clear.

Thousands of residents were also advised to keep doors and windows closed as a precautionary measure, although Strathclyde Police stressed there was no immediate risk to the public.

Figure 6.16   *Farmer spreading lime to neutralise acidic soil.*

The majority of plants grow best at pH 7. If the soil is acidic or alkaline the plant may not grow well. If the soil is too acidic – the most common complaint – it is treated by spreading lime (mainly calcium carbonate) onto their fields to neutralise the acid.

Waste from many factories is often **acidic.** If this acidic solution is not treated and enters rivers it can kill fish. Slaked lime (calcium hydroxide) is often used to neutralise the acid.

? Did you know ...?

Bee stings are acidic so can be treated by adding mild alkalis such as sodium hydrogen carbonate. The alkali neutralises the acid, which relieves the pain.

Figure 6.17 *Close-up of a bee stinger. Bee stings are acidic and they can be neutralised by mild alkali.*

GO! Activity 6.4

☺ 1. Read the Chemistry in action section 'Using neutralisation' on pages 85 and 86 and summarise the main points in a table like the one below. The first situation has been done for you.

Situation	Neutralised by
indigestion	antacids (magnesium hydroxide)

☺☺ 2. A group of students carried out some experiments to compare how well three different antacids neutralised stomach acid.

This is what they did.

- Step 1: they added universal indicator to some hydrochloric acid.
- Step 2: they added the recommended dose of one of the antacids to the acid.
- Step 3: they noted any colour change and the time it took.

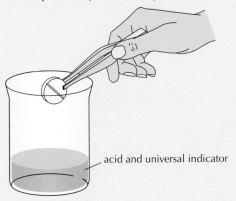

acid and universal indicator

They repeated the experiment for two other antacids.

(continued)

The table shows their results.

Name of antacid tablet	Colour of indicator at end	pH change	Time (min)
acid-ban	red → yellow	1 → 6	9
acid-ease	red → yellow	1 → 6	3
acid-cure	red → orange	1 → 4	5

(a) Suggest why:

    (i) they used hydrochloric acid

    (ii) the starting pH of the acid has to be 1.

(b) (i) State which tablet was the most effective.

    (ii) Explain your answer to part **(b) (i)**.

(c) One of the students suggested that the experiments should be repeated using crushed tablets. Suggest how this would affect:

    (i) the time taken to see a change in pH

    (ii) the final colour of the indicator.

National 3

Curriculum level 3

Planet Earth: Processes of the planet SCN 3-05b; Materials: Chemical changes SCN 3-18a

★ You need to know

Soluble non-metal oxides form acid solutions when they dissolve in water.

The chemical formulae of some acidic gases are:

carbon dioxide: CO_2

sulfur dioxide: SO_2

oxides of nitrogen: nitrogen oxide (NO) and nitrogen dioxide (NO_2), often shown as NO_x

☄ Make the link

See Chapter 11 for more information on sources of carbon dioxide.

Acidic gases and the environment

Learning intentions

In this section you will:

- learn the names of the non-metal oxide gases that have an impact on the environment
- learn the source of the non-metal oxide gases in the atmosphere
- learn what impact non-metal oxide gases have on the environment
- learn ways of reducing the volume of non-metal oxide gases released into the atmosphere
- learn ways of reducing the environmental impact of non-metal oxide gases in the atmosphere.

Burning fossil fuels such as petrol and diesel (in vehicles) and coal and gas (in power stations and homes) produces the acidic gases carbon dioxide, sulfur dioxide and oxides of nitrogen. These gases contribute to environmental issues such as acid rain and acidification of the oceans.

| N3 | L3 | N4 | L4 |

Acid rain

Normal rainwater is slightly acidic, with a pH between 5 and 6. This is mainly due to naturally occurring carbon dioxide dissolving in the rainwater. However, the other acidic gases produced when fossil fuels are burned are very soluble in water. When they dissolve in droplets of rainwater its pH drops to 4 and below. This results in **acid rain**.

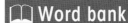

📖 **Word bank**

• **Acid rain**
Acid rain is formed when high concentrations of acidic gases such as sulfur dioxide and oxides of nitrogen dissolve in rainwater.

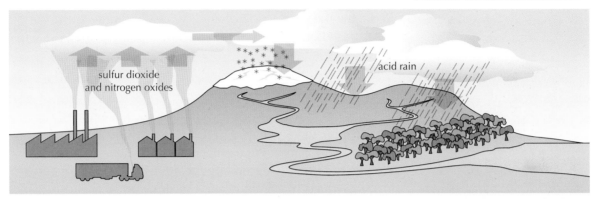

sulfur dioxide and nitrogen oxides

acid rain

Figure 6.18 *Acidic gases produced by burning fossil fuels cause acid rain.*

The effects of acid rain

Acid rain feels and looks like normal rain. However, it can cause serious harm to plants and animals:

* trees and even forests can be killed because acid rain dissolves important plant nutrients that get washed away
* damaged leaves can't photosynthesise efficiently
* the pH of lochs and rivers are lowered, causing fish to die.

Acid rain reacts with metal structures, such as bridges, and can weaken them over years.

The stonework of buildings is damaged by the weather, but acid rain makes the problem worse. Some types of rock used in building are easily weathered by chemicals. For example, limestone and marble are made of calcium carbonate. When acid rain falls on calcium carbonate the acid reacts with it. Some of the compounds formed are soluble and are washed away. Carbon dioxide gas and water are also produced.

Word equation:

$$\text{calcium carbonate} + \text{acid rain} \rightarrow \text{soluble salts} + \text{water} + \text{carbon dioxide}$$

Chemical weathering can hollow out caves and make cliffs fall away. The white cliffs of Dover in the south of England are made of chalk, a form of calcium carbonate. In 2012 a section of the eight-mile stretch of cliff collapsed. One of the reasons is thought to be the cliff weakening due to attack from acid rain.

Figure 6.19 *Trees in this forest have been damaged by acid rain.*

Figure 6.20 *The white cliffs of Dover are made of chalk, a form of calcium carbonate.*

| N3 | L3 | N4 | L4 |

Some types of rock, such as granite, used to construct a lot of the old buildings in Aberdeen, are much more resistant to chemical weathering. Granite contains little or no calcium carbonate.

Treatment and prevention of acid rain

Acidity in lochs and farmers' fields can be treated with lime (mainly calcium carbonate) which neutralises the acid rain. Reducing the production of acidic gases and preventing the gases from being released into the atmosphere is the best answer (see page 91).

Ocean acidification

Seawater is able to absorb carbon dioxide. It is naturally alkaline, having a pH around 8. Since the **Industrial Revolution** began in the eighteenth century, more carbon dioxide has been released into the atmosphere, from where it has subsequently been absorbed by the world's oceans, making them less alkaline (i.e. more acidic).

Seawater absorbs about one-quarter of the carbon dioxide currently released by human activities. This is good news because it reduces the amount of carbon dioxide in the atmosphere but bad news because it is increasing the acidity of the oceans. Most of that carbon dioxide combines with water to form carbonic acid.

Ocean acidity has increased roughly 30% since pre-industrial times. Rising ocean acidity creates major risks for marine organisms because it makes it more difficult for them to grow shells, making them more vulnerable to predators, among other effects.

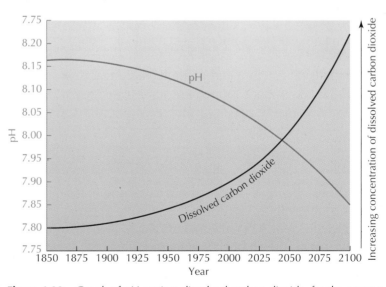

Figure 6.21 *Helicopters can be used to drop buckets of lime into lochs to neutralise acidic water.*

★ You need to know

Over the last 20 years the reduction in the amount of acidic gases being released into the atmosphere has led to a reduction in the acidity of rainwater and in the effects of acid rain on the environment.

Figure 6.22 *Graph of pH against dissolved carbon dioxide, for the oceans.*

N3   L3   N4   L4

The Great Barrier Reef in Australia is an underwater structure made from calcium carbonate produced by billions of marine organisms called coral. It is the world's largest coral reef and covers such a big area that it can be seen from outer space. It is home to a wide variety of life, including many endangered species, some of which can only be found in the reef system. There is concern among a number of researchers that the acidification of the oceans is making it more difficult for corals and other marine animals to form their calcium carbonate shell.

Some scientists believe that if carbon dioxide levels in the atmosphere continue to increase, the Great Barrier Reef coral could disappear within 50 years.

The good news is that if less carbon dioxide is released into the atmosphere, the acidification of the seas will be reduced and the corals will grow again.

Figure 6.23 *Aerial view of coral formations at the Great Barrier Reef near the Whitsunday Islands in Queensland, Australia.*

? Did you know ...?

Thirty species of whales, dolphins, and porpoises have been recorded in the Great Barrier Reef. More than 1500 fish species live on the reef and 49 species produce eggs only at the reef.

? Did you know ...?

The **Industrial Revolution** was the change to new manufacturing processes which took place in the UK in the eighteenth and nineteenth centuries. This change included going from hand production methods to machines, new chemical manufacturing and iron production processes, the increasing use of steam power, and the start of the factory system. Textiles were the main industry of the Industrial Revolution. These changes involved the use of more coal as a fuel, resulting in the release of acidic gases, such as carbon dioxide and sulfur dioxide, into the atmosphere.

Figure 6.24 *The volume of acidic gases released into the atmosphere has increased since the world became industrialised.*

Reducing acid gas emissions

Preventing acidic gases from reaching the atmosphere is better than having to treat the effects of acid rain and ocean acidification.

One obvious way is to burn less fossil fuels. Some countries, like Scotland, have invested a lot of money in developing renewable non-carbon-based energy sources like wind power to produce electricity (see Chapter 12).

Power stations that burn coal and gas remove the sulfur dioxide gas by scrubbing (cleaning) the waste gases. Alkali is often used to scrub (neutralise) the acidic gases (see Dunbar Cement Works, pages 104 and 105).

Cars that burn petrol can be fitted with catalytic converters to remove oxides of nitrogen. Electric cars are being developed and becoming more widely used.

Figure 6.25 *The use of electric cars is cutting down acid gas emissions.*

Figure 6.26 *Fertilisers are a source of oxides of nitrogen in the atmosphere.*

Scientists at Berkeley University, California, have evidence to suggest that fertilisers are a major source of oxides of nitrogen in the air. Microbes in the soil act on the fertiliser and oxides of nitrogen are produced. The use of fertilisers is so important for world food production that it is difficult to reduce their use by much. However, when and how they are applied could reduce the production of oxides of nitrogen. One approach, for example, is to time fertiliser application to avoid rain, because wet soil microbes can produce sudden bursts of nitrous oxide. Changes in the way fields are ploughed, when they are fertilised and how much is used can also reduce production of oxides of nitrogen.

GO! Activity 6.5

☺ 1. Acidic gases released into the atmosphere by human activity have an impact on the environment.
   (a) (i) Name three acidic gases released into the atmosphere.
       (ii) State the main source of the gases you named in part (a) (i).
   (b) (i) Give an example of how we can reduce the amount of acidic gases being produced.
       (ii) State how power stations can reduce the amount of acidic gases being released into the atmosphere.

2. Acid rain can cause lochs to become acidic.
   (a) Describe how acid rain is formed.
   (b) Give one other example of the effect of acid rain on the environment.
   (c) State how the acidification of a loch can be treated.

3. For millions of years the seas and oceans have been dissolving carbon dioxide from the atmosphere.
   (a) State one of the benefits of the oceans dissolving carbon dioxide.
   (b) (i) State one of the problems of the oceans dissolving carbon dioxide.
       (ii) Describe the effect more dissolved carbon dioxide in the oceans is having on marine life.

4. Look at Figure 6.22. Describe what happens to the pH of the oceans as the amount of carbon dioxide dissolving in it increases.

☺☺ 5. Write a letter to a politician or make up a poster or prepare a presentation about the effects acidic gases in the atmosphere have on the environment.
   Include a description of the effect and what we can do about it.
   Discuss the contents of your letter/poster/presentation with your teacher.

Greenhouse gases and global warming

National 3

Curriculum level 3

Planet Earth: Processes of the planet SCN 3-05b

Learning intentions

In this section you will:

- learn the names of some greenhouse gases
- learn what the greenhouse effect is
- learn how human activity affects concentrations of greenhouse gases
- learn about the link between greenhouse gases and global warming.

Gases such as carbon dioxide, methane, chlorofluorocarbons (CFCs), water vapour and nitrous oxide, which exist in Earth's atmosphere, are known as **greenhouse gases**. These greenhouse gases stop heat from Earth escaping into space and help to maintain our climate.

This is known as the **greenhouse effect**.

Human activities have caused higher concentrations of greenhouse gases. Many scientists think that this is causing the atmosphere to trap too much heat, increasing the greenhouse effect, and causing the temperature of Earth to rise.

This is known as **global warming**.

📖 Word bank

- **Greenhouse gases**

  Gases such as carbon dioxide, methane, CFCs, water vapour and nitrous oxide, which exist in Earth's atmosphere.

- **Greenhouse effect**

  The effect of greenhouse gases which stop heat escaping from Earth, helping to maintain our climate.

- **Global warming**

  The increase in the temperature of Earth due to the increasing greenhouse effect.

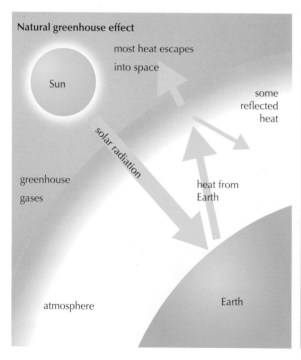

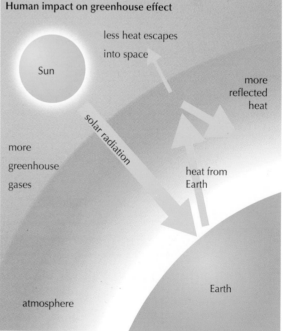

Figure 6.27 *Human activities may have increased the greenhouse effect.*

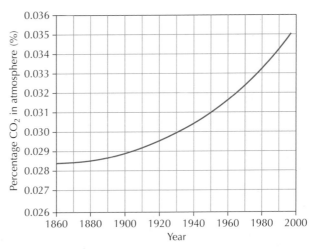

Figure 6.28   *The graph shows how the percentage of carbon dioxide in the atmosphere is increasing.*

Many scientists think that a rise in the average global temperature will cause climate change. Climate change is a large-scale, long-term shift in the planet's weather patterns or average temperatures. Earth's average temperature has risen by about 0.5 °C in the last 100 years. Some scientists have predicted a rise of between 1.4 and 5.8 °C in the next 100 years.

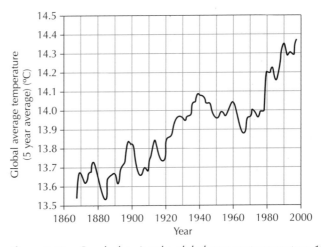

Figure 6.29   *Graph showing the global average temperature 1860–2000.*

↻ Keep up to date!

You can keep up to date with the hottest months of the year by searching online for 'hottest recorded months'.

❓ Did you know ...?

July and August 2016 were Earth's hottest months since records began in 1880. According to the North American Space Agency (NASA), the six hottest months on record have occurred since 2009. Scientists mainly blame man-made climate change from burning fossil fuels.

Such an increase in temperature could result in:

- melting of polar ice sheets, causing sea levels to rise and low-lying areas to be flooded

- some areas having less water for drinking and for growing food

- tropical diseases like malaria spreading over a wider area

- more frequent severe weather events

- bleaching of coral reefs like the Great Barrier Reef (see page 91).

Healthy coral is usually coloured but scientists have noted that since 1982 areas of coral have been losing their colour. This is known as bleaching. Bleaching can lead to the coral dying.

★ You need to know

Not all scientists agree that global warming is the cause of climate change. Some think that climate scientists may not be interpreting data correctly. Others think that climate change is not caused by human activity but is caused by natural processes like **El Niño**. Earth has had tropical climates and ice ages many times in its 4.5 billion years.

📖 Word bank

- **El Niño**

This is a band of warm ocean water that develops at certain times in the central Pacific. It is responsible for severe weather conditions in certain parts of the world. It is thought that there have been at least 30 El Niño events since 1900.

The United Nations Paris Climate Change Conference, 2015

At the Paris Climate Change Conference (COP21) in December 2015, 195 countries adopted the first-ever universal, legally binding global climate deal.

The agreement sets out a global action plan to put the world on track to avoid dangerous climate change by limiting global warming to well below 2 °C.

The agreement is due to enter into force in 2020.

At the summit governments agreed:

- a long-term aim of keeping the increase in global average temperature to well below 2 °C above pre-industrial levels

- on the need for global emissions to peak as soon as possible, recognising that this will take longer for developing countries

- to undertake rapid reductions once emissions have peaked.

Figure 6.30 *The Eiffel Tower was lit up with forest images to highlight the Paris Climate Change Conference held in Paris in December 2015.*

Make the link

See Chapter 11 for more about the Paris Climate Change Conference and also for the Scottish Government's commitments to reducing greenhouse gas emissions.

See Chapter 12 to learn about alternative (low-carbon) ways to meet energy needs in the future.

Before and during the Paris conference, countries submitted detailed national climate action plans.

The European Union was the first major economy to submit its intended contribution to the new agreement in March 2015. It is already taking steps to implement its target to reduce emissions by at least 40% by 2030. In September 2016 China and the USA ratified (agreed to sign) the agreement.

Keep up to date!

Governments that agreed the action plan had until 22 April 2017 to sign the agreement. It needs 55 countries that account for 55% of the greenhouse gas emissions to sign the agreement before it can enter into force.

Use the internet to find out how many countries have signed the agreement to date.

GO! Activity 6.6

1. Gases released into the atmosphere by human activity are thought to contribute to an increased greenhouse effect.

    (a) (i)   Name an acidic gas that contributes to the greenhouse effect.

    (ii) State how the increased greenhouse effect is thought to affect the temperature of Earth.

    (iii) What term is used to describe the effect caused by the change in Earth's average temperature?

    (b) Many scientists think there is a connection between global warming and climate change.

    (i)   State what the connection between global warming and climate change is thought to be.

    (ii) Describe one possible effect of climate change.

2. (a) (i)   Look at the graph in Figure 6.28. Describe the trend in carbon dioxide levels in the atmosphere since the 1860s.

    (ii) Suggest why the percentage of carbon dioxide in the atmosphere didn't change much before 1860.

    (b) Look at the graph in Figure 6.29. Describe the trend in the average global temperature since the 1860s.

N3 | L3 | N4 | L4

Learning checklist

After reading this chapter and completing the activities, I can:

N3 L3 N4 L4

- state that vinegar and lemon juice are examples of acids found in the home. **Activity 6.1 Q1** ○ ○ ○

- state that bicarbonate of soda and milk of magnesia are examples of alkalis found in the home. **Activity 6.1 Q1** ○ ○ ○

- state that acids turn universal indicator shades of red and have a pH less than 7. **Activity 6.1 Q2(a), (b)** ○ ○ ○

- state that alkalis turn universal indicator shades of blue and have a pH greater than 7. **Activity 6.1 Q2(a), (c)** ○ ○ ○

- state that pure water is neutral, turns universal indicator lime green and has a pH of 7. **Activity 6.1 Q2(d)** ○ ○ ○

- state that adding water to an acid dilutes it and moves the pH up towards 7. **Activity 6.1 Q5(a), (b)** ○ ○ ○

- state that adding water to an alkali dilutes it and moves the pH down towards 7. **Activity 6.1 Q5(a), (c)** ○ ○ ○

- give one example of an acid found naturally in a food or drink. **Activity 6.2 Q1** ○ ○ ○

- give an example of an acid found naturally in a food or drink which can cause health problems. **Activity 6.2 Q1** ○ ○ ○

- give an example of an acid added to foods to act as a preservative. **Activity 6.2 Q1** ○ ○ ○

- state that carbon dioxide is used in the drinks industry to make fizzy drinks. **Activity 6.2 Q2** ○ ○ ○

- state that carbon dioxide forms an acidic solution that can contribute to tooth decay. **Activity 6.2 Q2** ○ ○ ○

- state that an acid is neutralised when it reacts with an alkali. **Activity 6.3 Q1(a)** ○ ○ ○

- describe how an indicator can be used to follow the pH changes that take place during neutralisation. **Activity 6.3 Q1(b)** ○ ○ ○

N3 L3 N4 L4

- state that the pH of an acid goes up when it is neutralised. **Activity 6.3 Q1(a)** ○ ○ ○

- state that the pH of an alkali goes down when it is neutralised. **Activity 6.3 Q1(d)** ○ ○ ○

- state that when an acid is neutralised by an alkali water and a salt are formed. **Activity 6.3 Q2** ○ ○ ○

- name the salt formed when an acid reacts with an alkali. **Activity 6.3 Q3** ○ ○ ○

- give examples of common neutralisation reactions, including mouth acid, treating acid waste from factories, indigestion, acid spillages, acid soil, and bee stings. **Activity 6.4 Q1** ○ ○ ○

- state that carbon dioxide, sulfur dioxide and oxides of nitrogen are acidic oxides produced when fossil fuels are burned. **Activity 6.5 Q1(a)** ○ ○ ○

- state that we can reduce acidic gas production by using renewable energy sources such as wind power as alternatives to fossil fuels. **Activity 6.5 Q1(b)(i)** ○ ○ ○

- state that power stations reduce acid gas emissions by scrubbing (cleaning) waste gases so acidic gases don't reach the atmosphere. **Activity 6.5 Q1(b)(ii)** ○ ○ ○

- state that acid rain is formed when acidic gases dissolve in rainwater. **Activity 6.5 Q2(a)** ○ ○ ○

- state that acid rain has an effect on the environment, including: damaging leaves and killing trees; reacting with stone and metal structures; and killing fish and other animals in lochs and rivers. **Activity 6.5 Q2(b)** ○ ○ ○

- state that acidified water in lochs can be neutralised by adding lime. **Activity 6.5 Q2(c)** ○ ○ ○

- state that carbon dioxide dissolving in the oceans reduces the amount in the atmosphere but is making the oceans more acidic. **Activity 6.5 Q3** ○ ○ ○

N3 L3 N4 L4

- state that acidification of the oceans makes it difficult for sea creatures to grow shells. **Activity 6.5 Q3(b)(ii)**   ○ ○ ○

- state that carbon dioxide contributes to the greenhouse effect. **Activity 6.6 Q1(a)(i)**   ○ ○ ○

- state that the greenhouse effect is thought to be causing the average temperature of Earth to rise. **Activity 6.6 Q1(a)(ii)**   ○ ○ ○

- state that the increase in Earth's average temperature is known as global warming. **Activity 6.6 Q1(a)(iii)**   ○ ○ ○

- state that global warming is thought to be contributing to climate change. **Activity 6.6 Q1(b)(i)**   ○ ○ ○

- state that climate change could lead to environmental changes, such as: polar ice caps melting causing sea levels to rise; less freshwater available; tropical diseases spreading. **Activity 6.6 Q1(b)(ii)**   ○ ○ ○

- *present information in the form of a table.* **Activity 6.4 Q1**   ○ ○ ○

- *interpret information presented in a graph.* **Activity 6.1 Q3, Activity 6.5 Q4, Activity 6.6 Q2**   ○ ○ ○

- *apply my scientific knowledge in an unfamiliar situation.* **Activity 6.1 Q4, Activity 6.3 Q3(c), Activity 6.4 Q2(a)**   ○ ○ ○

- *make predictions and generalisations.* **Activity 6.4 Q2(c)**   ○ ○ ○

- *work with others to plan and carry out a practical activity.* **Activity 6.1 Q1**   ○ ○ ○

- *draw conclusions.* **Activity 6.4 Q2(b)**   ○ ○ ○

7 Acids and bases 2

This chapter includes coverage of:

N4 Acids and bases • Materials SCN 4-18a

You should already know:

- how to use indicators and the pH scale to identify acids, alkalis and neutral solutions
- that acids can be neutralised by alkalis
- how to name salts formed in a neutralisation reaction with an alkali
- that acids are present in various foods
- that acids in food can have an impact on human health
- that the acidic gases carbon dioxide, sulfur dioxide and oxides of nitrogen are produced by burning fossil fuels
- that acidic gases are linked to acid rain and acidification of the oceans (pages 88–91)
- that increased levels of greenhouse gases, including carbon dioxide, are linked to an increase in Earth's temperature, causing global warming (pages 93–96).

National 4

Bases

Learning intentions

In this section you will:
- learn to identify chemicals known as bases
- learn that soluble bases are called alkalis
- learn how to work out the names of the products of neutralisation using bases
- learn to write word equations for neutralisation involving bases
- learn how to follow neutralisation by titrating an acid with a soluble base
- learn how to neutralise an acid using an insoluble base.

A **base** is a compound that can react with an acid to form water and a salt. This is called **neutralisation** (see page 83). Metal oxides, metal hydroxides and metal carbonates are all bases. Soluble bases dissolve in water to form alkaline solutions. The pH of the water changes from 7 to above 7.

Table 7.1 gives examples of bases and their solubility in water. The example highlighted in blue in the table shows that sodium oxide is very soluble so will form an alkaline solution. Insoluble bases like nickel hydroxide (highlighted in green) are insoluble so have no effect on the pH of water.

Word bank

- **Base**

A base is a compound which reacts with an acid to form water and a salt.

Table 7.1 *Some bases and their solubility in water.*

	carbonate	oxide	hydroxide
barium	i	vs	vs
calcium	i	s	s
lithium	vs	vs	vs
magnesium	i	i	i
nickel	i	i	i
potassium	vs	vs	vs
sodium	vs	vs	vs
zinc	i	i	i

vs = very soluble; s = soluble; i = insoluble

Word equations can be written for neutralisation reactions. The first part of the name of a salt comes from the metal, the second part from the acid (see page 84).

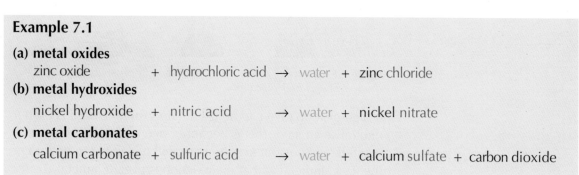

Example 7.1

(a) metal oxides
 zinc oxide   + hydrochloric acid → water + zinc chloride
(b) metal hydroxides
 nickel hydroxide + nitric acid   → water + nickel nitrate
(c) metal carbonates
 calcium carbonate + sulfuric acid   → water + calcium sulfate + carbon dioxide

Notice that when the base is a carbonate, carbon dioxide is produced along with water and a salt.

Given the name of the salt, the name of the acid from which it was formed can be worked out from the name ending, as shown in Example 7.2.

Example 7.2

barium chloride: hydrochloric acid

sodium sulfate: sulfuric acid

potassium nitrate: nitric acid

Indicators can be used to follow the course of a neutralisation reaction. If universal indicator is added to an alkali (soluble base) it turns purple. If acid is added the colour of the indicator changes through shades of blue and dark green. At the neutralisation point the indicator will be lime green. If the acid is added using a burette the volume of acid needed to neutralise the alkali can be measured accurately. This is shown in Figure 7.1.

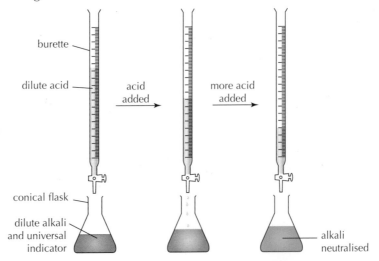

Figure 7.1   *Following a neutralisation reaction by observing the colour changes of universal indicator.*

Indicators don't have to be used if an insoluble base is used to neutralise an acid. This is because once all the acid has reacted, any unreacted solid base will be left in the bottom of the flask, indicating that the reaction is over. If a metal carbonate is used to neutralise the acid, bubbles of carbon dioxide gas will be seen. When the bubbles stop, this indicates that the acid has been neutralised.

GO! Activity 7.1

☻ 1. (a) Select the bases in the following list:

(i) potassium hydroxide (ii) sodium oxide

(iii) magnesium chloride (iv) calcium carbonate

(v) nickel fluoride (vi) barium iodide.

(b) From your answers to part (a), select the compounds that would form alkalis and say why.

(c) Use Table 7.1 to help you select:

(i) a soluble metal carbonate (ii) an insoluble metal oxide.

2. Complete the following word equations:

(a) potassium oxide + sulfuric acid → _____ + _____ _____

(b) lithium hydroxide + hydrochloric acid → _____ + _____ _____

(c) nickel carbonate + nitric acid → _____ + _____ _____ + _____ _____

(d) _____ _____ + _____ acid → _____ + calcium sulfate + carbon dioxide

☻☻3. Write out a plan describing how you could follow the neutralisation of sodium hydroxide solution with hydrochloric acid, by titration.

You should include: a list of chemicals you will need; a diagram of the arrangement you will use; a brief description of how you would carry out the experiment; how you would know the neutralisation point had been reached.

Ask your teacher to check your plan.

☻ 4. State **two** things you would see that indicate that zinc carbonate had neutralised nitric acid.

Carbon dioxide and the cement industry

Learning intentions

In this section you will:

- learn how carbon dioxide is produced during cement manufacture
- learn ways in which the cement industry is reducing the amount of carbon dioxide released into the atmosphere
- learn one way of measuring the impact that all your activities have on the environment.

Cement is one of the important materials used in the building and construction industries. It is used to make mortar for sticking bricks together and concrete used to make structures

National 4

Curriculum level 4

Materials: Chemical changes SCN 4-18a

★ You need to know

Acidic gases are linked to acid rain and the acidification of the oceans (pages 88–91).

Increased levels of greenhouse gases, including carbon dioxide, are linked to an increase in the Earth's temperature, causing global warming (pages 93–96).

like the pillars supporting the Queensferry Crossing, the new bridge across the Forth.

The main raw material used to make cement is limestone (calcium carbonate). It is mixed with clay and is heated. Carbon dioxide is produced as a waste product. If fossil fuels are used in the heating process, then carbon dioxide is again produced as a waste product. The cement industry is thought to produce about 5% of global man-made carbon dioxide emissions.

(a)   **(b)**

Figure 7.2   **(a)** *Cement is used to make concrete, which is essential for the construction industry.* **(b)** *The pillars supporting the new road bridge over the Forth are made of concrete.*

In Europe, the energy used for the production of cement has been reduced by approximately 30% since the 1970s. This reduction is equivalent to approximately 11 million tonnes less coal per year being burned, and has resulted in a big decrease in the amount of carbon dioxide produced. In addition, newly developed types of cement can absorb carbon dioxide from the air during hardening.

Chemistry in action: Dunbar Cement Works

Dunbar Cement Works is situated on the coast of East Lothian, Scotland. It was built there in 1963 because of the high-quality limestone in the area. Limestone is the main material used to make cement.

The works have been improved over the last 50 years to make the process more energy efficient and to reduce the impact of cement-making on the environment.

These improvements include:

- replacing fossil fuels with waste-derived fuels

- installing a gas scrubber (cleaner).

Figure 7.3   *Dunbar Cement Works. What looks like smoke coming from the chimney is mainly water vapour.*

Waste-derived fuels

Fossil fuels have been replaced by waste-derived fuels (WDF) such as:

- used car and lorry tyres, which can no longer be dumped in landfill sites; they produce as much energy as coal and contain **biomass**, which helps reduce the work's carbon footprint (see page 106)

- solid recovered fuels (SRF) – waste material that would normally go to **landfill sites**

Figure 7.4 *Dunbar Cement Works burns used tyres, which can no longer go to landfill sites.*

- recycled liquid fuel (RLF) – the byproduct from the manufacture of household items such as paints

- processed sewage pellets (PSP), a carbon-neutral fuel made from the leftovers of sewage treatment, which have a similar energy value to coal.

Gas scrubber

Sulfur dioxide is a very soluble gas so it is easily removed by passing it through water. Sulfur dioxide is acidic, so lime (calcium hydroxide), which is alkaline, can also be used to neutralise it. A salt and water are produced.

lime + sulfur dioxide → calcium sulfite + water

Since the installation of the gas scrubber at the Dunbar Cement Works in 2007, scrubbing (or cleaning) the sulfur dioxide in this way has halved the amount of sulfur dioxide being released into the atmosphere.

The scrubber also removes dust particles and has halved the amount of dust released into the atmosphere. Before 2007 the fields and buildings surrounding the works could often be seen to be covered with a thin layer of dust.

❓ Did you know ...?

As part of regulations to deliver its Zero Waste plan, the Scottish Government introduced a ban on both biodegradable and key recyclable materials being dumped in landfill sites.

Scotland's plan for Zero Waste sets out the Scottish Government's vision for a zero waste society. This vision describes a Scotland where waste is:
- minimised
- treated as a valuable resource not to be disposed of in landfill
- sorted, leaving only limited amounts to be treated.

📖 Word bank

- **Biomass**

Biomass is biological material made from living, or recently living, organisms.

- **Landfill site**

This is an area of land that is used to dump rubbish.

Carbon footprint

A **carbon footprint** is a measure of the impact that all our activities have on the environment. It calculates the total amount of greenhouse gases that you might expect to produce and expresses it as the equivalent amount of carbon dioxide. The footprint reflects the amount of carbon-based natural resources used by a person, company, community or country over a given period of time. More-developed countries tend to have higher carbon footprints than developing countries. The world average is about 4 tonnes of carbon dioxide per person per year. Table 7.2 shows the average carbon footprints for a person in some countries, as estimated in 2012.

Table 7.2   *The average carbon footprint of a person in some countries.*

Country	Carbon dioxide emissions (tonnes/person/year)
Australia	18.8
USA	16.4
Canada	16.0
UK	7.7
China	7.1
Brazil	2.3
India	1.6

Source: European Commission and Netherlands Environmental Assessment Agency

Everyone has a personal carbon footprint. It is based on a number of things including:

- the amount and type of energy we use in our homes, especially for heating – using fossil fuels for energy produces most carbon dioxide

- the food we eat – producing meat creates more carbon dioxide than producing vegetables

- where the food comes from – food imported from abroad uses more energy than locally produced food

- the amount of travelling we do and the distance travelled – for example, flying abroad for holidays.

There are some simple things you can do to help reduce your carbon footprint, for example switching off lights when you leave a room and completely switching off electrical appliances when you are not using them.

Word bank

- **Carbon footprint**

A carbon footprint measures the impact of all our activities on the environment by calculating the total amount of greenhouse gases produced and describing this as an equivalent amount of carbon dioxide.

? Did you know ...?

It has been estimated that the energy used to produce, deliver and dispose of junk mail produces more greenhouse gases than 2.8 million cars do.

? Did you know ...?

You can get an idea of what your carbon footprint is by doing an online survey. Type 'carbon footprint calculator' into a search engine to find an online calculator. There is a good one on the WWF website.

Governments are making some headway in reducing carbon footprints by, for example, promoting the use of alternative (non-carbon-based) energy sources (see the Paris Climate Change Conference agreement on page 95). The newspaper article below gives an indication of how using non-carbon sources of energy is helping reduce Scotland's carbon footprint.

Some countries argue that calculating carbon footprints per person doesn't give an accurate picture of which countries are the biggest greenhouse gas producers. Although Australia has a much higher carbon footprint per person than China, it has a much smaller population and so, as a country, produces much less carbon dioxide than China.

Figure 7.5 *Making electricity from wind power is one example of a renewable energy source that can reduce a country's carbon footprint.*

Drop in greenhouse gas emissions welcomed

Scotland has achieved the second-highest reduction in greenhouse gas emissions in Western Europe over a quarter of a century, Scottish government figures reveal.

The country managed a reduction of 39.5 per cent from 1990 to 2014.

This compared with drops of around 34 per cent for England and 33 per cent for the UK as a whole.

In Western Europe, Scotland's figures were bettered only by Sweden, which saw a decline of 54.5 per cent over the same period.

WWF Scotland director Lang Banks said: 'It's great to see more evidence that Scotland is in the vanguard when it comes to tackling climate change in Europe.

'Thanks to strong government leadership over the years we've embraced renewables,

helping to de-carbonise our power sector. However, looking ahead there is no room for complacency if Scotland is to maintain its position as a leader on climate change and to capture the many social, health and economic benefits of moving to a zero-carbon future.

'Outside of the electricity and waste sectors, progress to cut carbon has been far too slow.' Ministers plan to introduce a bill which will set a target to reduce emissions by more than 50 per cent by 2020.

Environment secretary Roseanna Cunningham said: 'Scotland is a world-leader in tackling climate change, and these figures reaffirm that Scotland continues to outperform the rest of the UK and punch above its weight in international efforts to cut greenhouse gas emissions.'

Source: *Metro, 1 August 2016*

GO! Activity 7.2

☺ 1. Give the **two** main sources of carbon dioxide produced by human activity.

2. Write a report of 50–70 words about the effect cement manufacture has on the environment and what is being done to reduce the effect.

   Discuss your report with a partner and your teacher.

   OR

   Prepare a presentation to deliver to the rest of the class on the effect cement manufacture has on the environment and what is being done to reduce the effect. You may wish to use PowerPoint for your presentation.

   Discuss your presentation with your teacher.

3. (a) Explain what a carbon footprint is.

   (b) Look at Table 7.2. Suggest why developed counties such as the USA and the UK have a larger carbon footprint than so-called developing countries such as India and Brazil.

4. Read the article 'Drop in greenhouse gas emissions welcomed' on page 107.

   (a) Part of the report says, 'we've embraced renewables, helping to de-carbonise our power sector'.

      (i) Give an example of a renewable source of energy.

      (ii) Explain how using renewable sources of energy helps to 'de-carbonise the power sector'.

   (b) Draw up a table to show the percentage reduction in carbon emissions between 1990 and 2014, for Scotland, England, the UK and Sweden.

Learning checklist

After reading this chapter and completing the activities, I can:

N3 L3 **N4 L4**

- state that metal oxides, hydroxides and carbonates are bases. **Activity 7.1 Q1(a)**   ○ ○ ○

- state that soluble bases form alkaline solutions. **Activity 7.1 Q1(b)**   ○ ○ ○

- use a table of solubilities to identify soluble and insoluble bases. **Activity 7.1 Q1(c)**   ○ ○ ○

- state that metal oxides and hydroxides react with acids to produce water and a salt. **Activity 7.1 Q2(a), (b)**   ○ ○ ○

N3 | L3 | **N4** | **L4**

- state that metal carbonates react with acids to form water, a salt and carbon dioxide. **Activity 7.1 Q2(c), (d)**   ◯ ◯ ◯

- name the salt formed in a neutralisation reaction. **Activity 7.1 Q2(a), (b), (c)**   ◯ ◯ ◯

- write a word equation for a neutralisation reaction. **Activity 7.1 Q2**   ◯ ◯ ◯

- state universal indicator turns lime green at the neutralisation point. **Activity 7.1 Q3**   ◯ ◯ ◯

- identify when an acid has been neutralised by an insoluble base because unreacted base remains when all the acid is reacted, or, if a carbonate, bubbles of gas are no longer produced. **Activity 7.1 Q4**   ◯ ◯ ◯

- state that the main sources of carbon dioxide produced by human activity are burning fossil fuels and making cement. **Activity 7.2 Q1**   ◯ ◯ ◯

- state that the cement industry is reducing the amount of carbon dioxide produced by: replacing fossil fuels with waste-derived fuels; using gas scrubbers. **Activity 7.2 Q2**   ◯ ◯ ◯

- state that a carbon footprint is a way of measuring the impact of our activities on the environment. **Activity 7.2 Q3(a)**   ◯ ◯ ◯

- *present information in the form of a table.* **Activity 7.2 Q4(b)**   ◯ ◯ ◯

- *extract scientific information from a report.* **Activity 7.2 Q4(a)**   ◯ ◯ ◯

- *obtain information from tables of data.* **Activity 7.1 Q1(c)**   ◯ ◯ ◯

- *make predictions and generalisations.* **Activity 7.2 Q3(b)**   ◯ ◯ ◯

- *work with others to plan and carry out a practical activity.* **Activity 7.1 Q3**   ◯ ◯ ◯

Unit 1 practice assessment

Total: 21 marks

N3 Rates of reaction

1. Sodium is a soft grey metal that can be cut with a knife. When a piece of sodium was dropped into a gas jar filled with green chlorine gas, a bright flame was seen. At the end of the experiment a white solid was seen at the bottom of the gas jar.

   State **one** piece of evidence from the report above that a chemical reaction had taken place.   1

2. State **one** way of speeding up a chemical reaction.   1

3. Explain why boiling water and paint drying are **not** chemical reactions.   1

N3 Chemical structure

4. Choose the correct word.

   When sodium chloride dissolves in water, the sodium chloride is the **solute / solvent / solution**.   1

5. Select your answers to the questions from the grid below. You can use a periodic table to help you.

A	B
sodium	calcium
C	**D**
magnesium	nitrogen

   (a) Identify the **two** elements with similar chemical properties.   1

   (b) Identify the non-metal element.   1

   (c) Identify the element that is an alkali metal.   1

6. The circuit diagram shows the arrangement which can be used to investigate electrical conductivity.

Experiment	Substance X
A	graphite
B	iodine
C	sulfur

   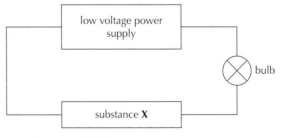

   Identify the experiment in which the bulb would light.   1

7. When aluminium reacts with iodine, aluminium iodide is formed.

   (a) Write the formula for aluminium iodide. 1

   (b) Write the word equation for the formation of aluminium iodide. 1

   (c) When aluminium iodide solution and lead nitrate solution are mixed together a chemical reaction takes place.

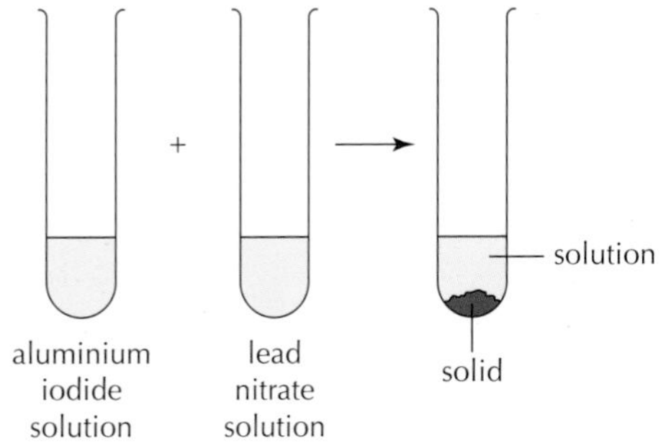

   How could the solid be separated from the solution? 1

N3 Acids and bases

8. Hydrochloric acid reacts with sodium hydroxide to form water and a salt.

   (a) Suggest what the pH of the acid might be. 1

   (b) What colour would the acid turn universal indicator? 1

   (c) Name the salt formed. 1

9. Some acid was spilled in the laboratory. The teacher poured water on the acid then added alkali.

   (a) What happens to the pH of the acid when water is added? 1

   (b) Name the type of reaction happening between the acid and the alkali. 1

10. Carbon dioxide is added to some drinks to make them fizzy.

    (a) What effect does dissolving carbon dioxide have on the pH of the water? 1

    (b) Suggest why dentists are concerned about carbon dioxide in drinks. 1

11. Non-metal oxide gases are released into the atmosphere because of human activity. Some of these gases cause acid rain and acidification of the oceans.

    (a) Name a source of these gases. 1

    (b) Describe how acid rain is formed. 1

    (c) State one effect that acidification of the oceans is causing. 1

Total 21

A score of 11 or more is a pass!

Section 2 National 4 Outcomes

Total: 29 marks

N4 Rates of reaction

1. The diagrams show two experiments set up to monitor rates of reaction. The mass of marble chips is the same in each experiment.

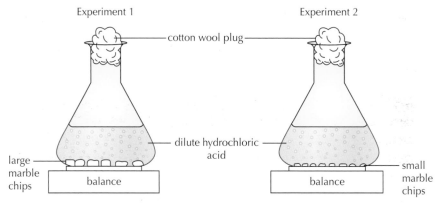

(a) (i) In which experiment would the reading on the balance go down more quickly? **1**

   (ii) Explain your answer to part **(a) (i)**. **1**

(b) The results of the two experiments are shown below.

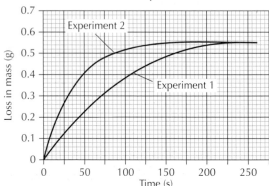

Using the graph:

(i) State how many seconds it took for experiment 1 to finish. **1**

(ii) State the total loss in mass in both experiments. **1**

2. Two experiments involving zinc reacting with hydrochloric acid were carried out. The volume of hydrogen gas produced was measured in each experiment. The results of the two experiments are shown below.

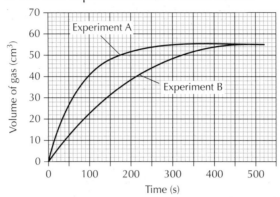

(a) (i) A group of students looked at the graph and stated that experiment A was faster.

What feature of the graph would lead them to their conclusion? **1**

(ii) Suggest what change could be made to the hydrochloric acid in experiment A in order to get the results for experiment B. **1**

(b) Describe how the gas in each experiment could be collected. **1**

N4 Atomic structure and bonding related to properties of materials

3. The diagram represents an atom.

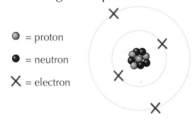

● = proton

● = neutron

✗ = electron

(a) Name the part of the atom which contains the protons and neutrons. **1**

(b) Using the information in the diagram:

(i) State the mass number of the atom. **1**

(ii) Explain why the atom is electrically neutral. **1**

(c) Name the element to which this atom belongs. **1**

4. The diagram shows the outer electrons in a chlorine atom.

   (a) Copy the diagram of the chlorine atom and add another diagram to show how two chlorine atoms join to form a chlorine molecule. 1

   (b) When chlorine gas is bubbled through calcium hydroxide, one of the products is calcium chlorate.

   Name the elements present in calcium chlorate. 1

5. The electrical conductivity of three substances was tested. The results are shown in the table.

Substance	Electrical conductivity	
	Solid	Liquid
X	no	yes
Y	no	no
Z	yes	yes

   Which substance is covalent? 1

6. (a) Calculate the formula mass of calcium chloride ($CaCl_2$).

   (Use the information in Table 4.14 to help you.) 1

   (b) State the type of bonding present in calcium chloride. 1

   (c) Write the formula equation for calcium reacting with chlorine to form calcium chloride ($CaCl_2$). 1

N4 Energy changes of chemical reactions

7. Combustion is an example of an exothermic reaction.

   State what is meant by exothermic. 1

8. A student mixed two solutions together in a beaker and noted a change in temperature as the chemicals reacted. He concluded, correctly, that the reaction was endothermic.

   What must have happened to the temperature of the mixture during the reaction? 1

N4 Acids and bases

9. The table shows the effect some metal and non-metal oxides have on the pH of pure water (pH 7).

Oxide	Solubility	Effect on pH of pure water
potassium oxide	soluble	increases
calcium oxide	soluble	(i)
phosphorous oxide	soluble	decreases
sulfur dioxide	soluble	decreases
oxide X	soluble	decreases
zinc oxide	insoluble	(ii)

   (a) Complete the table by giving answers for (i) and (ii). 2

   (b) Oxide X is a non-metal oxide.

      What information in the table supports this? 1

   (c) Potassium oxide is an example of a base.

      Give the name of one other potassium compound which is a base. 1

10. Acids are often added to foods to give them a tangy or sour taste.

   Give **one** other reason that acids are added to foods. 1

11. Copper carbonate is insoluble in water but reacts with acids.

   (a) State the type of reaction taking place. 1

   (b) The word equation for the reaction between copper carbonate and sulfuric acid is shown.

      copper carbonate + sulfuric acid → water + Z + carbon dioxide

      (i) Name compound Z. 1

      (ii) What type of compound is Z? 1

   (c) How would you be able to tell when all the acid had reacted? 1

   (d) A major source of carbon dioxide in the atmosphere comes from the burning of fossil fuels.

      State another major source of carbon dioxide. 1

Total 29

A score of 15 or more is a pass!

8   Fuels and energy 1: Fuels

What are fuels?   •   Fossil fuels   •   Hydrocarbons   •   Fuels that give our bodies energy   •   Alcohol as a fuel

9   Fuels and energy 2: Controlling fires

The fire triangle   •   Extinguishing fires safely

10 Fuels 1: Fossil fuels

Trapping the Sun's energy   •   Formation of fossil fuels   •   Fossil fuels and new technologies   •   Burning fuels   •   The law of conservation of mass

11 Fuels and energy 3: The problems with fossil fuels

The problem with using fossil fuels   •   Climate change targets

12 Fuels and energy 4: Meeting energy needs in the future

Sustainable sources of energy   •   Biomass and biofuels   •   Alternative technologies in Scotland   •   Advantages and disadvantages of different sustainable sources of energy

13 Fuels 2: Solutions to fossil fuel problems

Combustion of fossil fuels and the carbon cycle   •   Reducing carbon dioxide emissions   •   Sustainable sources of energy

14 Fuels 3: Hydrocarbons

Fractional distillation   •   The alkanes   •   The alkenes   •   Meeting market demand: Cracking

15 Everyday consumer products 1: Plants and food

Plants for food   •   Diet and disease   •   Plants and alcohol   •   Improving plant growth with fertilisers

16 Everyday consumer products 2: Cosmetic products

Plants and cosmetics   •   Using essential oils

17 Everyday consumer products 3: Plants for energy

Plants for food and energy   •   Carbohydrates – What's in a name?   •   Digestion   •   Alcohol from plants   •   Effect of temperature and pH on enzyme activity

18 Plants to products

Products from plants   •   Medicines from plants   •   Labelling of medicines

Unit 2 practice assessment

UNIT 2
Nature's chemistry

8 Fuels and energy 1: Fuels

This chapter includes coverage of:
N3 Fuels and energy • Planet Earth SCN 3-05b

What are fuels?

Learning intentions

In this section you will:
- explain what is meant by the term 'fuel'
- explain how energy can be obtained from fuels.

We live in an energy-dependent world. As the world's population grows and countries develop, more and more energy is required. We need energy to fuel all forms of transport. We need energy to manufacture all the goods we use in our daily lives.

Modern living would be impossible without huge supplies of energy.

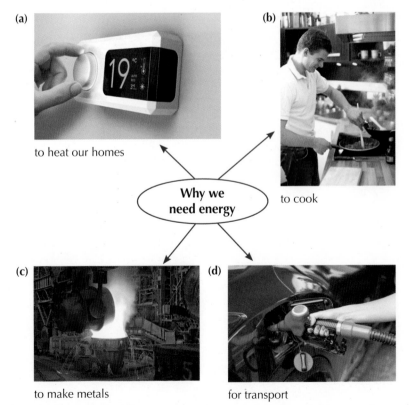

(a) to heat our homes

(b) to cook

Why we need energy

(c) to make metals

(d) for transport

Figure 8.1 *We need energy* **(a)** *to heat our homes and* **(b)** *to cook our food. Large amounts of energy are needed* **(c)** *to make metals such as steel and* **(d)** *for transport.*

Where do we get energy from?

From earliest times, humans have obtained energy by burning things. Early humans would burn wood to provide heat and light. As civilisations developed, energy was obtained by burning other substances such as animal fats, coal, oil and natural gas.

We refer to anything we can burn to give energy as a **fuel**.

What happens when fuels burn?

Fuels have chemical energy stored in the compounds that make up the fuel. When the fuel burns, the compounds that make up the fuel combine with oxygen in the air and this releases the stored energy as heat and light.

These are the energy changes that happen when a fuel burns:

chemical energy → heat energy + light energy

Fuels can be solids, liquids or gases. They range from the gas in a camping stove and the wood we put on an open fire to keep us warm, to kerosene from oil that can be used to propel rockets into space.

Figure 8.2 *Cave paintings show that early humans burned wood.*

❓ Did you know ...?

Modern jet aircraft like the Airbus A380 are as fuel efficient as many family cars and only use about 4 litres of fuel for every 100 passenger-kilometres travelled. This still means that 2100 litres of fuel are used to carry 535 passengers 100 kilometres!

Jet aircraft burn about 500 000 tons of fuel every day.

Figure 8.3 *When a fuel burns, compounds in the fuel combine with oxygen, releasing energy.*

Figure 8.4 *Camping gas stoves usually burn a mixture of the gases propane and butane.*

Figure 8.5 *Wood being burned on a fire.*

Figure 8.6 *Rockets burn kerosene, a liquid fuel, to propel them into space.*

National 3

Fossil fuels

Learning intention

In this section you will:

• find out about fossil fuels.

Word bank

• **Fossil fuels**

Fuels consisting of the remains of animals and plants that lived and died many millions of years ago.

Most of the energy used in the world today is obtained by burning **fossil fuels**. Fossil fuels are the remains of animals and plants that lived and died many millions of years ago.

Types of fossil fuel

Fossil fuels are coal, natural gas, crude oil and peat.

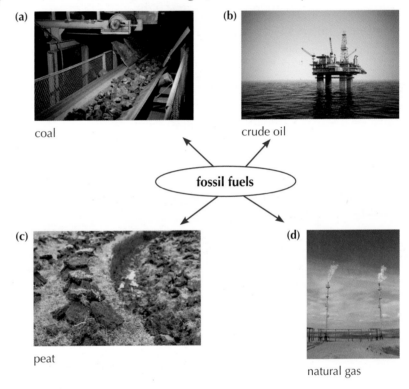

(a)

coal

(b)

crude oil

fossil fuels

(c)

peat

(d)

natural gas

Figure 8.7 *Different types of fossil fuel:* **(a)** *coal,* **(b)** *crude oil,* **(c)** *peat,* **(d)** *natural gas.*

Coal

Coal has been used as a fuel for many thousands of years. The first known use of coal as a fuel was in China in about 3500 BCE (Before Common Era). In the Americas, there is evidence of the Aztecs using coal as a fuel in the fourteenth to sixteenth centuries.

In Roman Britain, coal was used to heat public baths and the homes of important officials.

Figure 8.8 *Coal stores have been found at forts along Hadrian's Wall.*

Natural gas

Natural gas has also been known about for thousands of years. In some areas, it seeps out of the ground and can catch fire.

Figure 8.9 *Natural gas burning as it seeps out of the ground.*

The Ancient Greeks even built a temple to their god Apollo at Delphi where natural gas was burning as it seeped out of the ground.

As early as 500 BCE, the Chinese used crude bamboo piping to transport natural gas, so that they could boil sea water to produce drinking water. Today, natural gas is burned in our homes for heating and cooking, and is burned in many power stations to produce electricity.

Figure 8.10 *The ruins of the temple to Apollo at Delphi in Greece.*

> **? Did you know ...?**
>
> The earliest known oil wells were drilled in China in 347 CE. Drill bits attached to bamboo poles were used to drill as deep as 240 m to reach the oil.

Crude oil

Crude oil is another naturally occurring fossil fuel. It has been used as a fuel for thousands of years. Petrol and diesel are obtained from crude oil.

It is only since 1847, when a Scottish chemist, James 'Paraffin' Young, took some oil that was seeping into a colliery in Derbyshire, heated it and condensed the vapours that were given off, that crude oil has been **distilled** to give products like petrol and diesel.

Figure 8.11 *James 'Paraffin' Young, the Scottish chemist.*

> **? Did you know ...?**
>
> In 2013, 35 million vehicles were licensed to drive on British roads. That number is increasing every year. In 2016, the number of cars on English roads rose by 600 000. Such a large number of cars needs a lot of fuel, placing great demands on the supply of crude oil.

> **📖 Word bank**
>
> • **Distil**
>
> To boil a liquid, collect and condense (change from gas to liquid) the vapours that are given off.

Figure 8.12 *Hand-cut peats stacked to allow them to dry, ready for use.*

Peat

Peat is 'modern' fossil fuel, since it can be thought of as the first stage in the transformation of decaying vegetation into coal! It has been the fuel used by common people in Northern Europe since pre-Roman times. The countries that are the biggest consumers of peat as a fuel source today are Finland, Ireland and Scotland. Traditionally, the peats – they are also called turfs – were cut by hand in the summer and stacked to allow them to dry. In some areas, this is now done mechanically, with tractors scraping the surface of peat bogs. The peat is then squeezed and heated to remove the water.

National 3

Curriculum level 3

Planet Earth: Processes of the planet SCN 3-05b

Hydrocarbons

Learning intentions

In this section you will:

- learn what is meant by the term 'hydrocarbon'
- learn what is produced when hydrocarbons burn
- learn about tests used to identify water and carbon dioxide.

📖 Word bank

- **Hydrocarbon**
  A compound made up of only hydrogen and carbon atoms.

Fossil fuels are mainly composed of compounds known as **hydrocarbons**. These are compounds that contain only carbon and hydrogen atoms. There are many different hydrocarbons.

Natural gas is mainly made up of the simplest hydrocarbon, methane.

hydrogen

carbon

Figure 8.13 *A molecule of methane is made up of four hydrogen atoms joined to a carbon atom.*

Octane is another hydrocarbon. It is found in petrol, which is found in crude oil.

Figure 8.14 *Octane is a hydrocarbon with eight carbons and is just one of the compounds obtained from crude oil.*

Burning hydrocarbons

When a hydrocarbon burns, the atoms of the hydrocarbon combine with oxygen in the air.

Figure 8.15 shows the experiment used to identify the products when hydrocarbons are burned. The pump draws the gases given off when the hydrocarbon burns through the apparatus.

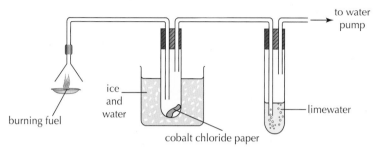

Figure 8.15 *Apparatus for testing products made when burning a hydrocarbon.*

In class, your teacher may demonstrate the burning of different hydrocarbons.

When a hydrocarbon is burned, the cobalt chloride paper turns from blue to pink. This shows that water is produced. You may also see droplets of water in the tube. The iced water bath causes any water vapour to condense. You can test the pH of the droplets. Water has a pH of 7.

The limewater in the second tube turns milky. This is because burning hydrocarbons produces carbon dioxide. Any carbon dioxide given off will be drawn through the limewater, turning it milky. The word equation for burning hydrocarbons is:

hydrocarbon + oxygen → carbon dioxide + water

> ### Make the link
> You can learn more about the pH of water in Chapter 6.

GO! Activity 8.1 | Answers to all activity questions and assessments in this Unit are available online at: www.leckieandleckie.co.uk/page/Resources

☺ **1.** Kerosene is a fuel that is obtained from crude oil.

   (a) State the term used to describe fuels such as crude oil.

   (b) The molecules in kerosene are made from only carbon and hydrogen atoms.
   State the term used to describe molecules that are made from only carbon and hydrogen atoms.

   (c) Name the gas in the air that is used up when fuels such as kerosene burn.

   (d) When kerosene is burned, a gas that turns limewater milky is produced.
   Name the gas.

   (e) Burning kerosene also produces water.
   State **two** tests which can be used to show that the liquid produced by burning kerosene is water. Give the results of each test.

Fuels that give our bodies energy

Learning intention

In this section you will:

* explain how our bodies obtain energy.

Figure 8.16 *Food and drink are the fuel for our bodies.*

Just as vehicles need energy to move, so do animals. Just as we burn fuel to heat our homes, so our bodies burn fuel to keep us warm.

The fuel for our bodies is the food and drink we take in. When combined with oxygen in the cells of our bodies, these foods provide energy for heat and movement.

A healthy diet should contain foods that are rich in carbohydrates (which include starch and sugars), proteins, and fats and oils.

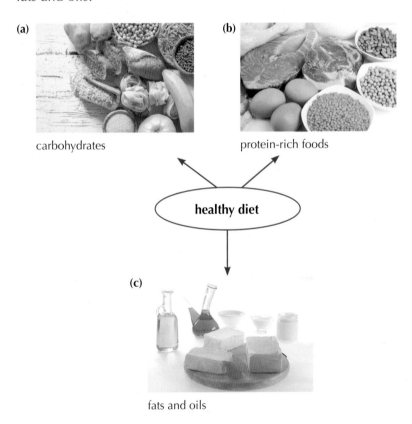

Figure 8.17 **(a)** *Carbohydrates.* **(b)** *Protein-rich foods.* **(c)** *Fats and oils.*

All of the compounds in these foods contain carbon and hydrogen atoms. Some of the foods contain atoms of other elements such as oxygen and nitrogen.

When foods burn in our bodies, they produce carbon dioxide and water, and energy for our bodies to use. The carbon atoms in food molecules combine with oxygen to produce carbon dioxide and the hydrogen atoms combine with oxygen to produce water.

Word equation: food + oxygen → carbon dioxide + water

If we blow through limewater, the limewater turns milky. This shows that the air we breathe out contains carbon dioxide.

We can show that water is also produced if we breathe onto a cold mirror. We see the mirror mist up as water vapour condenses. This is also the reason we can see our breath on a cold morning.

GO! Activity 8.2

☺ 1. Copy and complete the sentences below using the following words.

coal   crude   fuels   water   burned   carbon   energy   fossil   proteins   vegetable

Fuels are substances which when **(a)** _____ produce **(b)** _____. **(c)** _____ oil, natural gas, **(d)** _____ and peat are examples of **(e)** _____ fuels.

Many **(f)**_____ contain hydrocarbon molecules. Hydrocarbon molecules contain only hydrogen and **(g)** _____ atoms. When hydrocarbon fuels burn, **(h)** _____ and carbon dioxide are produced.

Starch, sugars, **(i)** _____ oil and **(j)** _____ are foods that provide the body with energy.

2. Many of the compounds in peanuts are made up of molecules that contain carbon and hydrogen atoms.

   (a) Name the substance that foods combine with in the body to produce energy.

   (b) The apparatus below was used to show what is produced when a peanut is burned.

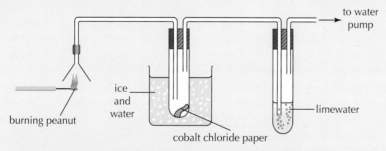

   Apparatus for burning a peanut and testing the products.

   (i) Suggest why the first tube is placed in a beaker containing ice and water.

   (ii) Describe what you would see hapenning in the tube containing limewater.

   (c) Write a word equation to show what happens when foods burn.

(continued)

3. Experiment

Use the apparatus shown in Question 2 (your teacher or a technician will need to set it up for you) to test different foods and show that they produce carbon dioxide and water when they burn.

You can try burning foods such as biscuit, bread, peanuts, potato crisps and sugar.

You will need: cobalt chloride paper, limewater, crushed ice, solid foods (such as peanuts, digestive or other biscuits, bread, breakfast cereal, sugar).

 Be careful when burning foods. Foods often need to be held in a Bunsen flame for a short time before they catch fire. DO NOT use cooking oils as these can spurt when burning and create a risk of burns and fires.

4. Experiment

Use the apparatus below to compare the energy given out by burning different foodstuffs.

You will need: boiling tube, deflagrating (burning) spoon, thermometer, weighing scales, solid foods (such as sugar, peanuts, biscuit, potato crisps, bread).

Be careful when burning foods. Foods often need to be held in a Bunsen flame for a short time before they catch fire. DO NOT use cooking oils as these can spurt when burning and create a risk of burns and fires.

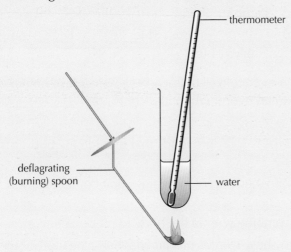

thermometer

deflagrating (burning) spoon

water

Apparatus for measuring the energy produced when food is burned.

Method

(a) Measure $20\,cm^3$ of water into a boiling tube and clamp the boiling tube in a stand.

(b) Record the temperature of the water.

(c) Weigh a sample of food (e.g. a peanut) and place it on the burning spoon.

(d) Hold the burning spoon in a Bunsen flame until the peanut catches fire.

(e) Hold the burning peanut below the boiling tube as shown in the diagram.

(f) When the flame from the burning peanut dies away, record the highest temperature the water reaches, while stirring with the thermometer.

(g) Repeat the experiment using the same mass of different foods. (You could use the same foods as in Question 3.)

Keep your experiment fair

To make your experiment fair, there are three things you should **keep the same** when carrying out your experiments:

- the mass (weight) of food being burned
- the volume of water being heated
- the distance between the burning food and the boiling tube.

The starting temperature of the water in the boiling tube doesn't have to be the same each time as you are measuring the temperature change.

Results

Record your results in a table like this one.

Food being burned	Temperature of water at the start (°C)	Highest temperature of water reached (°C)	Temperature rise (°C)*

*Temperature rise = highest temperature – lowest temperature

The rise in temperature is a measure of the amount of energy produced when the food is burned. Write a list of the foods you tested in order of the amount of energy produced, from most to least.

Alcohol as a fuel

National 3

Learning intention

In this section you will:

- learn that alcohol is a fuel and burns to give carbon dioxide and water.

Alcohol is also a fuel. It is made by fermenting sugars. In schools it is used in spirit burners and can be used as a fuel for model steam engines.

Make the link

For more about fermentation, see Chapter 15.

(a)

(b)

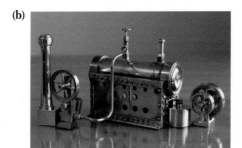

Figure 8.18 **(a)** *Spirit burners and* **(b)** *model steam engines use alcohol as fuel.*

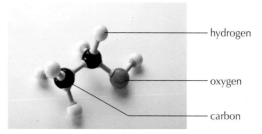

hydrogen

oxygen

carbon

Figure 8.19 *A model of an alcohol molecule.*

Alcohol molecules are made up of carbon, hydrogen and oxygen atoms.

Alcohol is not a hydrocarbon because the molecules contain oxygen in addition to carbon and hydrogen. However, like hydrocarbons, alcohols burn to give carbon dioxide and water:

alcohol + oxygen → carbon dioxide + water

Make the link

You can learn more about alcohol as a fuel in Chapter 13.

GO! Activity 8.3

☺ 1. (a) Alcohol molecules contain hydrogen and carbon. Why aren't they hydrocarbons?

   (b) Write the word equation to show the products from burning alcohols.

Learning checklist

After reading this chapter and completing the activities, I can:

N3 L3 N4 L4

- state that coal, natural gas, crude oil and peat are examples of fossil fuels. **Activity 8.1 Q1(a), Activity 8.2 Q1** ○ ○ ○

- state that hydrocarbons are compounds that are made up of only the elements hydrogen and carbon. **Activity 8.1 Q1(b), Activity 8.2 Q1** ○ ○ ○

- state that when hydrocarbons burn they combine with oxygen from the air. **Activity 8.1 Q1(c)** ○ ○ ○

- state that when hydrocarbons burn carbon dioxide and water are produced. **Activity 8.1 Q1(d), (e)** ○ ○ ○

N3 L3 N4 L4

- state that limewater is used to test for carbon dioxide because carbon dioxide turns limewater milky.
Activity 8.1(e), Activity 8.2 Q3   ○ ○ ○

- state that tests for water are that water has a pH of 7 and will turn blue cobalt chloride paper pink.
Activity 8.1 Q1(e), Activity 8.2 Q3   ○ ○ ○

- state that fuels are substances which can be burned to produce energy. **Activity 8.2 Q1**   ○ ○ ○

- state that coal, natural gas, peat and substances that can be obtained from crude oil, such as petrol and diesel, are fuels that can be burned to produce energy. **Activity 8.2 Q1**   ○ ○ ○

- state that fossil fuels are mainly hydrocarbons.
Activity 8.2 Q1   ○ ○ ○

- state that foods are fuels and combine with oxygen in the body to provide the body with energy.
Activity 8.2 Q2(a), Q4   ○ ○ ○

- state that starch, sugars, proteins and oils provide the body with energy. **Activity 8.2 Q1, Q4**   ○ ○ ○

- state that when foods burn in our body carbon dioxide and water are produced. **Activity 8.2 Q2, Q3**   ○ ○ ○

9 Fuels and energy 2: Controlling fires

This chapter includes coverage of:

N3 Fuels and energy

National 3

The fire triangle

Learning intention

In this section you will:

* learn about the fire triangle and the conditions needed for fire.

Figure 9.1 *Once the fuel is ready, all that's needed to get a fire going in the open air is a match.*

Figure 9.2 *A fire triangle shows us the conditions needed for burning. Take away any one of the three sides and the fire will go out.*

📖 Word bank

* **Fire triangle**

A helpful image that summarises the three conditions needed for fire: fuel, heat and oxygen.

What is needed for a fire?

When fuels burn they combine with oxygen in the air. A fuel and oxygen do not guarantee that burning will take place. A third condition is needed for burning to start. We need something to start the fire. This is normally a source of heat, such as a match or an electrical spark.

When lighting a campfire, a match can be used to supply the heat to start the burning. The campfire will continue to burn as long as more wood is added as fuel. This is summarised in Figure 9.2 and is known as the **fire triangle**.

If you want to put out a campfire there are different ways you can do it.

* You can shovel earth over the fire. This will prevent oxygen reaching the fuel. Since there is no oxygen, the fire will die out.

* You can throw water over the fire. This removes heat, putting the fire out.

* You can stop adding fuel. When all the fuel is used up, the fire will go out.

Because three things – a fuel, oxygen and heat – are required to get a fire started, we can think about putting fires out by taking away one of the conditions shown in the fire triangle.

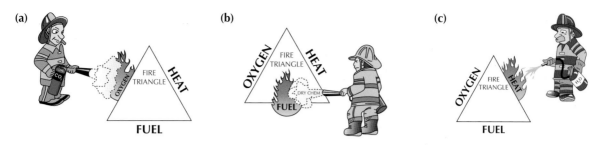

(a) **(b)** **(c)**

Figure 9.3 *Take away any one of (a) oxygen, (b) fuel or (c) heat, and a fire will go out.*

> **Activity 9.1**
>
> 😊 **1. (a)** Name the **three** conditions that are needed for a fire.
>
> **(b)** Name and draw the symbol used to represent the three conditions needed for a fire.

Extinguishing fires safely

National 3

Learning intentions

In this section you will:
- learn about different types of fire extinguisher
- explain how different types of fire can be extinguished safely.

Controlling different types of fires

Dealing with fires can be tricky. If a fire occurs, the first priority is to raise the alarm and make sure everyone is safe by evacuating the area. If you find a fire:

- alert everyone else in the building (use the fire alarm if there is one)

- get yourself out

- close doors as you leave to contain the fire

- call the fire brigade.

Remember that fire can spread quickly and people can easily be overcome by smoke and poisonous fumes.

Figure 9.4 *In the event of a fire, the first priority is to raise the alarm and evacuate safely.*

Word bank

- **Extinguish**

To put out, to stop burning.

Types of fire extinguisher

There are several different types of fire **extinguisher**. Each type is appropriate for use with a particular type of fire. There are situations in which it is **wrong** to use particular types of extinguisher. Table 9.1 lists different types of extinguisher and situations in which they can and shouldn't be used.

Table 9.1 *Types of fire extinguisher.*

Extinguisher		Type of fire					
Colour	Type	Solids (wood, paper, cloth, etc)	Flammable liquids	Flammable gases	Electrical equipment	Cooking oils & fats	Special notes
	Water	✓ Yes	✗ No	✗ No	✗ No	✗ No	Dangerous if used on 'liquid fires' or live electricity.
	Foam	✓ Yes	✓ Yes	✗ No	✗ No	✓ Yes	Not practical for home use.
	Dry powder	✓ Yes	✓ Yes	✓ Yes	✓ Yes	✗ No	Safe use up to 1000v.
	Carbon dioxide (CO_2)	✗ No	✓ Yes	✗ No	✓ Yes	✓ Yes	Safe on high and low voltages.

⟳ Keep up to date!

Search online for up-to-date information about recommendations on the types of extinguisher for different types of fire. Fire service websites have information relating to extinguishers.

How do fire extinguishers work?

Fire extinguishers are pressurised. You need to remove the safety pin in the handle before using an extinguisher. Squeezing the handles of the extinguisher together opens a valve which lets you spray the contents of the extinguisher on the fire.

The fire triangle can be used to explain how the different types of extinguisher work.

Water can be used to cool burning solid material. Heat is taken away from the fire triangle.

Dry powder extinguishers put a layer of powder on the burning material. This stops oxygen from reaching the burning material, smothering the fire.

Carbon dioxide and foam extinguishers work in two ways. The carbon dioxide or foam stops oxygen getting to the burning surface, and it cools the fuel (the foam contains water and the carbon dioxide jet is very cold).

Figure 9.5 *How different fire extinguishers are used to put out fires.*

Even if you think you can deal with a small fire, remember that using the wrong type of fire extinguisher could put your life at risk.

For example, **never** use a water extinguisher or a foam one on an electrical fire. This could lead to you, or anyone else nearby, being electrocuted because water is an electrical conductor. Instead, carbon dioxide or dry powder should be used.

The safest way to deal with a chip-pan fire is to cover the pan with a fire blanket (or a damp towel) and to switch off the heat source. The fire blanket or towel will stop oxygen getting to the burning oil. Putting water on an oil fire will make the situation much worse. If you use a carbon dioxide extinguisher, the powerful jet of carbon dioxide could force the burning oil out of the pan, spreading the fire.

Figure 9.6 *Extreme care needs to be taken when dealing with electrical fires.*

? Did you know ...?

In Scotland 1 in 10 household fires is started by unattended chip pans or by other hot food pans.

Figure 9.7 *The result when you try to put out chip-pan fire with water.*

✺ Chemistry in action: Putting out very big fires

Specially adapted aircraft are used to fight forest fires and other wildfires. An aircraft can drop a fire **retardant** on the fire and on the surrounding area to prevent the fire spreading.

Fire retardants are mainly water (85%). The fertiliser ammonium phosphate (10%) and colourants and thickeners such as clay (5%) are added. The ammonium phosphate makes the fire retardant sticky, so that the areas sprayed have a coating of water that doesn't run off. The water cools the burning material and also stops oxygen getting to the surface. The colourants are added so that the pilots can see the areas that have already been sprayed.

Figure 9.8 *An aircraft dropping fire retardant on a wildfire.*

📖 Word bank

• **Retardant**
A substance that slows the rate of burning.

GO! Activity 9.2

☻ 1. Use Table 9.1 to answer these questions.
 (a) (i) When is the only time you should use a water extinguisher?
 (ii) Which side of the fire triangle is removed when water is used to extinguish a fire?
 (b) Which types of extinguishers can be used safely on electrical fires?
 (c) Which **two** types of extinguisher can be used to extinguish burning oil?
2. Carbon dioxide extinguishers are used on fires involving flammable liquids.
 Name the gas that carbon dioxide prevents from getting to the fuel.
3. Firefighters use foam to extinguish vehicle fires. Draw the fire triangle to show the two ways in which foam will extinguish a fire.

🔍 Hint

Figure 9.5 will help you.

A firefighter extinguishing a car fire.

4. Describe the best way of dealing with a chip-pan fire.
5. Your television catches fire. Why shouldn't you use a water or foam extinguisher to put the fire out?

| N3 | L3 | N4 | L4 |

☺☺**6.** Design a home fire-safety leaflet.

Try to include:

- what to do in the event of a fire
- information relating to the use of different types of extinguisher
- statistics relating to fire in the home.

You could search online for up-to-date recommendations from fire services.

GO! Activity 9.3

 In these experiments you use dilute hydrochloric acid, which is harmful and corrosive. You should always wear safety goggles for experiments. If you get any acid on your skin or clothes wash with water.

☺☺**1. Experiment**

This is a simple demonstration of how a carbon dioxide extinguisher works.

You will need: some tea lights, a 400 cm³ beaker, measuring cylinder (50 or 100 cm³), sodium hydrogencarbonate powder (also called sodium bicarbonate or baking soda), dilute (1 mol/l) hydrochloric acid (you can use vinegar).

Method

(a) Light some tea lights and place them in a row on the bench.

(b) Put 5 g sodium hydrogencarbonate powder (sodium bicarbonate) in a 400 cm³ beaker.

(c) Pour 50 cm³ of 1 mol/l hydrochloric acid onto the powder and swirl the beaker.

(d) Now tip the beaker slightly and move along the row of tea lights.

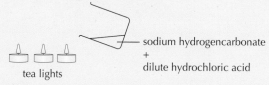

sodium hydrogencarbonate
+
dilute hydrochloric acid

tea lights

Tip the beaker slightly and pour the gas that is produced onto the tea lights.

Results

The carbon dioxide produced when the acid reacts with sodium hydrogencarbonate is heavier than air and stays in the beaker. When you tip the beaker, you pour the carbon dioxide onto the flames.

Question

Explain why tipping the beaker and pouring carbon dioxide onto the candles extinguishes the flames.

(continued)

2. Experiment

You can also make a model foam extinguisher by carrying out the reaction in a conical flask.

You will need: 250 cm³ conical flask, rubber bung with delivery tube, a small test tube (it needs to be small enough to lie in the bottom of the conical flask), thread, 5 g sodium hydrogencarbonate, 2 cm³ washing-up liquid, 50 cm³ 1 mol/l hydrochloric acid, a tea light, a heatproof mat.

Method

(a) Measure 50 cm³ of 1 mol/l hydrochloric acid into the conical flask.

(b) Add 2 cm³ of washing-up liquid to the acid.

(c) Measure 5 g of sodium hydrogencarbonate into a small test tube and tie a piece of the thread around the neck of the test tube.

(d) Suspend the test tube in the conical flask and stopper the flask with the rubber bung and delivery tube.

(e) Light a tea light and place it on a heatproof mat.

(f) Loosen the rubber bung slightly and lower the test tube of sodium hydrogencarbonate into the acid.

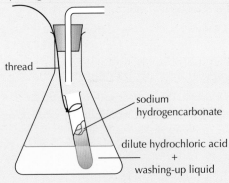

Set-up for making model foam extinguisher.

(g) Stopper the flask again and shake the flask gently.

(h) Direct the foam produced at the burning tea light.

Results

When the reactants mix, a foam will be produced that can be sprayed onto a burning candle.

Questions

(a) Name the gas produced when sodium hydrogencarbonate reacts with acid.

(b) Explain why the foam from your extinguisher has put out the burning tea light.

3. Carry out a survey of the fire safety precautions in your school. You may have to speak to a senior member of staff to get permission for this activity and to find out some of the information.

Try to find out the answers to these questions:

- How many extinguishers are there?
- How many types of extinguisher are there? Do certain departments such as Technology and Science have specific types?

- Is any other fire safety equipment available? Look out for sand buckets and fire blankets.
- Where are the fire alarms situated?
- Are there special fire doors?
- Are fire safety notices displayed?
- Are there special arrangements for evacuating wheelchair users or other people with a disability?
- How long does it take to evacuate the building?

Prepare a presentation on the topic 'Fire safety in our school'.

Learning checklist

After reading this chapter and completing the activities, I can:

N3 L3 N4 L4

- state that a fuel, oxygen and a source of heat are required for a fire. **Activity 9.1 Q1** ○ ○ ○

- state that in order to extinguish a fire oxygen, heat or the fuel must be removed. **Activity 9.2 Q1(a)(ii), Q3** ○ ○ ○

- state that water can be used to remove heat from a fire. **Activity 9.2 Q1(a)(ii)** ○ ○ ○

- state that carbon dioxide and dry powder extinguishers prevent oxygen reaching the fire. **Activity 9.2 Q2** ○ ○ ○

- state that different types of fire require specific methods in order to be extinguished safely, for example, ○ ○ ○

  - a damp towel or a fire blanket is the best way to extinguish a chip-pan fire. **Activity 9.2 Q4** ○ ○ ○

  - water or foam extinguishers should never be used on an electrical fire. **Activity 9.2 Q5** ○ ○ ○

- *extract information from tables.* **Activity 9.2 Q1(a)(i), (b), (c)** ○ ○ ○

- *work with others.* **Activity 9.2 Q6, Activity 9.3 Q1–3** ○ ○ ○

- *research information.* **Activity 9.2 Q6, Activity 9.3 Q3** ○ ○ ○

- *present information.* **Activity 9.2 Q6, Activity 9.3 Q3** ○ ○ ○

10 Fuels 1: Fossil fuels

This chapter includes coverage of:

N4 Fuels • Planet Earth SCN 4-04b • Materials SCN 4-16b • Topical science SCN 4-20b

You should already know:

- that fuels can be burned to produce energy
- that coal, natural gas, crude oil and peat are examples of fossil fuels
- that hydrocarbons are molecules made up of only carbon and hydrogen atoms
- that hydrocarbons burn to produce carbon dioxide and water
- about the conditions needed for fire.

National 4

Trapping the Sun's energy

Learning intention

In this section you will:
- explain how the Sun's energy is trapped by photosynthesis.

All of the energy we use in the world today can be traced back to the energy that has come from the Sun. The Sun's energy is trapped by plants and is passed on to animals through food chains.

Plant leaves contain chlorophyll, which gives plants their green colour and allows plants to capture the Sun's energy.

In sunlight, plants take in carbon dioxide from the atmosphere and absorb water through their roots. The chlorophyll in their leaves allows the plants to change the carbon dioxide and water into glucose. Plants can then convert the glucose into starch. This allows the plants to store the Sun's energy in the sugar and starch molecules.

This process is called **photosynthesis**.

Photosynthesis can be shown as a word equation:

$$\text{carbon dioxide} + \text{water} \xrightarrow[\text{Sun's energy}]{\text{chlorophyll}} \text{glucose} + \text{oxygen}$$

📖 Word bank

- **Photosynthesis**

The process by which green plants change carbon dioxide and water into sugars, starches and oxygen.

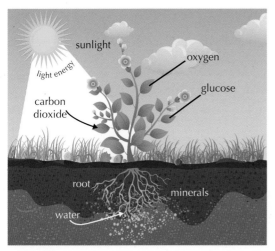

Figure 10.1 *Chlorophyll allows plants to capture the Sun's energy.*

All plants are stores of energy. The energy is stored as chemical energy in the bonds that join the atoms in plant molecules together.

It is not only living plants that are energy sources. The fossilised remains of trees that lived millions of years ago can be a source of energy. These remains are coal, a fossil fuel.

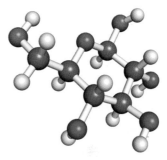

Figure 10.2 *A model of a glucose molecule – a sugar made by plants. The energy is stored in the bonds that hold the carbon, hydrogen and oxygen atoms together.*

🟢 Activity 10.1

☺ **1.** Green plants are able to store the Sun's energy using a process called photosynthesis.

   (a) Name the substance in green plants that allows carbon dioxide and water to react during photosynthesis.

   (b) Name the gas given out to the atmosphere by green plants during photosynthesis.

   (c) How is the Sun's energy stored in green plants?

🔗 Make the link

Photosynthesis and food chains are covered in *S1–N4 Biology Student Book*, Chapter 9.

Formation of fossil fuels

Learning intentions

In this section you will:
- explain what is meant by the term 'fossil fuel'
- explain how fossil fuels were formed.

Most of the energy used in the world today is obtained by burning **fossil fuels**.

How were fossil fuels formed? These fuels are the remains of animals and plants that lived and trapped the Sun's energy many millions of years ago.

National 4

Curriculum level 4

Planet Earth: Energy sources and sustainability SCN 4-04b

📖 Word bank

- **Fossil fuel**

A fuel that has been made from the decaying (rotting) remains of ancient plants and animals.

How coal and peat were formed

The acidic conditions in ancient forests meant that, when trees died, they did not completely decay (rot). Over time, partly decayed material was covered by mud and sand, compacting the material. This high pressure, and heat from within the Earth, turned the decaying material into coal.

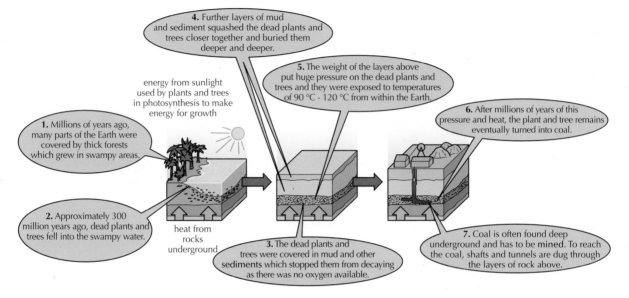

4. Further layers of mud and sediment squashed the dead plants and trees closer together and buried them deeper and deeper.

energy from sunlight used by plants and trees in photosynthesis to make energy for growth

5. The weight of the layers above put huge pressure on the dead plants and trees and they were exposed to temperatures of 90 °C - 120 °C from within the Earth.

6. After millions of years of this pressure and heat, the plant and tree remains eventually turned into coal.

1. Millions of years ago, many parts of the Earth were covered by thick forests which grew in swampy areas.

2. Approximately 300 million years ago, dead plants and trees fell into the swampy water.

heat from rocks underground

3. The dead plants and trees were covered in mud and other sediments which stopped them from decaying as there was no oxygen available.

7. Coal is often found deep underground and has to be mined. To reach the coal, shafts and tunnels are dug through the layers of rock above.

Figure 10.3 *How coal was formed.*

Sometimes you can find fossils of plant material in coal.

Figure 10.4 *A fossil fern, over 300 million years old.*

Today, similar processes to those of coal formation are leading to the formation of peat from mosses in acidic peat bogs.

How oil and natural gas were formed

Oil and natural gas were formed from microscopic sea creatures that lived in the oceans. As they died, they fell to the sea floor and were covered by sediments. As the sediments built up, they caused an increase in pressure. This, together with heat from within the Earth, helped to turn the dead remains into crude oil.

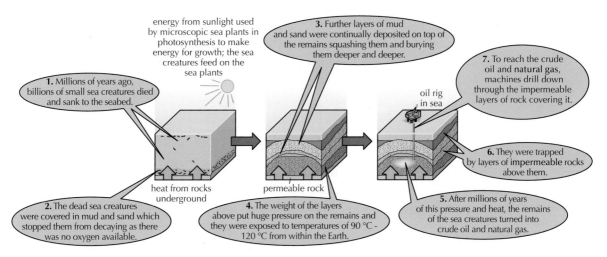

Figure 10.5 *How oil and natural gas were formed.*

Activity 10.2

1. Explain what is meant by the term 'fossil fuel'.
2. Look at Figures 10.3 and 10.5, showing how coal, oil and natural gas were formed.
   (a) In what ways were the processes similar?
   (b) In what way were the processes different?
3. Draw **two** flowcharts to show how:
   (a) coal was formed
   (b) oil and natural gas were formed.

Fossil fuels and new technologies

Learning intention

In this section you will:
- learn why new technologies are being developed to replace our dependence on fossil fuels.

Fossil fuels are a **finite resource**. This means supplies of fossil fuels will eventually run out. Once they are used up, they will be gone forever.

It was thought that the known oil and gas reserves would run out in the next 60 years, with coal reserves perhaps lasting for another 250 years.

National 4

Curriculum level 4

Planet Earth: Energy sources and sustainability SCN 4-04b; Topical science SCN 4-20b

📖 Word bank

- **Finite resource**

A resource which is non-renewable and will eventually be used up.

New discoveries

New discoveries of fossil fuel reserves in places like Africa and South America and the use of new technologies, such as fracking and deep-water drilling, to recover hard-to-reach reserves of oil and gas, could mean that supplies will last longer than is presently predicted.

Fracking

The term **fracking** is short for 'hydraulic fracturing'. This is a controversial process. It involves drilling deep down into shale rocks that contain natural gas. Explosives and a mixture of sand, water and chemicals are injected into the shale rocks to fracture (break) the rock and release the trapped gases.

> ### 📖 Word bank
>
> • **Fracking**
>
> Breaking shale rocks apart using pressure, sand and water to release trapped natural gas.

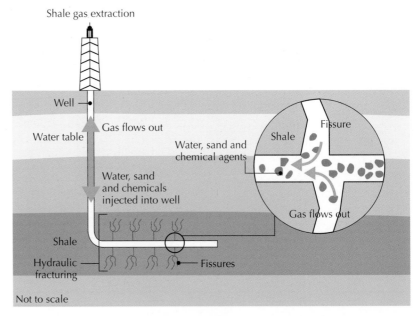

Figure 10.6 *Shale gas can be extracted by fracking shale rock.*

Steps in fracking:

1. Engineers drill down into the shale and form a well.

2. Water, sand and chemicals are injected at high pressure into the well up to 3000 m deep.

3. The pressurised mixture causes the natural fissures (cracks) and layers in the rock to crack.

4. Sand particles hold the fissures in the shale open, allowing natural gas to flow up the well.

5. Natural gas flows out of the well into storage.

It is thought that enough gas could be released from shale rocks in the north of England to supply the United Kingdom with gas for the next 500 years.

UK FRACKING GO-AHEAD BOOSTS SHALE GAS INDUSTRY

Britain's shale gas industry has won a significant victory after the government overturned local council objections to a fracking scheme in Lancashire, clearing the way for the first exploration since an earthquake halted drilling five years ago.

Sajid Javid's decision was the clearest sign yet of the government's willingness to push through shale gas development in the face of fierce opposition from local communities and environmental groups. The Communities Secretary said shale gas had "the potential to power economic growth, support 64,000 jobs, and provide a new domestic energy source, making us less reliant on imports". Mr Javid ruled that Cuadrilla Resources should be allowed to drill four horizontal wells at its Preston New Road site near Blackpool, the first time permission has been granted in the UK for a form of fracking that would extend beneath homes.

At the same time, the push for alternative sources of hydrocarbons took a blow in Scotland, where the Scottish National Party administration blocked a proposal to extract gas from under the Firth of Forth using a controversial coal-burning technique. There is already a moratorium on shale gas fracking in Scotland. Advocates of fracking said the English ruling was a breakthrough in the push to unlock Britain's shale gas resources after a near-standstill since exploratory drilling by Cuadrilla in Lancashire caused small earth tremors in 2011.

Campaigners say fracking – which involves pumping large volumes of water, sand and chemicals under the ground at high pressure to release gas and oil trapped in rock – pollutes groundwater, causes geological instability and, ultimately, undermines efforts to tackle climate change.

Source: *Financial Times 6 October 2016*

⟳ Keep up to date!

Oil reserves

The oil company BP is cautious in saying how big oil reserves are. On one of its web pages, the company states:

'Nobody knows or can know how much oil exists under the Earth's surface or how much it will be possible to produce in the future.'

BP estimates that, if known oil reserves are used at the same rate as now, they will run out in 53.3 years' time.

Use the internet to find out the latest predictions for how long oil and other fossil fuels will last.

Fracking

Fracking has caused a lot of controversy. Find out the latest developments relating to fracking by searching online using terms such as 'fracking' or 'Why is fracking controversial?'

In the USA fracking has been linked to an increased incidence of asthma and an increased frequency of attacks in those with the condition.

GO! Activity 10.3

☺ 1. The information on pages 142–143 presents arguments for and against the use of fracking to obtain natural gas.

Summarise reasons for and against the use of fracking. This will help you to decide whether fracking should or should not be used to help meet our future energy needs.

Figure 10.7 *The deep-water drilling rig, Deepwater Nautilus, a sister rig of the Deepwater Horizon, being transported on a heavy lift ship.*

Deep-water drilling

Until recently, it was not economical to try to exploit oil reserves that lay deep under the ocean. As oil reserves have diminished and drilling technology has improved, it has become more cost effective to try to exploit these resources.

In September 2009 the oil rig Deepwater Horizon drilled to a depth of 10 685 metres in water that was 1259 metres deep. This means it drilled nine and a half kilometres into the Earth's crust.

GO! Activity 10.4

☺ 1. (a) Why are fossil fuels described as finite resources?

(b) Name **two** technologies that are being used to recover hard-to-reach fossil fuel reserves.

2. The graph shows the known crude oil reserves in the USA.

(a) In 2015, how many billion barrels of crude oil were estimated to remain?

(b) In what year were the estimated reserves of crude oil lowest?

(c) Suggest why estimated reserves have increased during 2008 to 2014.

3. *You will need: internet access.*

The Deepwater Horizon drilling rig created controversy in 2010. Search online and find out what happened. Write a 50–100 word report of your findings.

Known crude oil reserves in the USA, 1965 to 2015
Source: *US Energy Information Administration*

4. The table shows the proven (known) reserves of natural gas in the USA.

   (a) Draw a line graph to show how the natural gas reserves have changed in the 50 years since 1965. (Your teacher may give you graph paper with the axes already drawn.)

   (b) Explain the trends that your graph shows from 1970 until 1990, and from 1995 until 2015, in terms of amount of gas used and discovery of new reserves.

Year	Gas reserves (trillion cubic feet)
1970	285
1975	220
1980	205
1985	200
1990	175
1995	175
2000	185
2005	200
2010	300
2015	320

Burning fuels

National 4

Learning intentions

In this section you will:

- describe the types of molecules that make up fossil fuels
- explain what happens when hydrocarbon fuels burn in a plentiful supply of oxygen
- learn what is formed when hydrocarbons burn in a limited supply of oxygen
- learn why combustion reactions can be described as oxidation reactions
- use the fire triangle to explain how fires can be extinguished.

Combustion

Fossil fuels are mainly composed of hydrocarbons. They also contain impurities, including sulfur compounds. These compounds create problems when the fossil fuel is burned and can be a major cause of acid rain.

Reactions in which fuels burn are **combustion** reactions.

> **:: Make the link**
>
> You can read more about acid rain in Chapter 6.

> **📖 Word bank**
>
> - **Combustion**
> Burning of a fuel.

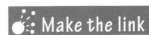

Figure 10.8 *The candle will go out when all of the oxygen in the air is used up.*

Candle wax obtained from crude oil is a mixture of hydrocarbons. If you float a burning candle in a glass trough and cover it with a bell jar, the candle will burn for a short while and then go out. You may also notice that the water will rise up the bell jar slightly. This is because oxygen in the air is used up as the candle burns.

When fuels burn, the fuel molecules combine with oxygen in the air. This releases energy, mainly in the form of heat. Since the fuel combines with oxygen, burning can also be referred to as **oxidation.**

Complete combustion

If there is sufficient air, this gives a good oxygen supply and a hydrocarbon fuel will burn completely, producing carbon dioxide and water.

The word equation for the **complete combustion** of a hydrocarbon fuel is:

hydrocarbon + oxygen → carbon dioxide + water

You should know how to test for carbon dioxide and water.

Incomplete combustion

Hydrocarbon fuels burn completely to give carbon dioxide and water only if enough oxygen is present. Sometimes there is not enough oxygen to burn a fuel completely.

When you heat a small beaker of water using the yellow flame of a Bunsen burner or using a candle, the bottom of the beaker becomes blackened. This is because there is not enough oxygen for the carbon atoms to become carbon dioxide and they form a sooty deposit (carbon) on the beaker. Carbon monoxide is also formed.

Burning fuels in a limited supply of oxygen is called **incomplete combustion**.

📖 Word bank

- **Oxidation**

The combining of a substance with oxygen in a chemical reaction.

- **Complete combustion**

When a hydrocarbon compound burns to give only carbon dioxide and water.

- **Incomplete combustion**

When a hydrocarbon fuel burns in a limited supply of air producing carbon and carbon monoxide.

Make the link

You can learn more about testing for carbon dioxide and water in Chapter 8.

Figure 10.9 *The yellow Bunsen flame on the left is due to incomplete combustion. The blue flame on the right produces complete combustion.*

Carbon monoxide

Carbon monoxide is an extremely poisonous gas, often referred to as a 'silent killer' as it has no colour and no smell. It is estimated that about 40 people in the UK die from accidental carbon monoxide poisoning each year. This can be due to poorly maintained gas appliances such as boilers and cookers.

All homes with gas appliances should be fitted with carbon monoxide detectors.

Figure 10.10 *A domestic carbon monoxide detector.*

Family killed in suspected carbon monoxide poisoning at caravan site 'died quickly'

Three members of the same family who died after suspected carbon monoxide poisoning at a caravan park would have been "unconscious within minutes", investigators said today.

It is also believed there was no working carbon monoxide detector in the static caravan where the bodies of John Cook, 90, his wife Audrey, 86, and their 46-year-old daughter Maureen were discovered on Saturday afternoon.

It is believed the family may have been using a heater to keep warm at Tremarle Home Park, Camborne, in Cornwall, as the region experienced freezing conditions overnight into the weekend.

Police and fire crews today continued their investigations at the mobile home.

The cause of death is still to be determined, although an early reading from experts at the site revealed a potentially lethal level of carbon monoxide present within the caravan at the time it was taken, shortly after the grim discovery.

Mark Pratten, crew manager in prevention with Cornwall Fire and Rescue Service, said: "Carbon monoxide was at an extremely high level. A significant dose such as this would have been fatal. The investigation is continuing, but it would appear that the people inside the home would have slipped into unconsciousness within a few minutes (of the leak)."

Source: Independent 25 February 2013

Car engines are a major source of carbon monoxide. This is why it is dangerous to run a car engine in an enclosed space, such as a garage.

In order to minimise emissions of carbon monoxide and the nitrogen oxide gases that cause acid rain, modern cars are fitted with catalytic converters.

In the catalytic converter, the harmful nitrogen oxide gases and carbon monoxide are changed to less harmful carbon dioxide and nitrogen. A simplified word equation for the reactions taking place is:

carbon monoxide + nitrogen oxides → carbon dioxide + nitrogen

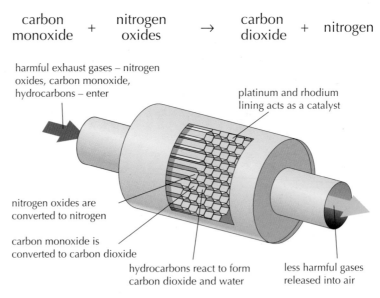

harmful exhaust gases – nitrogen oxides, carbon monoxide, hydrocarbons – enter

platinum and rhodium lining acts as a catalyst

nitrogen oxides are converted to nitrogen

carbon monoxide is converted to carbon dioxide

hydrocarbons react to form carbon dioxide and water

less harmful gases released into air

Figure 10.11 *A catalytic converter converts harmful gases into less harmful gases.*

Exothermic chemical reactions

Reactions are described as **exothermic** when they give out energy. All combustion reactions are examples of exothermic chemical reactions.

Sometimes the energy from an exothermic reaction is released very quickly (explosively). This can be demonstrated using a mixture of methane and oxygen gases.

 Your teacher might demonstrate this reaction. It is best carried out in a darkened room.

 Stand well back, and wear goggles and ear protection.

A small plastic bottle is filled with a mixture of methane and oxygen gases. If the stopper is removed from the bottle and a burning splint is held at the mouth of the bottle, the gas mixture explodes.

Word bank

• **Exothermic**
Giving out energy.

Make the link

You can read more about endothermic reactions and exothermic reactions in Chapter 5.

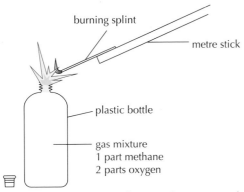

Figure 10.12 *Mixing methane and oxygen produces an exothermic reaction.*

Chemistry in action: Gas welding torches

Gas welding torches use a mixture of oxygen and a hydrocarbon called acetylene. Acetylene is the only hydrocarbon gas that gives a flame hot enough to cut through steel or to melt the edges of metal parts, welding them together.

Figure 10.13 *The temperature of an oxy-acetylene flame is hot enough to cut through steel.*

Controlling combustion reactions: the fire triangle

Combustion needs a fuel, oxygen and heat to get the reaction started. This is often represented by a fire triangle. Because combustion is an exothermic process, once the reaction starts, the energy given out keeps the reaction going until the fuel is used up or the supply of oxygen is removed.

Controlling combustion is a matter of removing one side of the fire triangle. There are, therefore, three ways of controlling fires:

- shutting off the fuel supply

- cooling the burning material, normally using water (water hoses and water fire extinguishers can be used)

- preventing oxygen from reaching the burning surface (carbon dioxide, foam and dry powder extinguishers create a barrier between the burning material and the oxygen of the air).

Figure 10.14 *The fire triangle.*

Make the link

You can read more about the fire triangle and dealing with fires in Chapter 9.

| N3 | L3 | **N4** | L4 |

⊙ Activity 10.5

☺ **1.** The molecules of many of the compounds in fossil fuels only contain carbon and hydrogen atoms.

    (a) State the term used to describe compounds that consist of molecules that only contain carbon and hydrogen atoms.

    (b) Name the products formed when a hydrocarbon burns in a good supply of air.

    (c) (i) Name **two** substances that can be formed when a hydrocarbon burns in a limited supply of air.

       (ii) State the term used to describe burning of a hydrocarbon in a limited supply of air.

2. (a) Acetylene is a hydrocarbon. When acetylene reacts with oxygen, only carbon dioxide and water are produced.

    (i) State the term used to describe reactions in which a substance combines with oxygen.

    (ii) State the term used to describe the burning of a hydrocarbon to produce only carbon dioxide and water.

    (b) Burning acetylene produces a large amount of heat energy.

    State the term used to describe a reaction that produces a large amount of heat energy.

3. Explain why car exhaust systems contain catalytic converters.

National 4
Curriculum level 4
Materials: Properties and uses of substances SCN 4-16b

The law of conservation of mass

Learning intention

In this section you will:

- explain the law of conservation of mass.

During all chemical reactions, including combustion reactions, none of the atoms of the reactants are lost. They are all changed to products. The **law of conservation of mass** states that in a chemical reaction the mass of the products must equal the mass of the reactants.

❔ Did you know ...?

During the eighteenth century many chemists worked on experiments relating to mass changes during reactions.

A French chemist, Antoine Lavoisier, is credited with the discovery of the Law of conservation of mass. He carried out experiments in which tin and lead were reacted with oxygen in sealed vessels. He had powerful lenses that could focus the Sun's rays, creating the very high temperatures needed for the reactions to take place. By carrying out the reactions in sealed vessels, Lavoisier was able to show that mass remains constant during reactions.

Figure 10.15 *Antoine Lavoisier.*

Methane burning

Methane is the simplest hydrocarbon and is found in natural gas. It is made up of a carbon atom bonded to four hydrogen atoms, CH_4.

Figure 10.16 *A methane molecule, formula CH_4.*

When methane burns in oxygen, the carbon atom joins with oxygen atoms forming carbon dioxide, and the hydrogen atoms join with oxygen atoms to form water.

We can show this with a diagram and an equation.

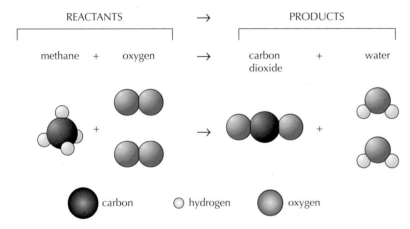

Figure 10.17 *The combustion of methane.*

$$CH_4 \quad + \quad 2O_2 \quad \rightarrow \quad CO_2 \quad + \quad 2H_2O$$
methane + oxygen → carbon dioxide + water

When you look at the diagram, you can see the same number of each type of atom on each side of the arrow.

- The methane molecule has **1** carbon atom; the carbon dioxide molecule has **1** carbon atom.

- The methane molecule has **4** hydrogen atoms; the **2** water molecules each have **2** hydrogen atoms, making **4** in total.

- The **2** oxygen molecules each have **2** oxygen atoms; the carbon dioxide has **2** oxygen atoms and both water molecules have **1**, making **4** in total.

Since the same number of each type of atom are on both sides of the equation, the mass of the products will be the same as the mass of the reactants. Nothing has been lost, confirming the law of conservation of mass.

GO! Activity 10.6

☺ **1.** Natural gas contains small amounts of propane. Each propane molecules contains three carbon atoms and eight hydrogen atoms.

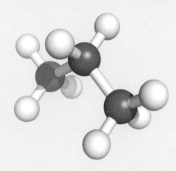

A model of a propane molecule.

(a) Write the formula for propane.

(b) How many molecules of carbon dioxide will be produced when a molecule of propane burns?

(c) How many molecules of water will be formed?

(d) How many molecules of oxygen will be needed to burn the propane molecule completely?

(e) Draw a diagram for the burning of propane and explain how this demonstrates the law of conservation of mass.

> 🔍 **Hint**
>
> In part **(d)**, count up the atoms of oxygen in the carbon dioxide and water molecules and divide by 2.

Learning checklist

After reading this chapter and completing the activities, I can:

N3 L3 **N4 L4**

- state that the Sun's energy is trapped by green plants by a process called photosynthesis. **Activity 10.1 Q1(a), (b)** ○ ○ ○

- state that the Sun's energy is trapped in chemical bonds that hold the atoms of plant molecules together. **Activity 10.1 Q1(c)** ○ ○ ○

- state that fossil fuels are fuels which formed from the remains of dead animals and plants that lived millions of years ago. **Activity 10.2 Q1** ○ ○ ○

- explain that fossil fuels were formed by the effects of temperature and pressure on decaying plant and animal material over a long period of time. **Activity 10.2 Q2, Q3**   ○ ○ ○

- explain that fossil fuels are described as finite as they cannot be replaced when they are used up. **Activity 10.4 Q1(a)**   ○ ○ ○

- state that fossil fuels are made up of hydrocarbon molecules. **Activity 10.5 Q1(a)**   ○ ○ ○

- explain that hydrocarbon fuels burning in a plentiful supply of oxygen to produce carbon dioxide and water is termed complete combustion. **Activity 10.5 Q1(b), Q2(a)(ii)**   ○ ○ ○

- state that hydrocarbons burning in a limited supply of oxygen can produce carbon and carbon monoxide, and this is termed incomplete combustion. **Activity 10.5 Q1(c)**   ○ ○ ○

- state that catalytic converters are fitted in car exhaust systems to reduce carbon monoxide emissions. **Activity 10.5 Q3**   ○ ○ ○

- state that combustion reactions can be described as oxidation reactions because the fuels combine with oxygen. **Activity 10.5 Q2(a)(i)**   ○ ○ ○

- state that combustion reactions can be described as exothermic reactions as they give out energy. **Activity 10.5 Q2(b)**   ○ ○ ○

- explain that mass is not created or lost during a chemical reaction and that this is known as the law of conservation of mass. **Activity 10.6 Q1**   ○ ○ ○

- *draw line graphs.* **Activity 10.4 Q3(a)**   ○ ○ ○

- *read and interpret line graphs describing the trends shown by graphs.* **Activity 10.4 Q2, Q3(b)**   ○ ○ ○

- *express opinions for and against an issue.* **Activity 10.3**   ○ ○ ○

11 Fuels and energy 3: The problems with using fossil fuels

This chapter includes coverage of:

N3 Fuels and energy • Planet Earth SCN 3-05b

You should already know:

- that energy can be obtained by burning fuels
- that fossil fuels were formed from living things which died millions of years ago
- that hydrocarbons are compounds made only from carbon and hydrogen.

National 3

Curriculum level 3

Planet Earth: Processes of the planet SCN 3-05b

The problem with using fossil fuels

Learning intention

In this section you will:

- explain the disadvantages of using fossil fuels.

Figure 11.1 *The Swedish chemist, Svante Arrhenius, who predicted global warming due to the burning of fossil fuels.*

Fossil fuels have been used for thousands of years. Records show that the Chinese were using coal in 3490 BCE. But it is only since the Industrial Revolution in the eighteenth century that humans have used fossil fuels on a large scale (see page 91). Today, most of the energy used in the world is obtained by burning fossil fuels. However, there are problems with this reliance on fossil fuels.

In the 1890s, a Swedish chemist, Svante Arrhenius, concluded that burning of fossil fuels would increase carbon dioxide levels in the atmosphere and that this would lead to global warming. But it wasn't until the 1980s, with record temperatures and droughts in the USA and a major international conference in Toronto, that politicians began to recognise the seriousness of environmental issues caused by modern industrial activity.

The continuing use of fossil fuels is likely to have severe environmental and health consequences.

Burning fossil fuels and global warming

Fossil fuels are mainly made up of hydrocarbon compounds. Burning hydrocarbons produces carbon dioxide, which is a greenhouse gas. Increased levels of carbon dioxide in the atmosphere is the main reason that that the atmosphere is trapping more heat, causing global warming and causing the Earth's climate to change.

Impact of burning fossil fuels on health

Burning fossil fuels doesn't only affect the Earth's climate. It has serious effects on human health.

The Organisation for Economic Cooperation and Development (OECD) is an organisation set up by the governments of 35 countries to promote trade.

The OECD estimates that:

- around the world, 3.5 million people die every year as a result of air pollution. This is greater than the number of deaths caused by dirty water and poor sanitation.

- half of these deaths are linked to vehicle exhaust emissions. Emissions from diesel vehicles cause the greatest harm.

- the cost to the economies of OECD countries is estimated to be $1.6 trillion.

Smog formation

In many Chinese cities smog – a mixture of smoke particles and fog, which is poisonous and foul-smelling – causes many health problems. Weather conditions often prevent the air pollution from being blown away. Many residents are forced to stay in their homes and those that do go out wear masks to avoid breathing the poisonous air.

Figure 11.3 *People wearing masks as a protection against the smog.*

In Beijing, authorities are taking steps to try to reduce air pollution problems in the future by:

- reducing the amount of coal burned by 30%

- closing coal-fired power stations

- getting rid of 300 000 high-polluting vehicles

- closing the most polluting factories and upgrading 2000 other factories

- installing air purifiers in schools.

Diesel cars

Exhaust fumes from diesel vehicles are a major source of air pollution in towns and cities.

Diesel engines cause problems in two key ways – through the production of very fine soot particles and emission of oxides of nitrogen. Very fine soot particles can get into the lungs, causing breathing problems and contributing to heart disease.

Nitrogen oxides can help form a chemical called ozone, which also causes breathing difficulties, even for healthy people.

Figure 11.4 *Diesel exhaust fumes contain very fine soot particles and nitrogen oxide gases. Many cyclists wear protective masks to stop fine particles being breathed in.*

? Did you know ...?

The leaders of four major cities – Paris, Mexico City, Madrid and Athens – say they will ban the use of all diesel-powered cars and trucks by the middle of the next decade, in order to improve air quality in their cities. Campaigners are calling for London's Mayor to commit to phase out diesel vehicles from London by 2025.

Premature deaths from the effects of air pollution are 10 times greater than deaths caused by road traffic accidents, and nitrogen dioxide and fine particle pollution in many of our streets is greater than European safety levels.

According to research by Friends of the Earth Scotland:

- there are 2094 deaths due to fine particle air pollution each year in Scotland (and 29 000 in the UK and 430 000 in the European Union)

- in comparison, this is more than 10 times the number of deaths every year in Scotland from road traffic accidents and 80% more than the number of alcohol-related deaths every year

- the average annual nitrogen oxides emissions of eight streets in Scotland are regularly greater than the European legal limit.

GO! Activity 11.1

☻ **1. (a)** What is meant by the term 'smog'?

  (b) State **two** things Chinese authorities intend to do to reduce smog formation.

  (c) Why do some cyclists in cities wear face masks?

  (d) State **two** types of health problem caused by breathing in air pollution.

☻☻**2.** Create a poster encouraging cyclists in cities to wear face masks.

Climate change targets

Learning intention

In this section you will:
- learn about plans to address problems caused by burning fossil fuels.

Climate change due to the use of fossil fuels poses such a threat to the Earth and its inhabitants that governments met in Paris in 2015 to reach an agreement regarding global warming. The Paris Agreement came into force in November 2016. By January 2017, 125 of the 195 countries that attended the climate conference had signed up to the agreement.

Some countries, including Denmark and Sweden, intend to be free of fossil-fuel use by 2050. But for other countries, particularly developing countries, the climate change agreement creates a problem.

Kosovo

Kosovo, a country in southern Europe, is reliant on two old coal-fired power stations for 97% of its electricity. There are frequent power cuts and, if the economy is to develop, the country needs to be able to generate more electricity. The country has the fifth biggest reserves of coal on Earth and the World Bank has concluded that the country's energy requirements can only be met by coal.

Make the link

For more detail about the Paris Agreement, see pages 95 and 96.

★ You need to know

In June 2017, Donald Trump, the President of the USA, withdrew support for the Paris Agreement. He wants to continue to use fossil fuels to help industry in the USA. He believes that burning fossil fuels does not cause climate change.

Figure 11.5 *Kosovo is one of the countries that was formed by the break-up of the Yugoslavia.*

Because of the problems associated with burning fossil fuels, the World Bank and other development banks have decided to lend money for new coal-powered plants only in exceptional circumstances. Kosovo's need for a new coal-powered plant is so great, however, that the World Bank is considering financing such a project. Kosovo is seeking US$1.1 billion to build a modern, clean coal-powered facility to replace one of the older plants.

Figure 11.6 *The Scottish Government has set out climate change targets to be met by 2032.*

Table 11.1 *Aspects of individual and household living that are responsible for greenhouse gas emissions.*

Aspect	% of emissions
housing	32
transport	30
food	16
consumption (shopping)	11
other	11

Scotland

The Scottish Government is fully committed to reducing greenhouse gas emissions. In 2009, it set an interim target to reduce emissions by 42% by 2020. This was achieved in 2014, six years early. The Government has set a new target of reducing emissions by 66% by 2032.

To achieve these targets the Government plans include:

- ensuring all electricity is generated from non-carbon sources by 2030

- wherever possible, using electricity instead of natural gas or oil for heating and cooling all public and commercial properties, such as hospitals, schools, shops and offices

- encouraging individuals and households to use low-carbon heat sources to heat homes

- replacing existing vehicles with ultra-low emission vehicles.

The policy identifies aspects of our daily living that are responsible for greenhouse gas emissions by individuals and households, as shown in Table 11.1. It also lists 10 areas of changes in behaviour that will help reduce greenhouse gas emissions.

Ten key behaviour areas

Housing

1. Keep the heat in by insulating, draught proofing and double glazing.

2. Manage heating better by turning down the heating and hot-water thermostats, and reducing the time the heating is on.

3. Save electricity with energy-efficient appliances and light bulbs, and by washing clothes at low temperatures.

4. Use energy-efficient heating systems and microgeneration technologies (that is, the generation of heat and electric power on a small scale).

Transport

5. Use the car less, by walking, cycling, taking public transport and joining car-share schemes.

6. Drive more efficiently or swap to low-carbon vehicles.

7. Use alternatives to flying, such as trains or video conferencing for meetings.

Food

8. Avoid food waste.

9. Eat a healthy, sustainable and seasonal diet.

Other

10. Reduce and reuse, as well as recycle.

Low carbon-emission transport

The Government has also set targets for increasing the use of low-emission transport, including buses.

In 2015, Inverness introduced fully electric buses. They look identical to a conventional bus but have an electric power pack in place of the diesel engine. The buses take just two hours to reach full charge. In addition, the braking system generates electrical energy when the brakes are applied.

Figure 11.7 *Electric buses are contributing to cleaner air in Scottish cities.*

GO! Activity 11.2

☻ 1. Draw a bar chart to show the aspects of individual and household living that are responsible for greenhouse gas emissions.

2. Look at the 10 key behaviour areas.

   (a) In what ways can people reduce emissions caused by heating their homes?

   (b) State **one** thing that can be done to save electricity.

   (c) In what ways can people reduce car use?

   (d) Find out what is meant by 'sustainable' and 'seasonal'. Suggest one way that 'Eating a healthy, sustainable and seasonal diet' will reduce greenhouse gas emissions.

3. The graph shows how the percentage of buses that have low emissions will change to help meet the Government's climate change targets.

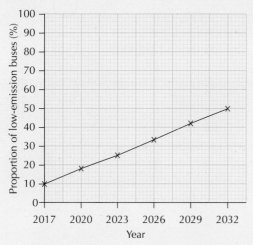

Targets for the proportion of low-emission buses in Scotland by 2032

   (a) What name is given to the type of gases that the Government wants to limit to prevent climate change?

   (b) State **one** way of making buses low emission.

   (c) What percentage of buses should be low emission by 2020?

   (d) (i) What is the Government's target for 2032?

   (ii) Use the shape of the graph to help you describe how the Government hopes the climate change target for 2032 will be achieved.

| N3 | L3 | N4 | L4 | **159** |

↻ Keep up to date!

Climate change is such an important issue that new technologies are being developed to help meet climate change targets. Search online to find out if countries are meeting their climate change targets and how they are achieving this.

Find out the latest predictions being made by climate change scientists.

Learning checklist

After reading this chapter and completing the activities, I can:

N3 L3 N4 L4

* state that burning fossil fuels has a significant impact on human health including: increased breathing difficulties, increased risk of cardiovascular (heart) disease, and increased risk of death. **Activity 11.1 Q1**   ◯ ◯ ◯

* describe ways in which the Government hopes to meet climate change targets. **Activity 11.2 Q2, Q3**   ◯ ◯ ◯

* *present information as a bar chart.* **Activity 11.2 Q1**   ◯ ◯ ◯

* *read a graph and use it to describe a trend.* **Activity 11.2 Q3 (d) (ii)**   ◯ ◯ ◯

12 Fuels and energy 4: Meeting energy needs in the future

This chapter includes coverage of:

N3 Fuels and energy • Planet Earth SCN 3-04b, SCN 3-05b

You should already know:

- that energy can be obtained by burning fuel
- that fossil fuels have been a source of energy for humans for a very long time
- that the use of fossil fuels has caused problems for the environment.

Sustainable sources of energy

National 3

Learning intentions

In this section you will:
- learn that fossil fuels are a limited resource
- learn what a sustainable source of energy is
- learn why the world needs sustainable sources of energy.

There are decreasing amounts of fossil fuels, which are a finite resource, meaning that eventually they will run out altogether.

To reduce carbon dioxide emissions and meet climate change targets, we need to develop ways of fuelling transport and generating electricity that do not rely on **finite energy sources**. Developing **sustainable energy sources** will be the key to meeting future energy needs and combating global warming.

Sustainable sources of energy are energy sources that future generations will continue to be able to use. They include **renewable energy sources**, such as wind power and solar power, as well as **non-renewable energy sources**, such as nuclear energy.

📖 Word bank

- **Finite energy sources**
Sources of energy that will run out, because they cannot be replaced on a human timescale.

- **Sustainable energy sources**
Energy sources whose present-day use will not prevent future use.

- **Renewable energy sources**
Energy sources that are naturally occurring, such as wind and solar energy, and sources that can be naturally replaced on a human timescale.

- **Non-renewable energy sources**
Energy sources that cannot be replaced on a human timescale, such as fossil fuels.

National 3

Curriculum level 3

Planet Earth: Energy sources and sustainability SCN 3-04b; Processes of the planet SCN 3-05b

Biomass and biofuels

Learning intentions

In this section you will:

- explain what biomass is
- explain what biofuels are
- give examples of biofuels
- learn how biofuels can be obtained from biomass.

Biomass

Approximately two billion people in developing countries still burn wood, dried animal dung or crop waste as fuel to provide energy. These fuels are referred to as **biomass** since they have been obtained from living things.

Biomass fuels release carbon dioxide into the atmosphere when they burn. But they only release the carbon dioxide that was absorbed by plants during their lifetime. Therefore, biomass fuels such as wood are **carbon neutral**, since they do not put any more carbon dioxide into the atmosphere than they absorbed during their lifetime.

Biofuels

As well as being burned directly, biomass can be processed to make fuels. For example, vegetable oils that are extracted from biomass can be used to make fuels. These **biofuels** include bioethanol, biodiesel and biogas.

Figure 12.1 *Girl carrying fuel made by mixing cow dung and straw.*

Word bank

- **Biomass**

Material obtained from living things that can be burned to produce energy.

- **Carbon neutral**

Not releasing more carbon dioxide than is used up.

- **Biofuel**

A fuel that is produced from biomass or from used biomass products such as reclaimed vegetable oil.

Figure 12.2 *Biofuels are produced from biomass.*

Bioethanol

Brazil was one of the first countries to use ethanol as a fuel for motor vehicles. The 'bioethanol' is made by fermenting sugar from sugarcane juices. In Brazil, this fuel is called 'Alcool'.

> **? Did you know ...?**
>
> There is a European Union (EU) requirement for 10% of transport fuel to come from renewable sources by 2020. One way to achieve this is by using petrol such as E20 (20% ethanol and 80% petrol), and by increasing the levels of biodiesel that are used in diesel fuel.

Figure 12.3   *Brazil pioneered the use of ethanol (Alcool) as fuel for cars.*

☀ Chemistry in action: Brazil

Brazil is the world's largest sugarcane ethanol producer. In 2015/16, Brazilian ethanol production reached 30.23 billion litres (8 billion gallons). It is sold as either pure ethanol fuel or mixed with petrol.

When bioethanol was first developed as a fuel for cars, people had to choose which type of fuel they would use. In 2003, Volkswagen in Brazil introduced a car that could run on a petrol–alcohol mixture or on pure ethanol. This is called flexible-fuel technology.

By 2014, 68% of the light commercial vehicles and 90% of the new cars sold in Brazil used flexible-fuel technology. Figure 12.4 shows how the growth in the use of flexible-fuel vehicles is expected to continue until 2020. The number of petrol and alcohol-only vehicles is predicted to fall.

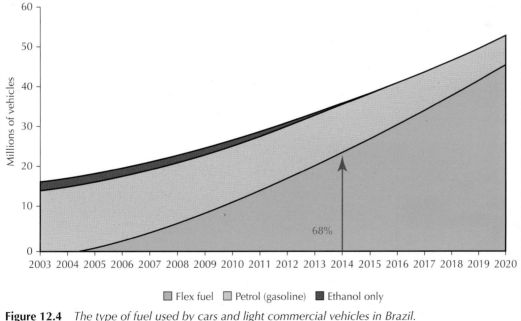

Figure 12.4   *The type of fuel used by cars and light commercial vehicles in Brazil.*

Figure 12.5 *Sugar beet looks like a fat parsnip.*

Figure 12.6 *The first British bioethanol plant at Wissington.*

Figure 12.7 *An old photograph of the sugar beet factory in Cupar, Fife.*

Figure 12.8 *Fields of rapeseed are easily recognised because of their vivid yellow colour. The rapeseed is used to produce cooking oil which, after it has been used, can be turned into biodiesel.*

Sugar beet as a source of sugar for bioethanol

In Brazil, bioethanol is produced from sugarcane, which is suited to Brazil's tropical climate. Sugar can also be obtained from a root crop called sugar beet, which can be grown in countries like the UK, which have a temperate climate. On average, 7.5 million tonnes of sugar beet are grown annually in the UK.

? Did you know ...?

A stalk of sugarcane contains about 12–14% sugar. A sugar beet contains 16–18% sugar. Although sugar beet contains more sugar, the same area of land produces more sugar from sugarcane than from sugar beet. Sugarcane gives a yield of 10 tonnes per hectare, whereas sugar beet gives a yield of about 7 tonnes per hectare. Overall, sugarcane is a better source of sugar.

In 2007, British Sugar opened the first bioethanol plant using sugar beet to produce ethanol at its Wissington factory. The plant can process up to 650 000 tonnes of sugar beet annually producing 55 000 tonnes (70 million litres) of bioethanol.

Scotland's sugar beet industry

The sugar beet industry in the UK really began during the First World War when the main sugar-beet producing areas in Northern France and Belgium became battlefields. Up until the 1970s there was a factory in Cupar in Fife processing sugar beet into sugar.

However, the industry was very labour intensive which made it uneconomic, and the factory closed in December 1971. Modern agricultural techniques and machinery could make growing sugar beet viable again in Scotland. Growing sugar beet and building a plant to process the sugar beet into bioethanol could be one way for Scotland to produce its own biofuels for transport use.

Biodiesel

Biodiesel can be used in place of ordinary diesel fuel.

It is more expensive to produce biodiesel from crops than to produce ordinary diesel from crude oil. So, in the UK and Europe, much of the biodiesel is produced from reclaimed (used) cooking oil and waste animal fats from the food-processing industry.

Biodiesel is biodegradable (it breaks down naturally), making it is safe for the environment, and it produces much less air pollution than diesel fuel. It even smells better than diesel fuel!

Using biodiesel reduces air pollution

Biodiesel reduces nearly all forms of air pollution, compared to diesel from crude oil. Using pure biodiesel (known as B100) reduces the toxic particles in vehicle emissions and can reduce cancer risks by 94%. Using B20 (a mixture of 20% biodiesel and 80% ordinary diesel) reduces cancer risks by 27%. Using biodiesel also reduces the risk of genetic mutations.

Biodiesel reduces greenhouse gas emissions

The plants used to make cooking oils absorb (take in) carbon dioxide while they grow. Burning biodiesel simply returns the carbon dioxide to the atmosphere. Using biodiesel is, therefore, considered to be carbon neutral. In fact, some studies have estimated that the use of 1 kg of biodiesel leads to the reduction of some 3 kg of CO_2.

Biogas

Biogas can be produced from animal manure, or municipal (town) waste at landfill sites. Many farms use biogas for heating and to produce electricity. Some companies that operate waste management sites clean and purify the gas producing **biomethane**. Biomethane can be liquefied and exported to the national gas grid network.

> **? Did you know ...?**
> Use of biodiesel in the mix for diesel fuel is expected to rise by 14% from 2016 to 2020 to help governments meet emission targets.

Figure 12.9 *Biogas storage facility on a farm.*

> **📖 Word bank**
> - **Biogas**
> A biofuel produced by decaying animal manure or municipal waste.
> - **Biomethane**
> Cleaned and purified biogas.

Biofuel production in the UK

Table 12.1 shows the different biofuel plants in the UK in 2013.

Table 12.1 *Biofuel plants in the UK.*

Company	Location	Year opened	Capacity* (million litres)	Fuel type	Feedstock (raw material)
Argent Energy	Motherwell, Scotland	2005	60	Biodiesel	Used cooking oil, animal fat, sewage grease
Harvest Energy	Seal Sands, Teesside	2006	284	Biodiesel	Primarily waste oils
British Sugar	Wissington, Norfolk	2007	70	Bioethanol	Sugar beet
Convert 2 Green	Middlewich, Cheshire	2007	20	Biodiesel	Used cooking oil
Greenergy	Immingham, Hull	2007	220	Biodiesel	Waste oils
Gasrec	Aldbury, Surrey	2008	5	Biomethane (liquid)	Municipal solid waste
Ensus	Wilton, Teesside	2010	400	Bioethanol	Wheat
Olleco	Bootle, Merseyside	2012	16	Biodiesel	Used cooking oil
Vivergo	Immingham, Hull	2013	420	Bioethanol	Wheat

* 'Capacity' means the amount which can be produced.

↻ Keep up to date!

1. Plans for two biofuel plants were put on hold in 2016 because it was not economical to build them. Search online to find out the latest in biofuel developments in the UK and elsewhere.

2. Search online for a map of the UK to show where biogas is being produced.

✳ Chemistry in action: Biofuels in jet aircraft

Biofuels for use in jet aircraft have even been developed.

Five bio-based jet fuels have been made for use for air travel.

The new fuel, known as Alcohol to Jet Synthetic Paraffinic Kerosene (ATJ-SPK), has been produced from an alcohol called isobutanol that is obtained from renewable feedstocks such as sugar, corn, or forest wastes.

Figure 12.10 *The development of biofuels will ensure that air travel continues to be possible after fossil fuels run out.*

GO! Activity 12.1

☻ 1. State the term used to describe material such as wood and crop waste that can be burned to produce energy.

2. Name **three** biofuels.

3. Explain why biofuels can be described as carbon neutral.

4. (a) Name **two** plants that can be used to produce bioethanol.

   (b) Name the biofuel made from used cooking oil.

   (c) Name the gas that is made by cleaning and purifying biogas.

5. (a) Copy and complete the table to show information about growing sugar beet and sugarcane.

	Type of climate required	Sugar content (%)	Yield (tonnes per hectare)
Sugarcane			
Sugar beet			

   (b) Until the 1970s sugar beet was grown and processed in Scotland.

   (i) Why did growing sugar beet stop in Scotland?

   (ii) Name the biofuel that can be made from sugar.

6. Use the information in Table 12.1 to complete the table below.

Biofuel	Number of plants	Total capacity (million litres)	Feedstocks
Biodiesel			
Bioethanol			
Biomethane			

Alternative technologies in Scotland

National 3

Learning intention

In this section you will:
- learn about sources of sustainable energy.

Scotland relies on a mix of energy sources, from fossil fuels and nuclear power to renewable energy sources, such as hydroelectricity, wind power, solar and even biomass.

Scotland is increasingly moving towards renewables for electricity generation.

Nuclear energy

Scotland has two nuclear power stations; one is at Hunterston in North Ayrshire and the other is at Torness in East Lothian.

Nuclear energy is a very dependable source of energy. It is sustainable but non-renewable, since our present-day use will not affect the potential to use it in the future. However, nuclear power stations are costly to build and to decommission (dismantle and clean up). There is also the problem of dealing with the highly radioactive waste produced by nuclear stations. There have been accidents at nuclear power stations, notably at Chernobyl in the Ukraine (1986) and Fukushima in Japan (2011), which resulted in loss of life for workers and people living near the power plants. Nuclear accidents can cause huge environmental damage – at the time of the accident and for a long time afterwards.

Figure 12.11 *Peterhead gas-fired power station is the last power station in Scotland to use fossil fuels.*

Figure 12.12 *Torness nuclear power station in East Lothian. Its life has been extended to 2030 to help meet energy demand.*

↻ Keep up to date!

1. The clean-up operations following the disaster at Fukushima will go on for many years. Find out what has happened in the area since the disaster.

2. Dounreay nuclear power plant has been decommissioned and the site is being restored (repaired to a state where it can be used for other purposes). Find out how the restoration project is progressing, and if any problems are being found.

⟳ Keep up to date!

Use the internet to find out the latest developments in fuel-cell technology and their use in transport.

Developing a hydrogen economy

One possible way to reduce carbon dioxide emissions is to use hydrogen as a fuel. At present, most hydrogen is made from fossil fuels but it can also be made by splitting water apart using electricity. The hydrogen can then be recombined with the oxygen in a fuel cell, providing energy. Fuel cells can be used to power vehicles and are more efficient than petrol or diesel engines. The great benefit is that the fuel cells only produce water, so no carbon dioxide is released to the atmosphere.

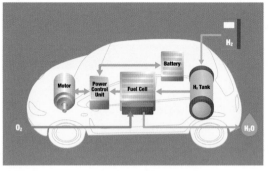

Figure 12.13   *How a fuel-cell car works.*

Scotland's main ferry operator is planning to develop the world's first ferry powered by hydrogen fuel cells. The zero-emissions vessel could run on hydrogen gas produced using excess electricity generated overnight from wind farms near to the ports it serves.

Figure 12.14   *Scottish ferry operator Caledonian MacBrayne hope to develop a car ferry powered by fuel cells.*

✴ Chemistry in action: Aberdeen Hydrogen Bus Project

On 11 March 2015, the UK's first hydrogen-production and bus-refuelling station opened in Aberdeen as part of a £19 million green transport project.

The project has support from Europe, the UK Government and the Scottish Government, and is designed to create a hydrogen economy in the city.

The 10 special buses that are to be used in the city will only emit (produce) water vapour, reducing carbon emissions and air pollution, as well as being quieter and smoother to run than diesel vehicles. There are plans for more hydrogen stations which could fuel other vehicles, including cars.

Aberdeen City Councillor Jenny Laing commented, 'We have a very clear Hydrogen Strategy for the future and are on the cusp [edge] of realising our aspiration [hope] of becoming a world-leading city for low-carbon technology, while maintaining our position as a leading world energy city.'

Derek Mackay, a Minister in the Scottish Government claimed, 'This project isn't just good news for transport – it also demonstrates how we can use hydrogen as energy from renewables, which integrates [blends] our energy and transport sectors, as well as making the most of Scotland's vast renewable energy resources.'

Wind farms

By March 2015 Scotland had 2683 wind turbines generating 5115 megawatts (MW) of electricity. Another 282 were under construction and planning permission had been granted for a further 2202 turbines. On one day in August 2016 it was reported that enough electricity had been generated by wind turbines to meet all of Scotland's energy needs for the first time.

But there are days when no electricity is generated. A major disadvantage of wind energy is simply that, if the wind doesn't blow or if the wind is too strong for the turbines to work safely, no electricity can be generated.

Figure 12.15 *A typical wind farm in the Moorfoot hills.*

Hydroelectricity

Hydroelectricity is the electricity generated by water flowing from a high level to a lower level.

The geography of Scotland – with mountains and fast-flowing rivers – makes it ideally suited to producing electricity from hydroelectric schemes.

Cruachan hydroelectric power station is an example of a pump storage system. When excess electricity is being generated in other ways, it can be used to pump water from Loch Awe into a reservoir over 1000 feet higher. Then, when electricity is needed, it is generated by allowing the water to flow back down through turbines located deep within the mountain.

In 2014 it was announced that Scottish Power would consider more than doubling the amount of electricity produced at Cruachan.

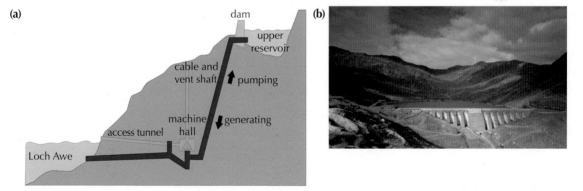

Figure 12.16 **(a)** *Schematic diagram of Cruachan pumped storage hydro scheme.* **(b)** *The dam above Cruachan power station.*

A major advantage of hydroelectric power plants is that, unlike most other types of power station, they can produce electricity almost immediately.

Figure 12.17 *The first turbine being lowered into position in the Pentland Firth in 2016. Eventually, turbines in this area will generate enough electricity to power 175 000 homes.*

Wave and tidal energy

Scotland is one of the leaders in producing renewable energy using the power of the waves and the tides. The tides in the Pentland Firth and around the Orkney Islands make this area ideal for developing energy from turbines on the sea bed.

In October 2016 the first of four turbines in a pilot scheme was installed on the seabed. By 15 November the BBC reported that the turbine had begun producing electricity.

It is thought that a combination of tidal and wave power from the area could be capable of producing up to 60 gigawatts (GW) of power. That is 10 times Scotland's annual electricity usage.

Solar panels

Scotland also generates electricity using solar (light) energy. Many homes, small businesses and public buildings, such as schools, are fitted with solar panels.

The Government promotes their use by making a payment to people who fit solar panels to their homes or businesses for the electricity they generate. The payment is called a 'feed-in tariff'. The electricity that is generated but not used by the home or business owner feeds back into the National Grid.

The disadvantage of relying on solar power is that in winter, when the light intensity is low and days are short, generation levels are low.

Figure 12.18 *Solar panels fitted to a house in Scotland.*

Geothermal energy

The Scottish Government is interested in the use of geothermal energy, which is heat produced from the Earth. Deep below the surface of the Earth, the rocks are much warmer than at the surface. This heat can be captured by drilling down into the rocks and pumping in water. When the water returns to the surface, it is warmer and the heat can be extracted.

❓ Did you know ...?

Scotland's largest solar-panel site was officially opened in June 2016. The site on Errol Estate in Perthshire consists of 55 000 solar panels. The scheme generates enough electricity for 3500 homes.

Figure 12.19 *Solar panels at Errol Estate.*

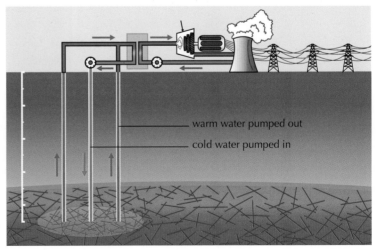

warm water pumped out

cold water pumped in

Figure 12.20 *Geothermal energy can be captured by pumping water down into the Earth.*

⊚ Activity 12.2

☺ **1.** Copy and complete the following paragraph.

Scotland relies on a mix of energy sources. These include n_ _ _ _ _ _ _,
h_ _ _ _ _ _ _ _ _ _ _ _, w_ _ _, and wave and t_ _ _ _ energy. Peterhead power station
is the only power station that still uses f_ _ _ _ _ f _ _ _ _.

2. (a) Complete the key by inserting the correct terms for **(i)**, **(ii)** and **(iii)**.

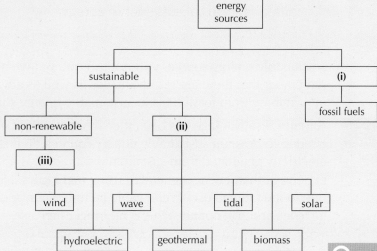

Types of energy sources.

> 🔍 **Hint**
>
> For part **(i)**, look at page 161.

(b) Which energy source would be unable to meet a
demand for electricity during the night?

(c) Burning wood pellets is an example of using which type of energy source?

(d) Which energy source uses the heat of the Earth to provide energy?

3. Aberdeen City has introduced buses powered by hydrogen fuel cells.

State **two** advantages of using hydrogen fuel cells rather than diesel engines to
fuel buses.

4. (a) What features of Scotland's geography make
Scotland suitable for the development of
hydroelectric schemes.

(b) The pumped storage hydroelectric system at Ben
Cruachan is unique in Scotland. Describe how the
system operates.

> 🔍 **Hint**
>
> You may want to refer back to
> Figure 12.16 to help you
> answer the question.

National 3

Curriculum level 3

Planet Earth: Energy sources and sustainability SCN 3-04b

Advantages and disadvantages of different sustainable sources of energy

Learning intention

In this section you will:

- learn about the advantages and disadvantages of different forms of sustainable energy.

Sustainable energy sources will not run out. But we must remember that there are disadvantages as well as advantages in switching from fossil fuels to sustainable energy sources.

A major disadvantage has been the large amount of money needed to develop sustainable energy sources. At first, electricity generated from sustainable sources was more expensive than energy generated using fossil fuels. However, due to the falling cost of renewable technologies, such as wind turbines and solar panels, this is no longer the case.

Table 12.2 summarises some of the advantages and disadvantages of different sustainable energy sources.

? Did you know ...?

In 2017, it was reported that in 30 countries, for the first time, generating electricity from solar or wind was as cheap or cheaper than generating electricity from fossil fuels.

Table 12.2 *Advantages and disadvantages of different sustainable energy sources.*

Energy source	Advantages	Disadvantages
Nuclear	• No greenhouse gas emissions • Dependable	• High cost of construction and decommissioning • Difficulties of dealing with highly radioactive nuclear waste • Risk of accident
Solar	• Energy from the Sun does not cost anything to produce • No greenhouse gases	• Solar-panel sites are expensive to build and take up a lot of space • Panels only work when Sun is shining and don't work at night • Generation levels in winter are lower when demand for electricity is high • Only certain places are suitable
Wind	• No greenhouse gases • Very safe • Can be sited offshore	• Large areas are needed for a wind farm to produce electricity • Turbines only work when it is windy • Some people think wind farms spoil the environment • Can affect wildlife, particularly birds

Hydroelectric	• No greenhouse gases • More reliable than wind and solar • Can be switched on and off quickly, therefore can respond quickly to changing generation needs	• Dams are expensive to build • Only certain areas are suitable • Land and wildlife habitat are lost when reservoirs are created
Wave and tidal	• No greenhouse gases • The amount of electricity produced can be calculated in advance	• Expensive to develop • Need to be sited near land • Not all sea areas are suitable • Energy production can be a long way from where it is needed
Biomass	• Biomass is carbon neutral; it only gives off as much carbon dioxide as it absorbed when growing • Can make use of materials that would otherwise be thrown away • Can be used to produce biofuels	• Land used to grow biomass crops could be used to grow food crops • Use of fertilisers to grow crops can cause pollution
Geothermal	• No greenhouse gases • Plentiful resource • Reliable • Low maintenance	• High cost • Possible emissions (of carbon dioxide, sulfur dioxide and nitrous oxide)

GO! Activity 12.3

😊 1. (a) (i) What is meant by a 'sustainable energy source'?

   (ii) What is the major advantage in terms of global warming and climate change of using sustainable energy sources?

   (b) State **two** disadvantages of relying on nuclear energy.

   (c) Explain why electricity generated by hydroelectric power stations is useful when there is a sudden demand for electricity.

   (d) Suggest why tidal power is more reliable than wind or solar power.

😊😊 2. Scotland has many options for generating the energy it will need to meet its future needs. Prepare a presentation to share with other members of your class to show how you think this should be done.

Learning checklist

After reading this chapter and completing the activities, I can:

N3 L3 N4 L4

- state that biomass is material obtained from living things that can be burned to produce energy. **Activity 12.1 Q1**   ○ ○ ○

- state that biomass can be processed to produce biofuels such as biodiesel, biogas and bioethanol.
  Activity 12.1 Q2   ○ ○ ○

- state that biodiesel can be obtained from used cooking oil. **Activity 12.1 Q4(b)**   ○ ○ ○

- state that bioethanol can be obtained from sugarcane and sugar beet. **Activity 12.1 Q4(a)**   ○ ○ ○

- state that biofuels are carbon neutral because they release only as much carbon dioxide as was absorbed by the plants that were used to make them.
  Activity 12.1 Q3   ○ ○ ○

- state that sustainable energy sources used for electricity generation include: nuclear, wind turbines, hydroelectricity, and wave and tidal. **Activity 12.2 Q1**   ○ ○ ○

- state that Scotland is suitable for the generation of hydroelectricity because it has mountains and rivers that flow quickly. **Activity 12.2 Q4**   ○ ○ ○

- state the major advantage of most sustainable energy sources used for electricity generation is that they do not produce greenhouse gases. **Activity 12.3 Q1(a)(ii)**   ○ ○ ○

- state that some sustainable energy sources such as solar and wind will only work in certain weather conditions.
  Activity 12.3 Q1(d)   ○ ○ ○

- *select information from a table.* **Activity 12.1 Q6**   ○ ○ ○

13 Fuels 2: Solutions to fossil fuel problems

This chapter includes coverage of:

N4 Fuels • Planet Earth SCN 4-04b, SCN 4-05b

You should already know:

- about fossil fuels
- why we need alternatives to fossil fuels as sources of energy
- what biomass is
- that biomass can be burned to produce energy
- that biomass can be processed to produce biofuels, including biodiesel, bioethanol and biogas
- that biofuels burn to produce carbon dioxide, but are considered carbon neutral
- about sources of sustainable energy, and their advantages and disadvantages
- about risks associated with the use of nuclear power.

Combustion of fossil fuels and the carbon cycle

Learning intentions

In this section you will:

- learn about the carbon cycle
- describe how burning fossil fuels contributes to the carbon cycle
- explain why carbon dioxide levels are increasing in the atmosphere.

All living organisms are made from carbon compounds. The carbon to make these compounds comes from the carbon dioxide in the atmosphere. Green plants remove carbon dioxide from the atmosphere and build carbon compounds by photosynthesis.

National 4

Curriculum level 4

Planet Earth: Energy sources and sustainability SCN 4-04b; Planet Earth: Processes of the planet SCN 4-05b

Make the link

For more about photosynthesis, look at Chapter 10.

The carbon in plants passes from organism to organism along food chains. Carbon dioxide passes back into the atmosphere from the decaying remains of living things and by respiration. Respiration happens in living things and involves the production of energy and carbon dioxide.

In certain conditions, when plants and animals die, their carbon compounds can become trapped in the Earth and are not released back into the atmosphere. This is what happened millions of years ago when fossil fuels were formed. The process continues today with the formation of peat in peat bogs.

This series of processes that recycles carbon between the atmosphere, the Earth and living organisms is known as the **carbon cycle**.

Word bank

- **Carbon cycle**

The series of processes, including photosynthesis, respiration and decomposition, by which carbon moves from the atmosphere into living organisms and back again.

Make the link

Photosynthesis and respiration are also covered in the *S1–N4 Biology Student Book*.

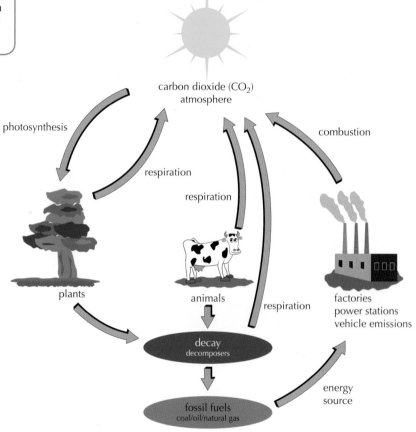

Figure 13.1 *The carbon cycle.*

The burning of fossil fuels is upsetting the natural balance between carbon in the atmosphere and carbon in living organisms. Fuels that have taken millions of years to form are being burned in just a few centuries, leading to an increased level of carbon dioxide in the atmosphere.

The activities of humans have caused the increase in the carbon dioxide level in the atmosphere.

At the Mauna Loa Observatory in Hawaii, scientists have been tracking the carbon dioxide level in the atmosphere for more than 50 years.

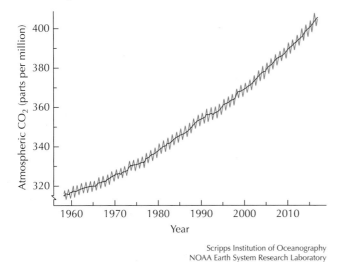

Figure 13.2 *Observations at Mauna Loa Observatory show that the carbon dioxide level in the atmosphere is increasing.*

Scientists have made measurements that show a steady increase in the level of carbon dioxide in the atmosphere. This is due to greater and greater demand for energy across the world, leading to the burning of more fossil fuels. The concern is that, if the carbon dioxide level continues to rise, this will lead to increased global warming and climate change. Governments are keen to stop this happening.

 **Make the link**

You can read more about global warming and the effect of fossil fuels in Chapter 6.

GO! Activity 13.1

☻ **1.** The diagram describes the carbon cycle.

   (a) Name processes **(i)**, **(ii)** and **(iii)**.

   (b) Describe the effect of burning fossil fuels on the level of carbon dioxide in the atmosphere.

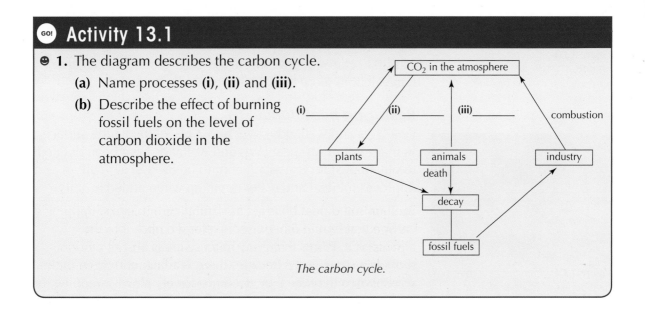

The carbon cycle.

National 4

Curriculum level 4

Planet Earth:
Energy sources and
sustainability SCN 4-04b

Reducing carbon dioxide emissions

Learning intentions

In this section you will:

- describe natural ways of reducing the carbon dioxide level in the atmosphere
- explain what is meant by carbon capture
- describe how captured carbon dioxide can be stored
- explain the role of biomass and biofuels in reducing carbon dioxide emissions.

In January 2017, the Scottish Government updated its climate change targets. Goals to be achieved by 2032 include:

- cutting greenhouse emissions by 66%

- generating all electricity from non-fossil fuel sources

- heating 80% of homes using low-carbon technologies

- increasing the use of ultra-low emission cars and vans

- increasing the amount of woodland

- restoring degraded (damaged) peatlands.

Replanting woodland and restoring peatlands

Replanting woodlands

Cutting down forests, particularly tropical rainforest to harvest the wood and to create farming land, means there are far fewer trees to absorb carbon dioxide from the atmosphere.

In order to help reverse the effects of cutting down woodland, the Scottish Government has set a target of planting at least 15 000 hectares of new woodland each year. This is an area equivalent to more than 20 000 football pitches.

Figure 13.3 *Cutting down rainforest: fewer trees means more atmospheric carbon dioxide.*

Restoring peat bogs

Peat bogs and wetlands cover about 6% of the Earth's surface. In many countries, peat bogs and wetlands have been drained to obtain peat for use as a fuel or to be used in agriculture as it improves mineral and water retention when added to soils.

Wetlands and peat bogs act as natural carbon sinks, trapping carbon that would otherwise be released back into the atmosphere. Plants that grow in these areas absorb carbon dioxide as they grow. Because there is a lack of oxygen in the water, when they die they decompose very slowly, trapping the carbon.

Figure 13.4 *Planting new trees will help Scotland meet its climate change targets.*

Figure 13.5 *A map of the world showing peatland areas.*

Peat bogs can be restored by blocking drainage channels that were created to drain the bogs, allowing the water content to increase again.

The Scottish Government's plan is to restore 250 000 hectares of peat bogs.

Planting new woodlands and restoring peatlands would be ways of lowering the carbon dioxide level in the atmosphere naturally.

↻ Keep up to date!

Try to find out if the Government is managing to achieve its targets. Search online using terms such as 'woodland creation in Scotland' or 'peatland regeneration in Scotland'.

GO! Activity 13.2

☺ 1. Name **two** natural ways to lower the carbon dioxide level in the atmosphere.

2. Explain why planting trees will help the Government meet its carbon emission targets.

3. Why do dead plants decompose very slowly in wet peat bogs?

Carbon capture and storage

Power stations, steel-producing plants and cement works all give off a lot of carbon dioxide (see pages 103–105). One possible way of reducing carbon dioxide emissions is to remove the carbon dioxide from the gases produced by these industrial plants and to store it.

(a)

(b)

(c)

Figure 13.6 (a) *Power stations,* (b) *steel works and* (c) *cement works are responsible for much of the carbon dioxide released into the atmosphere.*

? Did you know ...?

The cement industry accounts for 5% of all the carbon dioxide emissions into the atmosphere. Nearly 1 tonne of carbon dioxide is released for every tonne of cement made.

Cement is the most important ingredient in concrete. Concrete is the second most-used substance on Earth, after water. On average, each year, about 3 tonnes of concrete are used for every person on the planet.

📖 Word bank

- **Carbon capture and storage**

The process of trapping carbon dioxide produced by burning fossil fuels or other industrial processes and storing the liquefied carbon dioxide in underground rock formations.

One way to improve oil recovery from oil wells is to inject carbon dioxide into rock formations. This forces the oil out of the rock formation, increasing the amount of oil that can be recovered. The process is known as enhanced oil recovery. Attention is turning to this method, not only as a way of increasing output from oil wells, but also as a way of safely storing carbon dioxide. This process, known as **carbon capture and storage**, prevents the carbon dioxide from being released into the atmosphere.

In 2015, due to budget cuts, the UK government cancelled a £1 billion competition for companies to provide a suitable method for carbon capture and storage. This was just six months before the grant for the winning project was to be awarded.

? Did you know ...?

The United Nations' Intergovernmental Panel on Climate Change (IPCC) has concluded that carbon capture and storage (CCS) is hugely important for tackling climate change in the most cost-effective way. The IPCC has said that without CCS, the costs of halting global warming would double.

How will carbon capture and storage work?
Carbon-capture plants could be built next to any power stations that use fossil fuels, steel works and concrete production facilities.

There are three phases to the carbon-capture storage process.

1. The carbon dioxide produced from the power station is captured and compressed to turn it into a liquid.

2. The compressed carbon dioxide is transported to a storage site by pipeline, or by boat.

3. The compressed carbon dioxide is injected into suitable sedimentary rock formations deep underground, where it would be permanently stored.

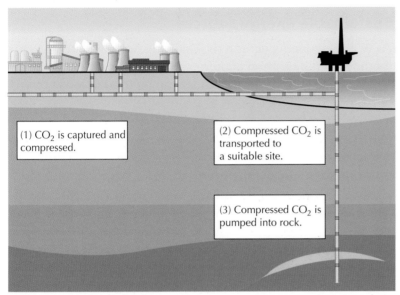

Figure 13.7 *How a carbon capture and storage scheme would work.*

Like any scheme, carbon capture and storage has advantages and disadvantages.

Advantages

- It could cut carbon dioxide emissions by up to 90%.

- If stored properly, in 1000 years' time, 99% of the stored carbon dioxide would still remain in the rock formations.

- North America alone has capacity to store 900 years of carbon dioxide emissions.

Disadvantages

- The technology needed is extremely expensive.

- The technology uses a lot of energy. As much as 40% of a power station's energy output would be needed to capture the carbon dioxide. Producing this energy would cancel out some of the gains that would be made.

- Injecting carbon dioxide into rock formations beneath the oceans could greatly increase the acidification of sea water. Scientists are already becoming aware of the problems created by acidification of the world's oceans (see pages 90–91).

> **Make the link**
>
> For more about different types of rock formation, see Chapter 17 in *S1–N4 Physics Student Book*.

> **? Did you know ...?**
>
> At the beginning of 2017 there were 21 carbon capture projects either operational or in development around the world.
>
> On 10 January 2017, the world's largest carbon capture and storage project at a power plant became operational. The Petra Nova Carbon Capture Project in the USA can remove 90% of the carbon dioxide (1.4 million tonnes every year) from the gases given off at the coal-fired power plant near Houston, Texas.

⬤GO! Activity 13.3

☻ 1. Carbon capture and storage is a process being used to reduce the amount of carbon dioxide released into the atmosphere by power stations or other industries using large amounts of fossil fuels.

Describe the **three** steps that occur during the carbon capture and storage process.

★ You need to know

You should already know about biomass and how it can be used to produce biofuels, which can be burned to give energy. See Chapter 12 for more detail.

Sustainable energy sources

As well as reducing carbon dioxide emissions by natural methods and by carbon capture and storage, sustainable sources of energy such as biomass can be used to lower emissions.

Biomass

Using biomass in place of fossil fuels reduces carbon dioxide emissions.

Figure 13.8 shows the main sources of biomass.

forestry crops and waste

agricultural crops and waste

sewage

biomass

industrial waste

animal waste

food waste

Figure 13.8 *Biomass is material obtained from a range of living things.*

Energy from biomass

The simplest way to get energy from biomass is to burn it. Burning biomass is described as being carbon neutral because it only releases the carbon dioxide that was absorbed by the biomass source when it was growing.

Sawdust and wood shavings from industrial wood processing are biomass that can be compressed into pellets, which can be burned in biomass boilers.

(a)

(b)

Figure 13.9 *(a) Wood shavings can be compressed into pellets. (b) The pellets can be burned in biomass boilers.*

? Did you know ...?

The UK's largest biomass-fuelled power station is in Scotland. Steven's Croft in Lockerbie uses 480 000 tonnes of biomass per year and can produce 44 megawatts (MW) of electricity, enough to supply the needs of 70 000 Scottish homes.

If fossil fuels were used to produce this amount of electricity it would result in an extra 140 000 tonnes of carbon dioxide being added to the atmosphere.

Figure 13.10 *Steven's Croft, Scotland's largest power station fuelled by biomass.*

Burning biomass is only one way to release its energy.

📖 Word bank

• **Biofuels**

Fuels that can be produced from biomass.

Figure 13.11 *Landfill gas being collected from decaying rubbish at a landfill site.*

Using biomass to produce biofuels

Some biomass materials, for example manure, are not suitable for burning but can be used to produce other fuels. These are called **biofuels**. Biofuels include biogas, bioethanol and biodiesel.

Biogas

Waste materials, such as our food waste, agricultural waste and even human waste, can be decomposed in containers called digesters to give a gas mixture called biogas that can be burned. Landfill gas can also be collected from decomposing materials in landfill sites. Biogas and landfill gases are mainly mixtures of methane (approximately 60%) and carbon dioxide (approximately 40%).

Biogas and landfill gas can be treated to remove the carbon dioxide and other gases that are present in smaller proportions to make them almost identical to natural gas. This purified biogas is called liquid biomethane (LBM) and can be added into the national gas grid.

Many farms are installing digesters to produce biogas from animal and crop waste. The biogas is used for heating and powering farm machinery and also provides fertilisers to be used on the farm. Other industries such as distilleries also use waste biomass materials to produce biogas.

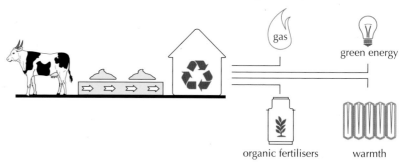

Figure 13.12 *Digesters on farms provide biogas that can be used for heating and producing energy, as well as providing fertiliser for use on the farm.*

🔄 Keep up to date!

Look out for the latest online information regarding biogas developments in your area.

🟢 Activity 13.4

☺ **1.** When biofuels burn, carbon dioxide is released into the atmosphere.

Explain why biofuels can be considered carbon neutral.

GO! Activity 13.5

1. Experiment

You can make biogas from biomass in the lab. It can be messy!

⚠ Wear disposable nitrile gloves!

Do not use glass bottles due to the risk of the glass shattering.

You will need: different types of biomass (e.g. fresh cow manure, vegetable peelings such as finely chopped carrot and potato, mashed banana or banana skins), soft drink bottles (nine identical 1 litre plastic bottles), disposable plastic cups for weighing, balloons (which must be the same size), marker pen, ruler, weighing scales, distilled water, rubber bands or tape.

Method

(a) Wash and dry the soft drink bottles.

(b) Label three of the bottles 'cow manure'.

(c) Label three of the bottles 'cow manure + vegetable peelings'.

(d) Label three of the bottles 'cow manure + mashed banana'.

(e) Use a permanent marker and ruler to draw a horizontal line about 2 cm from the top of each bottle. This is the level to which you will fill each bottle.

(f) Put on disposable gloves and weigh out 40 g of cow manure in a plastic cup and carefully transfer it to a bottle labelled 'cow manure'. Repeat this for the two other bottles labelled 'cow manure'.

(g) For the bottles labelled 'cow manure + vegetable peelings' add 20 g of cow manure and 20 g of vegetable peelings to each.

(h) For the bottles labelled 'cow manure + mashed banana' add 20 g of cow manure and 20 g of mashed banana to each.

(i) Carefully fill the bottles with deionised (distilled) water to the marks you have drawn and place a balloon over the neck of each bottle. (Make sure the balloons don't have holes but remember to squeeze out any air before fitting the balloons over the necks of the bottles.) Secure the balloon on the neck with a rubber band or tape.

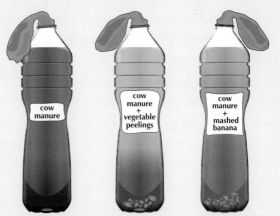

Apparatus to make and measure biogas in the lab.

Measuring and recording results

The experiment should be left for about 3 weeks.

The experiment will work better if the bottles are left somewhere warm and well ventilated, but NOT near a naked flame.

Results

Measure and record the circumference of the balloons throughout the 3-week period. Then draw graphs of your results. Plot 'Number of days' on the horizontal *x*-axis and 'Circumference of balloon' on the vertical *y*-axis of your graph.

Any equipment you use should be thoroughly cleaned at the end of the experiment.

Report

Write up the experiment. From the graph you will be able to draw conclusions about which type of biomass produces biogas most quickly and which type produces most biogas. In your report, state your conclusions, and also state two things you did to make the experiment fair.

More ideas

You could investigate the effect of temperature or the effect of pH on the rate of production of biogas.

2. Experimental design

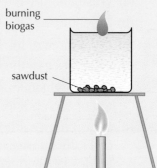

When sawdust is heated in the absence of air, biogas is given off. This can be shown by putting some sawdust in a syrup tin that has a small hole in the lid. When the tin is heated, as shown in the diagram, the gas given off can be burned. Your teacher may demonstrate this to you.

Sawdust heated in a can gives off biogas.

Design an experiment to compare the amount of gas given off from wood shavings and dried grass such as hay.

You will need to set up a clamp stand and boiling tube containing some of the biomass, as shown in the diagram. When the boiling tube is heated, biogas will be given off.

Prepare a plan for the experiment.

Think about what you need to keep the same if you want to be able to compare your results.

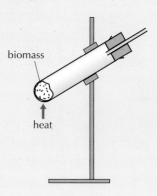

There are three ways you could decide how much biogas is given off.

- You could burn the biogas that is given off by each type of biomass, and time how long the flame lasts.

- You could collect the biogas and measure how much gas each type of biomass produced.

Comparing sources of biogas.

- You could weigh the boiling tube and biomass before heating and after the biogas is given off.

If you decide to weigh the boiling tube, make sure you let it cool before you attempt to remove it from the clamp stand.

Show your plan to your teacher, and carry out your experiment if your teacher permits.

Question

The results you would get from timing how long biogas burns for are likely to be less accurate than the other two methods.
Suggest a reason for this.

Bioethanol

Biomass can also be converted into fuels such as bioethanol or biodiesel, which are used as transport fuels. Bioethanol is an alcohol made by fermenting sugars.

$$sugar \rightarrow ethanol + carbon\ dioxide$$

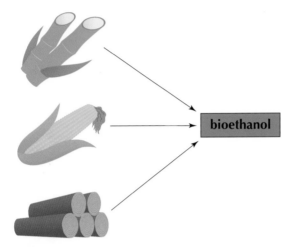

Figure 13.13 *Ethanol made from the sugars in plant material such as sugar cane, corn and wood and used as fuel is called bioethanol.*

Ethanol can also be made using compounds from crude oil fractions. Ethanol made in this way would not be classed as a biofuel.

Making bioethanol produces carbon dioxide, but the making and burning of bioethanol can be considered carbon neutral because the plants absorbed carbon dioxide from the atmosphere to make the sugar in the first place.

Biodiesel

Biodiesel is a biofuel that can be made from used cooking oil. It can also be produced from pure vegetable oils such as rapeseed oil but the cost of the oil makes this uneconomical.

Figure 13.14 *This bus runs on biodiesel made from used rapeseed oil.*

Activity 13.6

 1. Experiment

Biodiesel can be made in the lab by first extracting oil from plant material and then reacting the extracted oil with an alcohol called methanol to give biodiesel.

Part A: Extracting oil

You will need: mortars and pestles, plastic wash bottles, boiling tubes, filter funnel, stopper, test-tube rack, sweetcorn (preferably fresh), sunflower seeds or other oily nuts or seeds.

 Check to make sure that no-one has allergies to the nuts or seeds you are using.

Method

(a) Add a small amount of sweetcorn or sunflower seeds to the mortar.

(b) Grind the vegetable matter as finely as possible using the pestle. (You may need to add a small amount of water.)

(c) When your seeds are completely ground, use the wash bottle to add about 20 cm³ of water.

(d) Transfer the finely ground material and water to a boiling tube using the funnel.

(e) Rinse the mortar and pestle to remove any remaining traces of the seeds you are using and add the rinsings through the funnel to the tube.

(f) Stopper the boiling tube and place in a test-tube rack. Leave until the next time you come to class to allow time for the oils to separate.

Question

What do you need to do in order to compare which seeds or nuts give the most oil?

Part B: Changing oil to biodiesel

The oil you have made could be turned into biodiesel, but it is unlikely you will have enough. To make biodiesel you can use shop-bought cooking oil or sunflower oil instead.

You will need: test tubes and stoppers, small measuring cylinders, dropping pipette, cooking oil, 5% w/v (weight to volume) potassium hydroxide dissolved in methanol, disposable nitrile gloves.

 The potassium hydroxide in methanol solution is very corrosive and flammable. (It should only be made up by a teacher or technician.) It is made by dissolving 5 g potassium hydroxide pellets in 100 cm³ of methanol.

 The potassium hydroxide solution is toxic, corrosive and highly flammable. If carrying out this experiment you MUST be supervised by a teacher and take all the necessary safety precautions. Wear eye protection and disposable gloves and make sure there are no naked flames nearby.

Method

(a) Your teacher should give you 1.5 cm³ of the potassium hydroxide solution in a test tube.

(b) Add 10 cm³ cooking oil to the tube from a measuring cylinder.

(c) Stopper the tube.

(d) Slowly invert the tube (turn the tube upside down) to mix the oil and methanol layers.

(e) Repeat this at least 30 times.

(f) Set your test tube aside until the next day. This allows the biodiesel to separate out.

(g) You should see two layers in the test tube. The biodiesel is the top layer. When the layers have completely separated, use a dropping pipette to transfer the biodiesel to a clean test tube.

(h) Return the test tube containing the waste material to your teacher.

(i) The biodiesel can be 'washed' by adding an equal volume of water to the biodiesel and repeating the process of inverting the test tube and allowing the biodiesel to settle out. Washing with water removes impurities from the biodiesel.

(j) The washed biodiesel can then be removed by using a dropping pipette.

You can test the biodiesel to see how it burns.

2. Whoosh bottle experiment

This dramatic experiment shows the energy that can be released from ethanol. It is especially spectacular if carried out in a darkened room. Your teacher may demonstrate this to you, but you could also find a video online.

In the experiment, some ethanol is poured into a water-cooler bottle and the bottle is shaken to create ethanol vapours. Any extra ethanol is poured off. When a lit match is dropped into the bottle, the ethanol vapours ignite and a whoosh sound is heard.

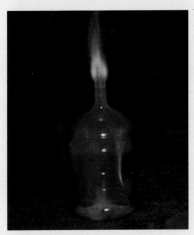

The 'whoosh bottle experiment' demonstrates that ethanol is a good energy source.

3. Copy and complete the passage below, using these words:

bioethanol   biomethane   burned   decomposing   biodiesel

Biomass is material that can be obtained from living things and _____ to produce energy. Three biofuels that can be obtained from biomass are biogas from _____ waste, _____ made by fermenting sugar, and _____ that can be made from used cooking oil. Purified biogas that is suitable for injecting into the national gas grid is known as liquid _____.

National 4

Sustainable sources of energy

Learning intentions

In this section you will:

- learn about some of the risks associated with sustainable sources of energy
- learn how switching to sustainable energy sources will affect the carbon cycle
- compare amounts of carbon dioxide produced by different sources of energy.

Risks of sustainable energy sources

Most of the risks associated with sustainable energy would appear to be financial for investors, but there are other risks as well.

Solar-panel arrays

Solar-panel sites are usually in remote rural areas. This increases the risk of vandalism and theft of the valuable panes. Sophisticated security measures including heat-seeking cameras and closed-circuit television are used to deter intruders.

Solar-panel plants also cover large areas of flat land, making them visually unattractive and reducing the amount of land available for other uses.

Figure 13.15 *The solar-panel arrays in commercial solar farms require large areas of flat land.*

Turbines

One of the biggest criticisms of wind farms is their visual impact. They are often sited on hills and can be seen for miles around. Many people feel that wind-farm developments have reached the point where they are spoiling the natural beauty of the landscape.

It is known that wind turbines pose a risk to birds and bats. In America, researchers have discovered that bats are most active when wind speeds are low. When wind speeds are low, generation levels from wind turbines are also low. Turbines, therefore, could be turned off without significant loss of energy production and lowering the risks to bats.

Tidal turbines could also damage marine life.

Figure 13.16 *Many people think that wind farms are spoiling the natural beauty of the landscape. They can also pose environmental risks, for example to wildlife.*

? Did you know ...?

In June 2013 a white-throated needle tail, the world's fastest flying bird, was spotted on the Isle of Harris. It was only the ninth time in 170 years that this species had been spotted in the UK.
Unfortunately, as birdwatchers looked on, the bird was hit and killed by the blade of a wind turbine.

Figure 13.17 *A white-throated needle tail, a rare visitor to the UK.*

Geothermal energy

Some geothermal facilities emit poisonous and greenhouse gases that have been trapped in rock formations underground. Drilling down into the rocks can release the trapped gases, including hydrogen sulfide, which smells like rotten eggs, and methane, ammonia and carbon dioxide, which are greenhouse gases.

Hydroelectricity

Building hydroelectric schemes can pose a risk to wildlife due to habitat loss. In tropical regions of the world, areas flooded to create reservoirs can release significant amounts of methane and carbon dioxide from decomposing plants and soil.

Nuclear power

The major catastrophes that took place at Chernobyl and Fukushima were well publicised by the media. But many smaller incidents that take place at nuclear facilities are not given so much media attention. Leaks from nuclear power facilities in the UK have polluted beaches and the seabed near to the facilities.

> **? Did you know ...?**
>
> The Scottish Environmental Protection Agency has said that radioactive contamination due to leaks over a 20-year period from the Dounreay nuclear plant will never be completely cleared up.
>
> Tens of thousands of particles, the size of a grain of sand, leaked from the plant, polluting beaches, the coastline and the seabed. The most dangerous particles could be deadly if swallowed. Since 1997, fishing within 2 miles of the plant has been banned.

GO! Activity 13.7

1. Name **two** types of sustainable energy development that can spoil the landscape.
2. State why the moving blades of wind and tidal turbines can be hazardous to wildlife.
3. Explain why flooding land for hydroelectric schemes can release greenhouse gases.
4. State **two** potential problems associated with the use of geothermal energy.

Sustainable sources of energy and the carbon cycle

Although using sustainable sources of energy is considered carbon neutral, carbon dioxide is given off into the atmosphere during the construction of sites and transport of components such as turbines. Over the lifetime of a sustainable source of energy, far less carbon dioxide is emitted into the atmosphere than is emitted by burning fossil fuels.

Switching to sustainable sources of energy will mean that the rate at which the carbon dioxide level in the atmosphere is rising will decrease. Eventually, the level may stabilise and even drop. This, however, is likely to take hundreds of years.

N3	L3	**N4**	L4

Table 13.1 shows the amount of carbon dioxide emitted by different energy systems over their lifetime.

Table 13.1 *The amount of carbon dioxide released to the atmosphere by different energy systems.*

Power generation source	Carbon dioxide emissions (kg of CO_2 per kilowatt hour of electricity produced)
Natural gas	0.3–1.0
Coal	0.7–1.8
Solar cells	0.04–0.1
Wind turbine	0.01–0.02
Geothermal	0.1
Wave/Tidal	0.025
Hydroelectricity	0.03*

*In tropical areas this can be much higher (0.25) due to the decay of plants and soils in the flooded areas.

The amount of carbon dioxide given off depends on the efficiency of the power plant. Burning fossil fuels, even if they are burned in the most efficient power plants, gives off much more carbon dioxide than any of the sustainable energy sources do over their lifetime.

● Activity 13.8

☻ 1. Name the power generation source that emits the least amount of carbon dioxide into the atmosphere over its lifetime.

2. **(a)** Draw a bar chart showing the carbon dioxide emissions for different power generation sources. Use the minimum values for each generation source given in Table 13.1. Label your horizontal (x-) axis 'Power generation source' and your vertical (y-) axis 'Carbon dioxide emissions'. The units for carbon dioxide emissions are kilograms of carbon dioxide per kilowatt hour (kg CO_2/kWh). (Your teacher may give you a prepared set of axes.)

(b) Explain how using sustainable energy sources instead of fossil fuels affects the carbon cycle.

Learning checklist

After reading this chapter and completing the activities, I can:

N3 L3 N4 L4

- describe the carbon cycle. **Activity 13.1 Q1(a)** ○ ○ ○

- state that burning fossil fuels upsets the natural balance of the carbon cycle by increasing the concentration of carbon dioxide in the atmosphere. **Activity 13.1 Q1(b)** ○ ○ ○

- state that natural ways of limiting the concentration of carbon dioxide in the atmosphere include planting forestry, and regenerating peat bogs and wetlands. **Activity 13.2 Q1** ○ ○ ○

- state that carbon dioxide can be removed from gases emitted from industrial plants including power stations. **Activity 13.3 Q1** ○ ○ ○

- describe how compressed carbon dioxide can be stored in rock formations deep below the Earth's surface. **Activity 13.3 Q1** ○ ○ ○

- state that biomass and biofuel burned for energy generation release carbon dioxide. **Activity 13.4 Q1** ○ ○ ○

- explain why biomass and biofuels can be considered carbon neutral. **Activity 13.4 Q1** ○ ○ ○

- describe risks associated with the use of sustainable energy sources. **Activity 13.7 Q2–Q4** ○ ○ ○

- *design a procedure for carrying out an investigation.* **Activity 13.5 Q2** ○ ○ ○

- *present information as a bar chart.* **Activity 13.8 Q2(a)** ○ ○ ○

14 Fuels 3: Hydrocarbons

This chapter includes coverage of:

N4 Hydrocarbons • Materials SCN 4-17a

You should already know:

- that hydrocarbons are molecules made up of only carbon and hydrogen atoms.

National 4

Curriculum level 4

Materials: Earth's materials SCN 4-17a

📖 Word bank

- **Fraction**

A group of compounds with similar boiling points that are obtained from crude oil.

- **Fractional distillation**

A method of separating crude oil into groups of compounds according to their boiling points.

❓ Did you know ...?

Carbon atoms have the ability to bond to each other and also to many of the other elements in the periodic table. There are so many carbon compounds that a whole branch of chemistry is devoted to them – organic chemistry.

Fractional distillation

Learning intentions

In this section you will:

- state the type of compounds that make up crude oil
- learn about fractional distillation
- describe how the compounds in crude oil can be separated into fractions
- compare the properties of crude oil fractions
- give examples of the uses of crude oil fractions
- explain why large numbers of carbon compounds exist.

As it comes out of the ground, crude oil is not very useful. It is a complicated mixture of compounds. To make it useful, crude oil is separated into smaller mixtures. This is described as 'refining' the oil and the first step in the process is known as **fractional distillation**.

The compounds that make up crude oil are mainly hydrocarbons, compounds made up of the elements carbon and hydrogen only.

In fractional distillation, the hydrocarbons in crude oil are separated into groups of compounds with similar boiling points. We refer to these groups of compounds as **fractions**.

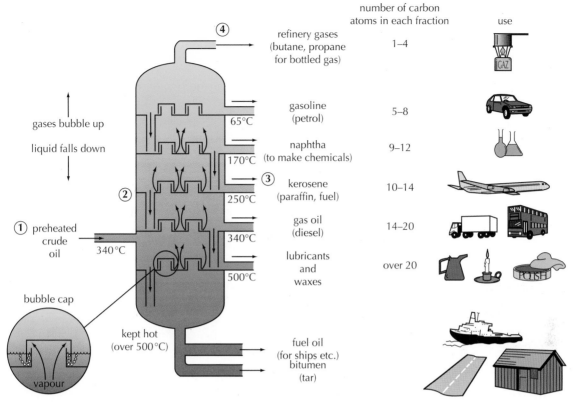

number of carbon atoms in each fraction | use

refinery gases (butane, propane for bottled gas) — 1–4

gasoline (petrol) — 5–8 — 65°C

naphtha (to make chemicals) — 9–12 — 170°C

kerosene (paraffin, fuel) — 10–14 — 250°C

gas oil (diesel) — 14–20 — 340°C

lubricants and waxes — over 20 — 500°C

fuel oil (for ships etc.) bitumen (tar)

gases bubble up

liquid falls down

① preheated crude oil — 340°C

② ③ ④

bubble cap

kept hot (over 500°C)

vapour

Figure 14.1 *Crude oil is separated into useful fractions in the fractionating column.*

1. Crude oil is heated in a furnace to a very high temperature to vaporise the oil, which then passes into the fractionating column. The fractionating column can be 60 metres tall.

2. The heated vapours rise up through the different levels of the fractionating column. At each level the vapours pass through a system of special trays which have holes that are covered by bubble caps. These force the rising vapours to pass through liquids that are already condensed in the tray.

3. At each level, the vapours cool and some of the vapours condense. The liquid fractions that are formed can then be piped off. The smaller molecules in the vapour rise furthest up the column before being cooled sufficiently to condense.

4. At the very top of the column, vapours that have not condensed are piped off. These contain hydrocarbons with low boiling points that are normally gases at room temperature (20°C).

Figure 14.2 *Grangemouth oil refinery on the banks of the River Forth.*

The properties of the fractions

The properties of the fractions obtained by distilling crude oil vary depending on the level at which they condense in the fractionating column. The fractions that come off at the top of the column have small molecules and are described as **light fractions**. Those that come off at the bottom have large molecules and are described as **heavy fractions**.

📖 Word bank

- **Light fractions**
Fractions obtained from crude oil that have low boiling points and contain small molecules.

- **Heavy fractions**
Fractions obtained from crude oil that have high boiling points and contain large molecules.

Word bank

- **Volatile**
Evaporates easily.

- **Viscous**
Thick and sticky.

The liquid fractions that come off at the top of the column are light coloured and **volatile**. This means they evaporate easily. The liquids which come off at the bottom of the column are very dark and sticky. Thick sticky liquids are described as **viscous**.

Figure 14.3 summarises the properties of the fractions obtained from crude oil.

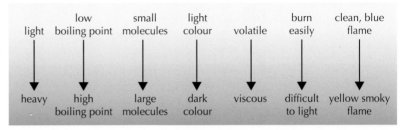

Figure 14.3 *The properties of the fractions from crude oil.*

GO! Activity 14.1

☺ **1.** At an oil refinery, crude oil is first separated into mixtures of compounds that boil in particular temperature ranges.

   (a) Name the process used to separate crude oil.

   (b) The compounds in crude oil are mainly hydrocarbons.

   State what is meant by a hydrocarbon.

   (c) State the name given to the mixtures of hydrocarbons that can be separated using their boiling points.

2. Two bottles contain two fractions from crude oil.

   Fraction A has a boiling range 20–120 °C.

   Fraction B has a boiling range 220–350 °C.

   Which fraction:

   (a) is lighter and is collected nearer the top of the refinery tower

   (b) burns with the smokier flame

   (c) has the lighter colour

   (d) contains the smaller molecules?

3. Look at Figure 14.2 and the number of carbon atoms in the molecules of the different fractions.

   • The gasoline fraction contains hydrocarbons with 5–8 carbon atoms.

   • The kerosene fraction contains hydrocarbons with 10–14 carbon atoms.

   • The gas-oil fraction contains hydrocarbons with 14–20 carbon atoms.

   (a) Which fraction will be most volatile?

   (b) Which fraction will burn with the cleanest flame?

   (c) Name the fuel that can be obtained from:

   (i) the gasoline fraction   **(ii)** the gas-oil fraction.

   (d) State a use for the kerosene fraction.

The alkanes

Learning intentions

In this section you will:
- learn about the family of hydrocarbons called alkanes
- learn how to name alkanes
- write molecular formulae for alkanes
- draw full structural formulae for alkanes.

The molecules that make up the fractions obtained from crude oil mostly belong to a family of hydrocarbon compounds called **alkanes**.

Every carbon atom can make four bonds to other atoms. Hydrogen can only make one bond to another atom. The simplest hydrocarbon has one carbon atom bonded to four hydrogen atoms, as shown in Figure 14.4. It is called methane.

Figure 14.4 *A model of one methane molecule.*

The **molecular formula** for methane is CH_4.

A molecular formula gives the number of different types of atom in a molecule, but it doesn't show how the atoms are joined together. To do this we draw a structure showing the bonds between atoms. For larger molecules, showing the three-dimensional shape of the molecule is complicated. A simplified two-dimensional representation, called a **full structural formula**, is used instead, as shown in Figure 14.5.

The next simplest hydrocarbon is ethane.

In ethane molecules there is a bond between two carbon atoms. Each of the carbon atoms can, therefore, bond to three hydrogen atoms. The molecular formula for ethane is C_2H_6. Figure 14.6 shows a model of an ethane molecule and its full structural formula.

National 4

Curriculum level 4

Materials: Earth's materials SCN 4-17a

★ You need to know

Carbon and hydrogen can form molecules of different sizes which results in many different hydrocarbon molecules.

📖 Word bank

- **Alkane**
A hydrocarbon in which the carbon atoms are joined by single carbon-to-carbon bonds.

- **Molecular formula**
A formula showing the number of atoms in a molecule.

- **Full structural formula**
A formula that shows the bonds between atoms in a molecule.

Figure 14.5 *The full structural formula of methane.*

(a)

(b)

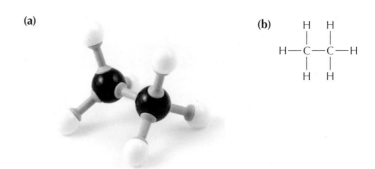

Figure 14.6 **(a)** *A model of an ethane molecule.* **(b)** *The structural formula of ethane.*

Because carbon atoms can bond to other carbon atoms, molecules with long chains of carbon atoms can form. Methane and ethane are the first two members of the alkanes. In alkanes the carbon atoms in the molecules are joined by single carbon-to-carbon bonds.

The next member of the group, propane, has three carbon atoms joined in a chain.

In propane, the carbon atoms on the ends are joined to three hydrogen atoms and the middle carbon is joined to two hydrogen atoms. The molecule therefore consists of three carbon atoms and eight hydrogen atoms and has the molecular formula C_3H_8. The full structural formula for propane is shown in Figure 14.7.

(a)

(b)

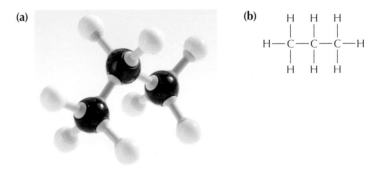

Figure 14.7 **(a)** *A model of a propane molecule.* **(b)** *The full structural formula for propane.*

★ You need to know

You need to know the names of the alkanes up to C_8 and be able to draw full structural formulae and write molecular formulae for them. You also need to be able to name and identify alkanes up to C_8 when given full structural formulae and molecular formulae.

So, the first three members of the group of alkanes are methane, ethane and propane. Notice that the names of the alkanes all end in '**-ane**'. This is true for all alkanes.

The fractions in crude oil contain lots of alkane molecules. Many have a lot more than eight carbons in the chain and the chains are not always a straight line. Many of the molecules have branched chains, where a carbon in the middle of the chain is attached to more than two carbons.

GO! Activity 14.2

☺ 1. Hydrocarbon molecules can be represented using molecular formulae or using full structural formulae.

State what is indicated by:

(a) a molecular formula

(b) a full structural formula.

☺☺2. The table below shows the names of the first eight members of the alkanes.

(a) If you have molecular mode kits, build models of the molecules of these members of the alkane family: butane, pentane, hexane, heptane and octane.

(b) For each molecule you build, count the number of hydrogens attached to the carbons.

(c) Copy and complete the table by writing the number of hydrogens, a molecular formula and drawing a full structural formula for a molecule of each alkane.

Alkane name	Number of carbons	Number of hydrogens	Molecular formula	Full structural formula
methane	1	4	CH_4	
ethane	2	6	C_2H_6	
propane	3	8	C_3H_8	
butane	4			
pentane	5			
hexane	6			
heptane	7			
octane	8			

(d) The name 'pentane' consists of two parts: pent- and -ane. What does each part of the name indicate?

(e) State the property of carbon that allows it to make so many different compounds.

National 4

The alkenes

Learning intentions

In this section you will:

- learn about the family of hydrocarbons called alkenes
- learn how to name alkenes
- write molecular formulae for alkenes
- distinguish between alkenes and alkanes chemically
- draw full structural formulae for alkenes.

Word bank

- **Alkene**

A hydrocarbon whose molecules contain a carbon-to-carbon double bond.

Alkenes are hydrocarbon molecules that contain a carbon-to-carbon double bond.

Alkenes are very important chemicals because small alkenes are the starting materials for the plastics industry.

The simplest alkene is ethene, which has two carbons – you need to have two carbons to have a carbon-to-carbon double bond!

The first part of the name indicates the number of carbons, just as in the alkanes. The ending '**-ene**' indicates that the molecule contains a carbon-to-carbon double bond.

The molecular formula for ethene is C_2H_4. Figure 14.8 shows a model of ethene and its full structural formula.

(a)

(b)

Figure 14.8 **(a)** *A model of an ethene molecule.* **(b)** *The structural formula for ethene.*

The alkene with three carbon atoms is propene, C_3H_6. The alkene with four carbon atoms is butene, C_4H_8. The full structural formula for butene can be drawn with the double bond in two different places (Figure 14.9).

Figure 14.9 *The double bond in butene can be drawn between different carbon atoms.*

Reactivity of alkenes

Alkenes are much more reactive than alkanes and are very useful as starting materials for making many other products.

The difference in reactivity of alkanes and alkenes can be shown using bromine solution.

When bromine solution is shaken with a liquid alkene (hexene) in a test tube, the bromine solution immediately decolourises. The bromine molecules have reacted with the hexene molecules.

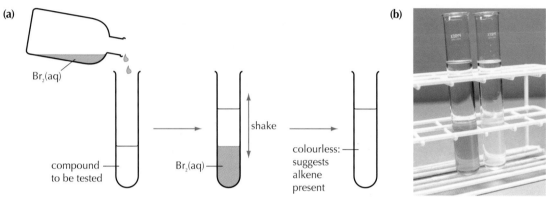

(a)

Br₂(aq)

compound to be tested

Br₂(aq)

shake

colourless: suggests alkene present

(b)

Figure 14.10 **(a)** *Bromine solution is decolourised immediately when shaken with hexene.* **(b)** *Bromine solution before (left) and after (right) being shaken with hexene.*

If the experiment is repeated with hexane, the colour goes into the hexane but remains. The colour goes into the hexane because bromine dissolves better in hexane than in water. The bromine colour only disappears when the mixture is left in sunlight or has another bright light source shining on it.

This reaction is used as a test to distinguish between alkanes and alkenes.

★ You need to know

You need to know the chemical test used to distinguish between alkanes and alkenes. A few drops of bromine solution is immediately decolourised when shaken with an alkene but not with an alkane.

GO! Activity 14.3

😊😊**1.** *You will need: a molecular model kit.*

   (a) Use a molecular model kit to build models for alkenes with up to eight carbon atoms.

(continued)

(b) Copy and complete the table below, using the naming rules for alkenes.

Alkene name	Number of carbons	Number of hydrogens	Molecular formula	Full structural formula
propene	3			(structural formula shown)
pentene		10		
			C_6H_{12}	(structural formula shown)
heptene	7			
			C_8H_{16}	

2. What do you notice about the numbers of carbon and hydrogen atoms in alkene molecules?

3. Ethene and propene belong to the same family of hydrocarbons.

   (a) State the name of the family which ethene and propene belong to.

   (b) State what -ene in the names of the compounds indicates about the bonds between carbon atoms.

4. You are given two test tubes, A and B, containing colourless liquids. You are told that one contains an alkane and the other an alkene. Describe a chemical test you could carry out to tell which is which.

National 4

Meeting market demand: Cracking

Learning intentions

In this section you will:

- describe how market demand for lighter hydrocarbon fractions can be met
- learn about the process of cracking
- describe what happens when hydrocarbon molecules are cracked.

When crude oil is separated into different fractions by fractional distillation, the fractions are not in the proportions that the market needs. Fractional distillation produces too many heavy fractions and not enough lighter fractions.

Table 14.1 *Fractions produced by fractional distillation.*

Fraction (in order of increasing boiling point)	Percentage produced by fractional distillation
liquefied petroleum gases (LPG)	3
gasoline	13
naphtha	9
kerosene	12
diesel	14
heavy oils and bitumen	49

The gasoline fraction is used to make petrol for motor vehicles. The market demand for gasoline is much higher than can be obtained by fractional distillation. This problem is solved by breaking down the larger molecules by passing them over a heated catalyst. This is known as **catalytic cracking**.

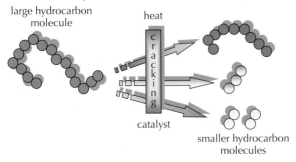

Figure 14.11 *Cracking produces smaller hydrocarbon molecules.*

In industry much larger molecules in heavy oils and bitumen are cracked to give more fractions such as gasoline (petrol) and gas oil (diesel). When large alkane molecules break apart a mixture of smaller alkane and alkene molecules is produced. This is complicated to show, so hexane is used here as an example.

A hexane molecule can break apart to give a molecule of butane (an alkane) and a molecule of ethene.

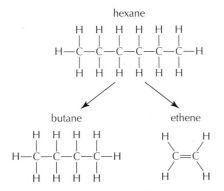

Figure 14.12 *Cracking of hexane can produce butane and ethene.*

An equation can be written to show the cracking of hexane:

$$C_6H_{14} \rightarrow C_4H_{10} + C_2H_4$$

hexane   butane   ethene

📖 **Word bank**

• **Catalytic cracking**

A process that splits long-chain hydrocarbons into short-chain hydrocarbons by passing them over a heated catalyst.

★ **You need to know**

Cracking of large hydrocarbon molecules produces a mixture of alkanes and alkenes.

Make the link

You can learn more about the cracking that takes place at Mossmorran in Fife on page 204.

☀ Chemistry in action: Mossmorran Natural Gas Liquids Plant

At the Natural Gas Liquids Plant at Mossmorran in Fife, ethane gas is cracked to give ethene.

The ethane used by the plant comes from the North Sea. Natural gas from the North Sea is piped ashore at St Fergus, north of Peterhead. Here, the methane is removed for distribution as domestic gas for use throughout the UK.

Figure 14.13 *The Natural Gas Liquids Plant at Mossmorran in Fife.*

The remaining gas liquids are sent to the Fife Natural Gas Liquids Plant through a 222 km underground pipeline. At the plant, the gas liquids are purified to produce ethane, which is then fed into furnaces where it is mixed with steam and heated to a temperature above 800°C.

This produces a mixture of gases from which ethene is extracted. The ethene is cooled and compressed, and either pumped to chemical plants throughout the UK using a 1000 km pipeline network or to a storage facility at Braefoot Bay, from where it is transported by tanker to Europe.

Each year the plant produces 830 000 tonnes of ethene.

Figure 14.14 *Distribution of ethene by pipeline across Europe.*

GO! Activity 14.4

😊😊**1.** *You will need: a molecular model kit.*

    (a) Use the kit to build a model of an octane molecule as shown below.

      An octane molecule.

    (b) Now break (crack) the molecule to give the two molecules shown here.

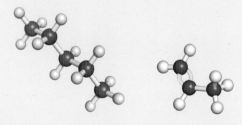

    (c) Draw full structural formulae to represent this cracking reaction.

    (d) Write an equation using molecular formulae for the reaction.

2. **(a)** Write the molecular formula for heptane.

(b) (i) Show, using full structural formulae, a heptane molecule breaking apart to give a butane molecule and a molecule containing a double bond.

(ii) Write an equation to show a heptane molecule breaking apart to give an alkane molecule and a molecule of ethene.

3. Explain why heavy fractions obtained by fractional distillation are cracked.

4. A long-chain hydrocarbon is an alkane and has the molecular formula $C_{13}H_{28}$. The hydrocarbon can be broken apart by passing it over a heated catalyst.

(a) Name the process that takes place when long-chain hydrocarbons are broken into smaller molecules.

(b) The equation for the breakdown is:

$$C_{13}H_{28} \rightarrow C_8H_{18} + C_2H_4 + X$$

(i) Work out the formula for compound X.

(ii) Name the compounds formed when $C_{13}H_{28}$ is broken down.

5. **Experiment**

Liquid paraffin is a mixture of alkanes that are made up of molecules of long-chain hydrocarbons. Liquid paraffin can be cracked in the laboratory as shown in the diagram below. (Your teacher may demonstrate this experiment to you.)

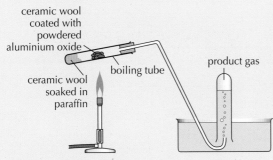

ceramic wool coated with powdered aluminium oxide

ceramic wool soaked in paraffin

boiling tube

product gas

Cracking liquid paraffin.

To crack the liquid paraffin, the aluminium oxide which acts as a catalyst is heated strongly. When the catalyst is hot the liquid paraffin is heated. The vapours produced pass over the hot catalyst and are broken down. The gas that is given off is collected by displacement of water. The first test tube of gas collected will contain the air displaced from the boiling tube and is discarded.

Before you stop heating the boiling tube, the delivery tube must be removed from the water. This prevents water being sucked back into the boiling tube as the tube cools, causing it to shatter.

(a) Liquid paraffin is a viscous liquid of mixed alkane molecules.

(i) Suggest why liquid paraffin is a viscous liquid.

(ii) State the result you would expect from shaking bromine solution with liquid paraffin.

(continued)

(b) The table shows the results when the gas collected from cracking liquid paraffin is tested.

Test	Burning	Testing with bromine solution
	gas — burning taper	bromine solution
Result	Gas burns with a smoky yellow flame	Bromine solution is immediately decolourised when shaken with the gas

(i) Explain why the gas burns.

(ii) Explain why we can conclude the gas contains small alkene molecules.

(c) State the safety precaution that must be taken before the heating of the boiling tube is stopped.

Learning checklist

After reading this chapter and completing the activities, I can:

N3 L3 **N4 L4**

* state that crude oil can be separated into fractions by fractional distillation. **Activity 14.1 Q1(a)** ◯ ◯ ◯

* state that a fraction is a group of hydrocarbons with boiling points in a given range. **Activity 14.1 Q1(c)** ◯ ◯ ◯

* state that the fractions in crude oil have different properties and different uses. **Activity 14.1 Q2, Q3** ◯ ◯ ◯

* state that, in a refinery, lighter fractions are obtained at the top of the fractionating column. **Activity 14.1 Q2(a)** ◯ ◯ ◯

* state that the heavier the fraction, the more smoky and yellow the flame when the fraction burns. **Activity 14.1 Q2 (b)** ◯ ◯ ◯

* state that lighter fractions contain smaller molecules than heavier fractions. **Activity 14.1 Q2(d)** ◯ ◯ ◯

N3 L3 N4 L4

- state that lighter fractions are more volatile and burn more easily than heavy fractions. **Activity 14.1 Q3(a)**

- state that molecular formulae show the numbers of each type of atom in a molecule. **Activity 14.2 Q1(a)**

- state that full structural formulae show how atoms in a molecule are joined together. **Activity 14.2 Q1(b)**

- name straight-chain alkanes with up to eight carbon atoms from full structural formulae and molecular formulae. **Activity 14.2 Q2**

- state that the names of alkane molecules end in '-ane' and the name ending can be used to identify molecules as alkanes. **Activity 14.2 Q2(d)**

- state that there are many different hydrocarbons because carbon atoms can bond together to form molecules with long chains of carbons. **Activity 14.2 Q2(e)**

- name straight-chain alkenes with up to eight carbon atoms from full structural formulae and molecular formulae. **Activity 14.3 Q1(b)**

- state that the '-ene' in the names of alkene molecules indicates that the molecules contain a carbon-to-carbon double bond. **Activity 14.3 Q3**

- state that alkanes and alkenes can be distinguished using bromine solution. **Activity 14.3 Q4**

- state that cracking is a process used to meet the demand for shorter chain alkanes and alkenes. **Activity 14.4 Q3, Q4**

15 Everyday consumer products 1: Plants for food

This chapter includes coverage of:

N3 Everyday consumer products • Planet Earth
SCN 3-02a

National 3

Curriculum level 3

Planet Earth: Biodiversity and interdependence SCN 3-02a

Plants for food

Learning intentions

In this section you will:

- learn that plants are a vital source of nutrients
- learn about the different food groups that are required for a healthy diet
- learn how foods can be tested to show different food groups.

Keeping our bodies healthy

What are our bodies made of?

Just four chemical elements make up 96% of our bodies. These are oxygen, carbon, hydrogen and nitrogen. Table 15.1 shows the percentage by mass (weight) of each.

Table 15.1 *Oxygen, carbon, hydrogen and nitrogen make up most of the human body.*

Element	Percentage (%) by mass
oxygen	65
carbon	18
hydrogen	10
nitrogen	3

These elements are present as compounds such as proteins and fats. The main compound present in our bodies is water. This accounts for over 60% of our body weight.

The importance of a balanced diet

Plants are a good source of the **nutrients** we need to keep our bodies healthy. If we are to maintain good health, it is

📖 Word bank

- **Nutrient**

A chemical element or compound that is essential for healthy growth and life.

important to eat the correct foods. A **balanced diet** provides all the essential elements and compounds that we need.

There are five main food groups that are part of a balanced diet and provide us with nutrients. These are: carbohydrates, proteins, fats and oils, vitamins, and minerals. Figure 15.1 summarises how the body uses each food group.

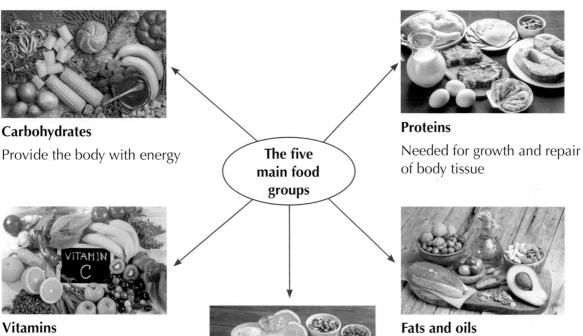

Carbohydrates

Provide the body with energy

The five main food groups

Proteins

Needed for growth and repair of body tissue

Vitamins

Needed by the body for normal growth and to help repair tissue; needed to fight infection and prevent diseases

Minerals and trace elements

Needed for healthy bones and teeth (calcium), and for healthy blood (iron); trace elements are needed in small amounts for special purposes in the body (e.g. iodine is needed in our thyroid glands and selenium is needed by our immune system)

Fats and oils

Provide the body with energy and act as an insulator against cold

Figure 15.1 *The five main food groups and why they are important for healthy bodies.*

Carbohydrates, fats and proteins provide us with the main chemical elements we need, as shown in Table 15.2.

Table 15.2 *The elements present in carbohydrates, fats and oils, and proteins.*

Food group	Elements present			
	oxygen	carbon	hydrogen	nitrogen
carbohydrates	✓	✓	✓	✗
fats and oils	✓	✓	✓	✗
proteins	✓	✓	✓	✓

Certain foods provide us with the vitamins, minerals and trace elements we need. For example, milk provides us with calcium for strong bones and teeth. Meat and leafy green vegetables provide us with iron needed for healthy blood.

It is important not to have too much of particular food groups, as overeating can lead to health problems. Too much carbohydrate, fats and oils in our diets can cause us to become overweight.

starchy foods – bread, potatoes, pasta

fruits and vegetables

protein foods – meat, fish, eggs

foods high in fat and sugar

milk and dairy products

Figure 15.2 *Fruit and vegetables, as well as foods derived from plants, are a source of the nutrients needed to keep our bodies healthy.*

Testing foods

Different food groups are made of a variety of compounds. Chemical tests can be done to check for the presence of particular types of compounds.

Testing for carbohydrates

Carbohydrates are used by our bodies to give us energy.

There are two different types of carbohydrate in our diet – sugars and starches. Sugars are made up of small molecules that dissolve in water. Starches are much larger molecules that are not very soluble.

Sugars in our diet include glucose, fructose, sucrose and maltose. Glucose is the sugar that our bodies use to give us energy.

Figure 15.3 *Foods with a high carbohydrate content are needed to give us energy.*

Figure 15.4 *Dried fruits such as raisins and dried apricots are particularly good sources of glucose.*

Testing for sugars

We can test for sugars in foods using Benedict's solution.

Benedict's solution is added to a sample of food in a test tube, which is then heated in a water bath. It helps to crush the food in a little water before testing. Benedict's solution is blue. If sugars are present in the food, the blue colour turns brick-red.

food sample + Benedict's solution

Figure 15.5 *Testing for sugars.*

Testing for starch

We can test for starch using iodine solution. A few drops of iodine solution can be placed directly onto the food, or the food can be ground up with some water and a few drops of iodine solution added. A blue–black colour indicates the presence of starch.

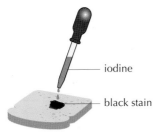

iodine

black stain

Figure 15.6 *Testing for starch.*

> **?** **Did you know ...?**
>
> When ancient Egyptian kings died, dried fruits were buried with their treasures in their tombs. The dried fruits were food for the afterlife.

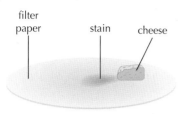

Figure 15.7 *Testing for fats.*

Testing for fats

Foods that contain fats and oils leave a greasy mark if rubbed on filter paper.

Testing for proteins

If foods containing proteins are heated with soda lime, a gas that turns red litmus paper or pH paper blue, is given off.

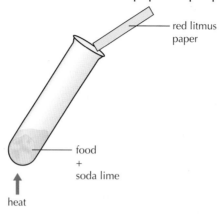

Figure 15.8 *Testing for proteins.*

GO! Activity 15.1

 Before using nuts you should ensure that no one in your group has a nut allergy.

 1. *You will need: foods (such as potato, bread, nuts, fruit, pasta, breakfast cereals), beaker, boiling tubes, Bunsen burner, stand, red litmus paper, Benedict's solution, filter paper, dropper.*

Carry out the food tests described on pages 211–212 on a number of different foods to find out which food group each food belongs to.

Record your results in a table like this one.

Food	Carbohydrate		Fats and oils	Protein
	Sugars	Starch		
bread				
Brazil nut				
apple				
breakfast cereal				
pasta				

Activity 15.2

1. Copy and complete the following passage.

Plants are a vital source of n_ _ _ _ _ _ _ _. The foods we obtain from plants include p_ _ _ _ _ _ _, c_ _ _ _ _ _ _ _ _ _ _, fats and o_ _ _, v_ _ _ _ _ _ _ and minerals.

2. Carbohydrates provide the body with energy. Name the **two** different types of carbohydrate.

3. Name the food group that is needed for growth and repair of body cells.

4. When iodine solution is dropped onto a biscuit, the biscuit turns blue–black. When the biscuit is crushed and heated in some Benedict's solution, the solution turns brick-red.

State the **two** types of compound in the biscuit.

5. When a piece of meat is heated with soda lime, a gas that turns pH paper blue is given off.

Which food group does meat belong to?

6. Name the **five** main food groups and state how they are used in the body.

Diet and disease

National 3

Learning intentions

In this section you will:

- learn that lifestyle can have an effect on the body
- learn that the body needs a range of different nutrients.

Diet and health

Our **diet** – what we eat and how much we eat – affects our health.

Lifestyle and healthy eating are key to good health. If we have a very active lifestyle, such as taking part in regular sporting activities, we need to consume more foods to give us energy than if we are less active. Many athletes follow strict diets in order to be in peak condition for events.

📖 Word bank

- **Diet**
  The kinds of foods that a person generally eats.

Figure 15.9 *Athletes follow special high-calorie diets to ensure they can meet the energy demands of their sports without tiring.*

? Did you know ...?

The NHS website offers advice on healthy eating and lifestyle choices. These eight tips can help people make healthier choices.

- Eat meals containing starchy carbohydrates
- Eat lots of fruit and vegetables
- Eat more fish, including oily fish
- Cut down on unhealthy (saturated) fats and sugar
- Eat less salt
- Get active and be a healthy weight
- Don't get thirsty
- Always eat breakfast

The key to a healthy diet is to:

- Eat the right amount of calories for your activity level, balancing the energy you consume with the energy you use.
- Eat a wide range of foods to have a balanced diet and make sure that your body is receiving all the nutrients it needs.

? Did you know ...?

A typical teenager would need to cycle for more than 2 hours to burn off the calories contained in three iced doughnuts.

Figure 15.11 *Foods such as iced doughnuts contain a lot of calories.*

? Did you know ...?

The average adult needs to consume about 2000–2500 calories every day. But mountaineers and polar explorers can consume up to 6000 calories a day. In an extremely cold climate your body needs lots of calories because of the energy you use just to keep warm. The daily diet of the members of Captain Scott's ill-fated expedition to the South Pole (1910–1913) was 2000–3000 calories short of what they needed to cope with the demands of their expedition.

Figure 15.10 *Captain Scott's expedition reached the South Pole but they didn't survive the return journey.*

Too much food

Like many other countries in the developed world, Scotland is experiencing an obesity epidemic. Fast food restaurants, food takeaways and ready meals mean many families don't cook at home as much as they did in the past. It's easy in these circumstances to eat too much of the wrong types of foods, especially carbohydrates and fats and oils, and to put on weight and become **overweight** or **obese**.

Only the United States and Mexico have a higher percentage of obese people than Scotland. Over 60% of adults in Scotland are classed as overweight with nearly 30% classed as obese.

📖 Word bank

- **Overweight**

Having a body mass index (BMI) of between 25 and 30.

- **Obese**

Having a body mass index (BMI) of above 30.

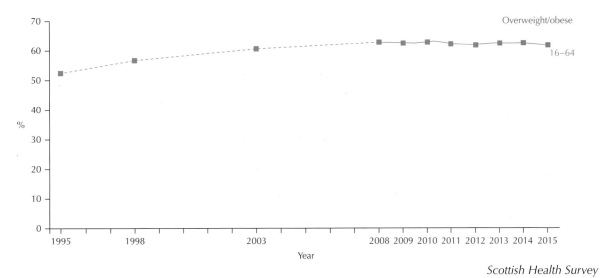

Scottish Health Survey

Figure 15.12 *Proportion of Scottish adults who are overweight and obese, 1955–2015.*

? Did you know ...?

Body mass index (BMI) is a measure of a person's body fat based on the person's height and weight.

A person with a BMI above 25 is classed as overweight and a person with a BMI above 30 is classed as obese. Generally, the higher a person's BMI, the more fatty tissue they have. Carrying this extra weight puts strain on the joints and also means the heart has to work harder to pump blood to all this extra tissue.

The World Health Organization reports that, each year, at least 2.8 million people die as a result of being overweight or obese.

People who are obese are at greater risk of some health problems than people who have a healthy weight. These problems include heart disease, high blood pressure and type 2 diabetes. People who are obese are also more likely to experience joint pain.

Figure 15.13 *An obese person is at greater risk of experiencing health problems.*

Not enough food

In many countries, particularly in parts of Africa, it isn't too much food but too little food that is the problem.

It is estimated that one person in every nine people on the planet does not get enough food to eat to keep them healthy. This is known as **starvation**. Starvation causes severe muscle loss and fatigue.

📖 Word bank

• **Starvation**
A lack of food causing suffering or death.

Figure 15.14 *Drought in African countries can mean that children do not get enough food and some starve.*

? Did you know ...?

Starving people do not take in enough vitamins and minerals to keep their bodies healthy. In 2015, 5.9 million children under the age of 5 died and 45% of these deaths were linked to starvation and **malnutrition**.

Figure 15.15 *This child is malnourished and starving. The swollen abdomen indicates a lack of protein.*

📖 Word bank

- **Malnutrition**

Having a diet that doesn't contain the right amount of nutrients to maintain good health.

- **Vitamin deficiency**

A lack of one or more vitamins in the diet.

The importance of vitamins and minerals

A balanced diet includes the vitamins and minerals that are necessary to prevent disease and maintain good health.

A lack of vitamins – **vitamin deficiency** – can cause severe health problems. Table 15.3 lists some diseases caused by vitamin and mineral deficiencies.

Table 15.3 *Diseases caused by a lack of vitamins or minerals.*

Disease	Vitamin or mineral deficiency in the diet
scurvy	vitamin C
rickets	vitamin D
anaemia	iron
beriberi	vitamin B1
pellagra	vitamin B3

Scurvy

Scurvy is a disease caused by a lack of vitamin C in the diet. It was known about in Ancient Greece. During the sixteenth to eighteenth centuries, many sailors on long sea voyages were affected by scurvy. It is estimated that at least 2 million sailors died from scurvy.

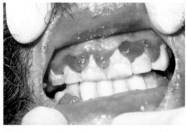

Figure 15.16 *Scurvy is caused by a lack of vitamin C.*

The sailors' diet tended to be based on dried meat and grain; there was little fresh fruit or vegetables, because these did not keep well on board ship. Fresh fruit and vegetables are important sources of vitamin C. Symptoms of a lack of vitamin C include swollen gums, loss of teeth, joint pain and general tiredness.

Rickets

People with rickets have bowed legs. The disease is caused by a lack of vitamin D. Vitamin D is needed for our bodies to absorb calcium, which we need to grow strong bones. A lack of vitamin D means bones are weaker because the body is unable to absorb calcium. This can cause the legs of sufferers to bow.

Figure 15.17 *A child with rickets caused by a lack of vitamin D.*

Anaemia

Anaemia is caused by a lack of iron in the diet, meaning the body can't make enough healthy red blood cells. The iron is needed to make haemoglobin, the compound that carries oxygen round our bodies. A lack of healthy cells leads to tiredness, breathlessness and increased susceptibility to infection. Young people with anaemia can have difficulty concentrating.

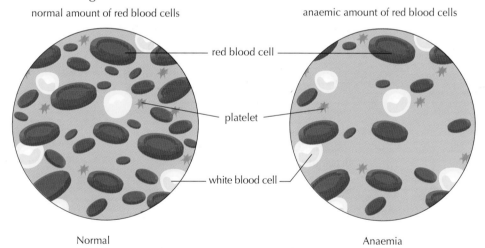

normal amount of red blood cells

anaemic amount of red blood cells

red blood cell

platelet

white blood cell

Normal

Anaemia

Figure 15.18 *People with anaemia have a reduced number of healthy red blood cells in their blood due to a lack of iron in their diet.*

Beriberi

Beriberi is due to having a lack of vitamin B1 (thiamine) in the diet. This causes heart and muscle strength problems. It became common in South-east Asia when people began to replace brown rice with polished white rice. Removing the outer rice coating removed most of the vitamin from the rice.

Pellagra

Pellagra is a condition caused by a lack of vitamin B3 (niacin) in the diet. It was common in rural South America among people who relied heavily on maize as an energy food. Symptoms include high sensitivity to sunlight, dermatitis (skin problems) and aggressive behaviour.

Plant sources of vitamins and minerals

Many of the diseases and conditions that are caused by a particular vitamin deficiency can be improved by including foods which contain the vital vitamins and minerals in people's diet.

Most of these essential vitamins and minerals can be obtained from plants. The exception is vitamin D. Our bodies make vitamin D when our skin is exposed to sunlight. This is why it is sometimes referred to as 'the sunshine vitamin'. Some vitamin D can be obtained from oily fish.

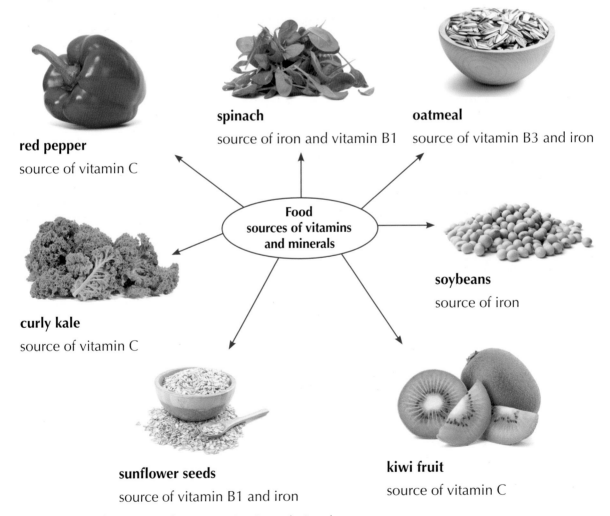

spinach
source of iron and vitamin B1

oatmeal
source of vitamin B3 and iron

red pepper
source of vitamin C

Food sources of vitamins and minerals

soybeans
source of iron

curly kale
source of vitamin C

sunflower seeds
source of vitamin B1 and iron

kiwi fruit
source of vitamin C

Figure 15.19 *Plant sources of important vitamins and minerals.*

Plants and essential fatty acids

Foods containing fat provide us with energy. Fats contain fatty acids. Special fatty acids called **omega-3 and omega-6 fatty acids** give health benefits when we include them in our diets.

If we don't get enough of these special fatty acids, research indicates that there is increased risk of conditions such as heart disease, stroke and breast cancer. We can also suffer from liver and kidney problems, and not be able to fight infections.

Our bodies can make most of the omega-3 and omega-6 fatty acids they need from the food we eat. But there are two fatty acids, linoleic acid and linolenic acid, which the body can't make. These are known as **essential fatty acids** and it is very important that we include foods that contain them in our diet.

The best sources of omega-3 and omega-6 fatty acids are plants.

Plants and proteins

Our diet needs to contain proteins.

The proteins we eat are broken down during digestion to provide the building blocks for the proteins that make up our bodies. About 16% of a human body is made up of proteins.

Proteins can be obtained from animal and plant sources. Studies suggest that moving to a diet containing more plant-based proteins can have health benefits.

People whose diets include higher levels of plant-based protein, such as vegetarians, tend to be healthier than those who get their protein from animal sources. They tend to have a lower body weight, lower cholesterol and lower blood pressure. They also have a lower risk of stroke, cancer and death from heart disease.

Word bank

- **Omega-3 and omega-6 fatty acids**

Special fatty acids that are thought to lower the risk of conditions such as heart disease, stroke and breast cancer.

- **Essential fatty acids**

Fatty acids that cannot be made by our bodies and that must be obtained from food in our diet.

Figure 15.20 *Walnuts, seeds and avocados are rich sources of omega-3 and omega-6 fatty acids.*

Figure 15.21 *Foods such as beans and lentils are a good source of plant proteins.*

🔵 Activity 15.3

☻ 1. **(a)** Name the condition which could result if you constantly eat too much food.

   (b) Name the **two** food groups mainly responsible for weight gain.

   (c) (i) A person weighs 55 kg and is 1.4 m tall. Calculate their body mass index (BMI) using the following equation:

$$\text{BMI} = \frac{\text{weight (kg)}}{\text{height (m)} \times \text{height (m)}}$$

     (ii) What term is used to describe someone with the BMI you calculated in part **(i)**?

2. Scurvy and rickets are diseases that can affect people with poor diets.

  Explain why people develop diseases like scurvy and rickets.

3. Our bodies can only make some of the fatty acids we require for good health.

  State the term used to describe fatty acids that can't be made by our bodies and that must be obtained from the food we eat.

(continued)

| N3 | L3 | N4 | L4 |

4. Describe **one** benefit of a diet rich in plant protein.

5. Complete the table to show the vitamin or mineral deficiency that is responsible for each disease and give a plant source for each vitamin or mineral.

🔍 Hint

Look at pages 216–218 to help complete Question 5.

Disease or condition	Vitamin or mineral deficiency	Plant source of vitamin or mineral
scurvy		
rickets		
anaemia		
beriberi		
pellagra		

National 3

Plants and alcohol

Learning intentions

In this section you will:

- learn that alcohol can be made from plants
- learn that fermentation and distillation are two processes used in the manufacture of alcoholic drinks
- learn that alcohol content in drinks is measured in units
- learn about government guidelines on safe drinking levels.

📖 **Word bank**

- **Alcohol**

A colourless flammable liquid made by the fermentation of sugars.

Making alcoholic drinks from plants

Since ancient times, plants have been used to make **alcohol**. Grapes, barley, potatoes, sugarcane and many other plants have been used.

(a)  (b)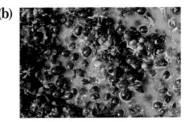

Figure 15.22 *Grapes are* **(a)** *harvested and* **(b)** *fermented to produce alcohol.*

Different plants are used to make different alcoholic drinks.

Table 15.4 *The plant sources of some alcoholic drinks.*

Alcoholic drink	Fruit or vegetable used
wine	grapes
whisky	barley
beer	barley, wheat
cider	apples
rum	sugarcane
vodka	potatoes, grain

Fermentation

When plants are fermented, the sugars they contain change to alcohol. Yeast is added to provide natural (biological) catalysts called **enzymes**. Carbon dioxide gas is also produced during the **fermentation** process.

$$\text{sugars in plants} \xrightarrow[\text{yeast}]{\text{fermentation}} \text{alcohol} + \text{carbon dioxide}$$

Wines and beers are made by fermentation. The alcohol content of drinks made by fermentation can be as high as 15%. When the alcohol concentration reaches this level, it kills the yeast and stops the fermentation process.

Distillation

To make alcoholic drinks with a higher alcohol content, a second process is needed. This process is called **distillation**.

Distillation is used to produce spirits such as whisky, brandy, rum, vodka and gin. In whisky production, barley is fermented to produce a beer. This beer is then heated in whisky stills, producing vapours with a high alcohol content. These gases are condensed and collected. The raw spirit, which has a much higher alcohol content than the beer, is matured in wooden barrels for at least three years before it can be sold as whisky.

Some alcoholic drinks such as port and sherry are made by adding brandy to wine to increase the alcohol concentration.

Alcohol content of drinks

The amount of alcohol in an alcoholic drink depends on the volume of the liquid and concentration of the alcohol. This is often measured in units and is shown on the bottles.

One unit of alcohol is approximately equal to the alcohol in a half pint of beer, a small measure (25 cm³) of spirits such as whisky, or a very small glass of wine.

> ### 📖 Word bank
>
> - **Fermentation**
>   The process that turns sugars into alcohol.
>
> - **Enzyme**
>   A natural (biological) catalyst which speeds up chemical reactions.
>
> - **Distillation**
>   The process of purifying a liquid by heating it and condensing the vapours that are given off.

Figure 15.23 *Copper stills at a distillery on Islay. A still can hold 9400 litres.*

Table 15.5 shows the alcohol content of different drinks.

Table 15.5 *The alcohol content of some drinks.*

Alcoholic drink	Alcohol content (units)
small measure (25 cm³) of whisky	1
pint of 4% beer (568 cm³)	2.3
175 cm³ glass of wine	2.0–2.5
330 cm³ bottle of strong (6%) beer	2
alcopop	1.1–1.5

The UK Government encourages responsible drinking. In the UK, the labels on all alcoholic drinks must show how many units of alcohol are contained and have information on the recommended maximum number of units to be drunk in a day. The maximum recommendations for men and women are different. The limit for women is lower than that for men.

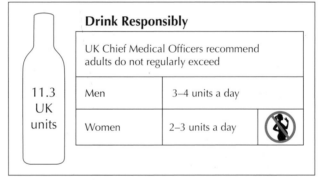

Drink Responsibly

11.3 UK units	UK Chief Medical Officers recommend adults do not regularly exceed	
	Men	3–4 units a day
	Women	2–3 units a day

Figure 15.24 *Labels on alcoholic drinks show the alcohol content in units and give advice about recommended maximum amounts of units per day for men and women.*

The Government also gives advice on its website. The UK Chief Medical Officer's guideline for both men and women is that, to keep health risks from alcohol to a low level, it is safest not to drink more than 14 units a week on a regular basis.

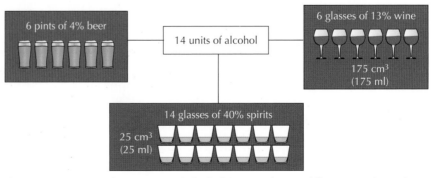

6 pints of 4% beer

14 units of alcohol

6 glasses of 13% wine
175 cm³ (175 ml)

14 glasses of 40% spirits
25 cm³ (25 ml)

Figure 15.25 *Note that 14 units of alcohol is equivalent to different numbers of glasses of different alcoholic drinks.*

ⓖⓞ! Activity 15.4

☺ 1. **(a)** Name the type of compound in plants that can be used to produce alcohol.

   (b) Name the process that produces alcohol from sugars.

   (c) Brandy has an alcohol content of 40%. It can be made from wine with an alcohol concentration of 10%. Describe how brandy could be made from wine.

2. Use Table 15.5 to help you calculate your answers to these questions.

   (a) A woman drinks two pints of beer and a whisky. How many units of alcohol will she have drunk?

   (b) Calculate which combination of drinks will contain less alcohol: a 330 cm³ bottle of strong beer and a whisky (25 cm³), or two 175 cm³ glasses of wine.

   (c) State **two** ways in which the Government promotes responsible drinking.

Improving plant growth with fertilisers

National 3

Learning intentions

In this section you will:

- learn that fertilisers are used to improve plant growth
- investigate fertilisers.

As well as absorbing carbon dioxide from the air and taking in water through their roots, plants need to take in elements for healthy growth. These elements are present in the soil and are taken in through the plant roots. If plants are unable to take in sufficient minerals, plant growth will be poor.

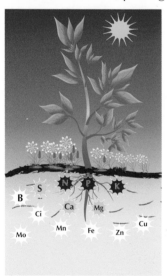

Figure 15.26 *Plants need to take in elements from the soil to grow well.*

📖 Word bank

- **Plant nutrients**

The chemical elements needed for healthy plant growth.

- **Fertiliser**

A substance that replaces nutrients removed from the soil.

Figure 15.27 *The elements nitrogen (N), phosphorus (P) and potassium (K) are the three main plant nutrients in the fertilisers added by farmers to fields.*

For instance, plants need magnesium to make chlorophyll, the green pigment that gives the leaves their colour. If plants cannot absorb enough magnesium, the leaves will be yellow.

Elements needed for healthy plant growth are called **plant nutrients**.

Cropping and fertilisers

If plants die and are left to decay, the nutrients they have absorbed return to the soil. However, when crops are harvested, the nutrients in the plants are taken away with the harvest and are lost from the soil. Fields can gradually lack the nutrients that plants need. To prevent this and to ensure that healthy crops can be grown, **fertilisers** are added to the soil.

The three main nutrients needed by plants and that are replaced by adding fertilisers are the elements nitrogen, potassium and phosphorus.

Farmers can replace the nutrients in fields either by adding natural fertilisers such as animal manure or by adding industrially produced fertilisers.

❓ Did you know ...?

Guano (dried seabird droppings) contains very high levels of nitrogen, phosphorus and potassium.

Figure 15.28 *Seabird droppings are a valuable natural fertiliser.*

The Chincha Islands off the coast of Peru were covered in guano 200 feet deep. In the 1830s, American sailors mined the guano and transported it back to the United States to be used as fertiliser.

☀ Chemistry in action: Different types of fertiliser

For centuries, farmers have added natural materials to land to improve the condition of the soil. These natural materials include:

- worm compost
- wood ash
- pond mud
- green manure
- animal manure

- limestone
- peat
- river sand
- chalk
- bone meal.

Bone meal is made by crushing animal bones. It releases phosphates, a source of phosphorus, into the soil. This happens very slowly. In the early nineteenth century, a German chemist, Justus von Liebig, found that treating the bones with sulfuric acid made them more effective than untreated bones. This led other chemists to experiment with rocks that contained phosphates. This was the beginning of the superphosphate fertiliser industry.

❓ Did you know ...?

Modern chemical fertilisers can double crop yields. This means twice as much food can be produced using fertilisers. Without modern fertilisers, the world would need the manure from up to 7 billion more cattle to maintain soil fertility.

The amount of fertiliser used world-wide is expected to reach 199 million tonnes per year by the end of 2019.

Figure 15.29 *Chemical fertilisers can double crop yields.*

GO! Activity 15.5

1. Experiment

You can investigate the effect of nutrients on plant growth by experimenting on mung beans. You can grow a mung bean in different culture solutions, and in distilled water as a control, so that you can compare the effect of the different nutrients.

You will need: germinated mung beans (the roots should be of similar lengths), test tubes, aluminium foil, a bright light source, distilled water, and a range of culture solutions containing nitrogen (N) and/or phosphorus (P) and/or potassium (K); if you do not use commercially available solutions, you can use:

- *complete solution containing N, P and K*
- *solution containing P and K but not N*
- *solution containing N and K but not P*
- *solution containing N and P but not K.*

(continued)

Method

(a) Label the test tubes 1–5 to show which culture solution is present.

(b) Fill each test tube with its corresponding culture solution.

(c) Cover each test tube with foil.

(d) Gently push down the centre of the foil in each test tube to make an upside down cone shape.

(e) Make a little hole in the centre of each cone. This will allow the culture solution to be topped up if needed.

(f) Carefully place a mung bean seed into each cone and gently push the root through the hole into the culture solution. The seed should be on top of the foil with the root dipping into the liquid below. Top up with culture solution if necessary.

Germinating mung bean.

(g) Place the test tubes in a rack and leave under bright light.

(h) Record your observations after 3, 7, 10 and 14 days, using a table like the one below.

Days elapsed	Plant appearance					
	Complete culture medium	Culture medium lacking N	Culture medium lacking P	Culture medium lacking K	Distilled water	
3						leaves
						roots
7						leaves
						roots
10						leaves
						roots
14						leaves
						roots

2. Experiment

You can carry out another experiment to investigate the effect of adding different amounts of fertiliser.

Do the experiment in the same way as Experiment 1. Use a liquid fertiliser such as Miracle-Gro, but this time vary the number of drops of fertiliser added to the water.

You can try adding 1 drop, 2 drops, 3 drops and 4 drops of fertiliser to your test tube of water. Remember to have one tube without fertiliser to use as a control.

GO! Activity 15.6

☻ 1. The symbols for the 16 elements that are plant nutrients are N, P, K, etc. Draw up a table and complete it by naming the elements that are plant nutrients.

Symbol	Element
N	nitrogen

Hint

Use Figure 15.26 to help.

2. (a) Name the **three** main elements in fertilisers.

(b) (i) Crop growth is improved by using fertilisers. State a reason for wanting to improve crop growth.

(ii) Explain why it is necessary for farmers to add fertilisers to fields from which crops have been taken.

Hint

Use the periodic table on pages 34–35 to help.

Learning checklist

After reading this chapter and completing the activities, I can:

N3 L3 N4 L4

- state that plants are a vital source of nutrients. **Activity 15.2 Q1**   ○ ○ ○

- state that nutrients from plants include proteins, carbohydrates, oils, vitamins and minerals. **Activity 15.2 Q1**   ○ ○ ○

- name the main food groups and say why they are important in the diet (carbohydrates and fats and oils are needed for energy, proteins are needed for growth and repair of tissue, vitamins and minerals are needed to prevent disease). **Activity 15.2 Q6**   ○ ○ ○

- work with others to carry out tests used to identify food groups. **Activity 15.1, Activity 15.2 Q4, Q5**   ○ ○ ○

- explain that lifestyle can have an effect on health. **Activity 15.3 Q1**   ○ ○ ○

- state that vitamins and minerals in the diet are required to keep the body healthy. **Activity 15.3 Q2, Q5**   ○ ○ ○

N3 L3 N4 L4

- state that essential fatty acids are fats that must be obtained from our diet. **Activity 15.3 Q3**

  ○ ○ ○

- state that health benefits of a diet high in plant-based proteins include: lower body weight, lower cholesterol and blood pressure levels, lower risk of stroke, cancer and death from heart disease. **Activity 15.3 Q4**

  ○ ○ ○

- state that alcohol is made by fermentation of sugars in fruits and other plant material. **Activity 15.4 Q1(a), (b)**

  ○ ○ ○

- state that distillation is a method used to increase the alcohol concentration in alcoholic drinks. **Activity 15.4 Q1(c)**

  ○ ○ ○

- state that the alcohol content in drinks is measured in units. **Activity 15.4 Q2(a), (b)**

  ○ ○ ○

- state that there are guidelines on safe drinking levels. **Activity 15.4 Q2(c)**

  ○ ○ ○

- work with others to carry out experiments to show that fertilisers promote plant growth. **Activity 15.5 Q1**

  ○ ○ ○

- state that fertilisers are needed for healthy plant growth. **Activity 15.6 Q2(b)(ii)**

  ○ ○ ○

- state that improved crop growth means more food can be grown. **Activity 15.6 Q2(b)(i)**

  ○ ○ ○

16 Everyday consumer products 2: Cosmetic products

This chapter includes coverage of:

N3 Everyday consumer products • Materials
SCN 3-17b

You should already know:

- that carbohydrates, plants and oils from plants are used as foods and for fuel.

Plants and cosmetics

National 3

Learning intention

In this section you will:

- learn that carbohydrates, and fats and oils from plants are used in the cosmetics industry.

📖 Word bank

- **Cosmetics**
Cosmetics are used to change the way the body looks or smells.

Cosmetic products have been used since ancient times. Perfume jars that had been filled with precious oils and ointments were found in King Tutankhamun's tomb and it is known that, in ancient Greece, athletes would use a mixture of sand and oil on their skin when competing. It is thought that this acted as a sunscreen. The oils and perfumes that were used were extracted from plants.

Many of the substances used in the cosmetics industry today are plant extracts. These include carbohydrates, and edible fats and oils. Essential oils, which give plants their various aromas, are also used.

Carbohydrates

Carbohydrates are energy foods, but they are also used in the cosmetics industry. Examples of such carbohydrates are agar and alginates, which are both obtained from seaweed.

Figure 16.1 *An alabaster jar from an Egyptian tomb. This contained essential oils.*

Figure 16.2 **(a)** *Agar, a carbohydrate extracted from red seaweed, is used to produce* **(b)** *cosmetic face masks.*

Figure 16.3 *Cocoa butter is used in chocolate making, body moisturising products and lip balms.*

Fats and oils

Cocoa butter

As well as being used in chocolate making, cocoa butter is included in many moisturising body creams. This fat is extracted from cocoa beans.

Argan oil

Argan oil is obtained from the nuts of the argan tree. For centuries, women from the Berber tribe in North Africa used the oil on their skin and hair to protect them from the harsh environment. The oil has a high vitamin E content, which protects against the damaging effects of ultra-violet rays in sunlight, and it is easily absorbed into the skin. Because the tree is only found in southwest Morocco, supplies are limited. For this reason, and because it is highly valued for its cosmetic effects, it is sometimes referred to as 'Moroccan liquid gold'.

Figure 16.4 *Argan oil is extracted from nuts from the argan tree.*

Castor oil

Castor oil has been used for medical purposes since ancient times. According to an ancient Egyptian medical text written in 1500 BCE, doctors used castor oil to treat eye irritations. But it was also used in many cosmetic preparations.

The oil is extracted from castor beans. Today, it is used in lipsticks as it forms a tough shiny film when it dries.

(a) **(b)**

Figure 16.5 **(a)** *Castor beans.* **(b)** *Half of the weight of a lipstick can be due to castor oil.*

Activity 16.1

☻ **1.** Copy and complete the paragraph below. Use the following words to help.

carbohydrates   fats   oils   cosmetics

Plants are a source of many compounds used in products such as shampoos and body lotions in the **(a)** _____ industry. Different compounds can be obtained from plants. Agar and alginates are examples of **(b)**_____ that can be obtained from red seaweeds. **(c)** ____ and **(d)** ____ are obtained from the nuts and seeds of plants.

Using essential oils

National 3
Curriculum level 3
Materials: Earth's materials SCN 3-17b

Learning intentions

In this section you will:

- learn how essential oils can be extracted from plant material
- learn about the uses of essential oils
- learn about the essential oils that give some plants their distinctive scent
- learn that different essential oils are blended to give perfume fragrances.

Essential oils are the oils that give different plants their characteristic, special aromas (smells). They are mixtures of many different compounds.

Essential oils are included in many products because of their aromas and also because of health benefits associated with their use. Many essential oils have antiseptic and antifungal properties.

Methods of extracting essential oils

Essential oils can be extracted from different parts of plants, including roots, leaves and flowers.

There are a number of different ways of extracting the oils from the plants:

- cold pressing
- using a solvent
- steam distillation.

Figure 16.6 *A steam distillation still in a factory producing essential oils.*

Cold pressing involves squeezing the plant material, so that plant cells burst and release their essential oils. You get the same effect when you dig your fingernails into the skin of an orange.

Another way to obtain the oil is to use a solvent. Using a solvent dissolves the essential oil out of the plant material.

The most common method of extracting essential oils is by a process called steam distillation.

Steam distillation can be carried out in the school laboratory.

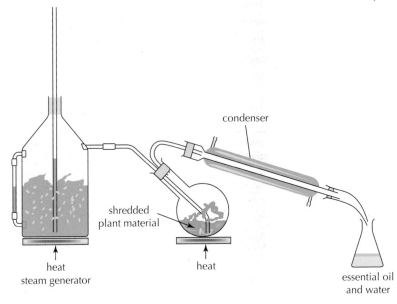

Figure 16.7 *Apparatus used to obtain essential oils by steam distillation.*

Steam from a generator is passed through shredded plant material in a heated flask. The compounds that make up the essential oils are very **volatile** (they vaporise easily). The vapours are carried with the steam into the condenser. The essential oils form an oily layer on top of the water in the flask that is used to collect the condensed liquids.

Word bank

• **Volatile**
Evaporates easily.

? Did you know ...?

In 1910 Otto Wallach was awarded the Nobel Prize in Chemistry. His work had a major influence on the branch of the chemical industry that processes essential oils.

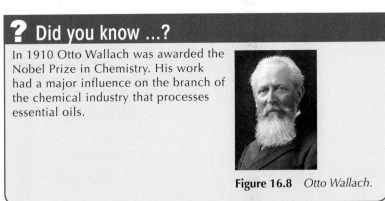

Figure 16.8 *Otto Wallach.*

Table 16.1 lists some essential oils, their plant sources and their uses.

Table 16.1 *Essential oils, their plant sources and uses.*

Essential oil	Part of plant used to extract the oil	Uses
lavender	flowers	sleep aid, skin creams, body lotions
lemon	peel	soaps and hair products
orange	peel	household cleaners
geranium	leaves and stalks	moisturisers
peppermint	stems and leaves	breath fresheners, toothpastes, indigestion relief
ginger	root	muscle pain relief
eucalyptus	leaves	antiseptics, sinus inhalants, muscle rubs
rosemary	leaves	shampoos and hair oils
spearmint	flowering tops	toothpaste
thyme	flowers and leaves	mouthwashes
sandalwood	powdered wood	skin creams, perfumes and aftershaves

Figure 16.9 *Oils such as lavender are popular in aromatherapy because of their relaxing effects.*

Effects on our bodies

Different essential oils can have different effects on our bodies. Some can have a calming effect, while others can make people feel energised. Table 16.2 lists the effects of different essential oils.

Table 16.2 *Effects of different essential oils.*

Effect of essential oil	Essential oil
calming	lavender, geranium, mandarin, bergamot, ylang ylang, neroli, jasmine, melissa, palmarosa, patchouli, petitgrain, sandalwood
energising	rosemary, clary sage, bergamot, lemongrass, eucalyptus, peppermint, spearmint, tea tree, cypress, pine, lemon, basil, grapefruit, ginger
memory boosting	basil, cypress, lemon, peppermint, rosemary

GO! Activity 16.2

☺ 1. Essential oils are responsible for the distinctive a_ _ _ _ of plants and can be extracted by s_ _ _ _ distillation.

> 🔍 **Hint**
>
> Use the information in Tables 16.1 and 16.2 to help you answer these questions.

2. Suggest why lavender is sometimes sprayed on bedroom pillows.

3. Name **two** essential oils used in toothpastes.

4. State a use for ginger essential oil.

5. An aftershave contains eucalyptus oil. How might someone feel after splashing on an aftershave containing eucalyptus oil?

6. Name an oil for use in a room diffuser to help you remember what you have been studying.

☺☺ 7. **Experiment**: Extracting an essential oil by steam distillation

You will need: 25 g finely diced orange peel, 250 cm³ conical flask, basket made from wire gauze, delivery tube, boiling tube, beaker, crushed ice.

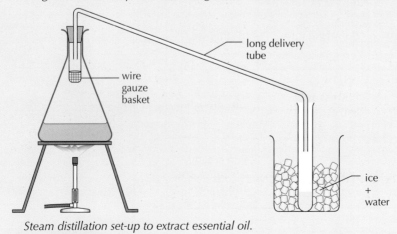

Steam distillation set-up to extract essential oil.

(a) Place 25 g of finely diced orange peel in a wire gauze basket.

(b) Hang the basket in a 250 cm³ conical flask containing 100 cm³ of water.

(c) Connect a long delivery tube to the conical flask.

(d) Place the open end of the delivery tube in a boiling tube that is sitting in an ice water mixture.

(e) Heat the conical flask to create steam.

(f) When the boiling tube is nearly full of condensed liquid, remove the delivery tube.

(g) Stop heating. Remember you must remove the delivery tube before stopping heating in order to prevent suck-back occurring.

(h) Write a brief description of how essential oils can be extracted.

Note: You are not likely to get much oil. If there is no sign of oil on the surface of the water in the boiling tube, smell the liquid to see if there is any indication that oil has been extracted.

Variation: You can extend the range of plant material to include other citrus fruits like lemon, grapefruit and mandarin. You could also try herbs such as rosemary, thyme, oregano and lavender.

? Did you know ...?

It takes 5 tonnes of roses to produce 1 kg of rose essential oil. That's equivalent to 1 drop of oil from 30 rose blossoms. It is one reason why rose essential oil is so expensive.

Blending essential oils to make perfumes

Perfumes are blends (mixtures) of different essential oils. It takes a lot of skill to produce a successful blend.

Perfumers can spend many months perfecting a perfume, adjusting the amounts of essential oils added to the mix in order to achieve a balanced fragrance.

Figure 16.10 *Perfumes are blends of many essential oils.*

? Did you know ...?

It takes up to 10 years to qualify as a perfumer. As well as gaining qualifications in chemistry, it is essential that perfumers have an excellent sense of smell.

Perfumers in France are referred to as *le nez* (the nose) as their sense of smell is so good.

It is not just expensive perfumes that perfumers work to produce. They are involved in the development of all sorts of products, even detergents and shampoos.

Figure 16.11 *It is essential that a perfumer has an excellent sense of smell.*

Essential oils can be divided into three categories: top notes (also known as head notes), middle notes (also known as heart notes) and base notes.

Figure 16.12 *The top notes of a perfume give you a first impression and have a strong influence on whether or not you like a perfume.*

An oil is classed as a top, middle or base note according to how long the scent lasts on the skin (Table 16.3).

Table 16.3 *Characteristics of top, middle and base notes.*

Note	Description
Top	These are the essential oils you smell immediately when you use a perfume. This is because the oils vaporise very easily. They tend to be the lightest molecules and fade away about 10 to 15 minutes after applying. They are very important as they give you the first impression of a perfume and determine whether or not you like a particular perfume.
Middle	As the top notes evaporate, the middle notes begin to take over. Combinations of floral, fruity and spicy scents tend to work well together. The middle notes tend to persist for 3 to 4 hours on the skin.
Base	In combination with the middle notes, these give the perfume its body. The scents from the base note essential oils can linger on the skin for many hours.

Table 16.4 lists essential oils by note.

Table 16.4 *Essential oils by note.*

Top (head) notes	Middle (heart) notes	Base notes
basil	black pepper	cinnamon
bergamot	chamomile	clove
eucalyptus	geranium	patchouli
grapefruit	ginger	sandalwood
lemon	jasmine	vanilla
mandarin	juniper	vetiver
orange	lavender	ylang ylang
peppermint	lemongrass	
pine	nutmeg	
spearmint	rose	
tea tree	rosemary	

To be perfectly balanced, a perfume needs to have a mixture of top, middle and base notes. Perfumers blend many different essential oils to obtain what they consider to be the correct balance.

Essential oils can also be classified according to how they smell. Table 16.5 shows a simplified classification of some essential oils by smell.

? Did you know ...?

In 1709, Giovanni Maria Farina, an Italian perfumer living in Cologne, created a fragrance he named 'Eau de Cologne'. He wrote to his brother: 'I have found a fragrance that reminds me of an Italian spring morning, of mountain daffodils and orange blossoms after the rain.'

Figure 16.13 *An Eau de Cologne called 4711 is still being produced more than 220 years after being first introduced.*

Table 16.5 *Essential oils by scent type.*

Scent type	Essential oil
citrus	bergamot, grapefruit, lemon, lemongrass, mandarin, orange
floral	geranium, jasmine, lavender, rose, vanilla, ylang ylang
herbal	basil, chamomile, eucalyptus, peppermint, pine, spearmint, tea tree, rosemary
spicy	black pepper, cinnamon, clove, ginger, nutmeg
woody	juniper, patchouli, sandalwood, vetiver

GO! Activity 16.3

1. Experiment: Making a simple Eau de Cologne

You will need: pure alcohol (you will need to be given this by your teacher), distilled or deionised water and a few drops of lemon and orange essence.

⚠ Pure essences can cause skin irritation and should only be applied to the skin dissolved in a carrier oil, alcohol or water.

Method

(a) Mix 7 cm³ of the alcohol and 3 cm³ of the water.

(b) Add 5 drops of lemon essence and 5 drops of orange essence.

You have made yourself a light fragrance in the style of Eau de Cologne!

(c) Rub a small amount on the back of your hand. Smell your hand every 5 minutes to see how long you can detect the smell. Orange and lemon are top notes in perfumes so will disappear fairly quickly.

(d) Record your results in a table like the one below.

Time since application (minutes)	How strongly is smell detected (strongly–faintly)
0 (on application)	
5	
10	
15	
20	
25	
30	

(e) **(i)** Describe how a simple Eau de Cologne-type fragrance is made.

(ii) Explain why the scent becomes more faint over time.

(continued)

2. Experiment

You will need: petri dishes, food essences such as lemon, orange, vanilla, almond, peppermint and lavender (you can also use essential oils from an aromatherapy kit to extend the range of scents), filter papers cut into strips.

This is a smelling game. One person (or the teacher) needs to set up the game for others to try.

To set up the game:

(a) Drop some food essence onto a strip of filter paper and place the filter paper in a petri dish. Label the petri dish with the number 1 and note the extract you have used.

(b) Repeat this with the other essences, labelling the petri dishes 2, 3, 4 … etc.

(c) Now give each person taking part an answer grid.

Petri dish	Essence or essential oil
1	
2	
3	
4	
5	

(d) Each person should lift the lid of the petri dish, smell the extract and write down the name of the extract they think they are smelling.

If anyone is unfamiliar with the scents, you can use labelled petri dishes containing filter paper strips with the essential oils for them to smell before taking part in the game.

You could also try dropping two different essential oils onto a filter paper to see if anyone can detect both.

☻ **3.** Collect labels from household products, such as cosmetics, toiletries, cleaning products and items from the medicine or first aid cabinet. Identify any labels that mention the names of essential oils; create a list of essential oils and the products that contain them.

4. The annual revenue from sales of cosmetics in the UK is approximately £9 billion. The pie chart shows the percentage that comes from each sector.

(a) Calculate the percentage of sales for haircare and skincare products combined.

(b) Calculate how much is spent on toiletries if £9 billion is spent on cosmetics in total.

Use the formula:

$$\text{amount} = \frac{\text{percentage}}{100} \times \text{total}$$

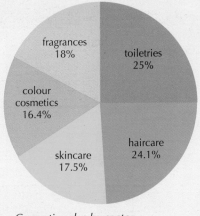

Cosmetic sales by sector.

5. A student was attempting to extract rosemary essential oil from chopped rosemary leaves using the following apparatus.

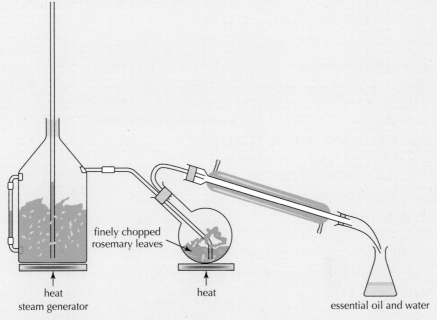

finely chopped
rosemary leaves

heat
steam generator

heat

essential oil and water

Set-up to extract essential oil from rosemary leaves.

(a) State the name of the technique that the student was using to extract the essential oil.

(b) What evidence would there be to show that rosemary essential oil was extracted from the finely chopped rosemary leaves?

(c) Rosemary leaves are used to extract essential oils. Name **two** other parts of plants from which essential oils can be extracted. Give examples.

6. Use the information in Tables 16.3, 16.4 and 16.5 to help you answer these questions.

Coco Mademoiselle is a popular female fragrance. The following description has been used to promote the fragrance.

> **An oriental fragrance with a strong personality, yet surprisingly fresh. Sparks of fresh and vibrant orange awaken the senses. A clear and sensual heart reveals the notes of jasmine and rose. The trail leaves behind the notes of patchouli and vetiver.**

(a) Suggest which essential oil is used as the top note for the fragrance.

(b) What type of scent will jasmine and rose give the fragrance?

(c) The base notes for the fragrance are patchouli and vetiver.

   (i) Explain why patchouli and vetiver essential oils are used as base notes in fragrances.

   (ii) What type of scent would be given by patchouli and vetiver essential oils?

(d) Why can Coco Mademoiselle be described as a balanced perfume?

Learning checklist

After reading this chapter and completing the activities, I can:

N3 L3 N4 L4

- state that carbohydrates, fats and oils from plants can be used to produce cosmetic products. **Activity 16.1**   ○ ○ ○

- state that essential oils are the oils that give plants their distinctive aroma. **Activity 16.2 Q1**   ○ ○ ○

- state that essential oils can be obtained using different parts of plants including roots, leaves and flowers. **Activity 16.3 Q5**   ○ ○ ○

- describe how essential oils can be extracted from plant material by steam distillation. **Activity 16.2 Q7, Activity 16.3 Q5**   ○ ○ ○

- identify essential oils by smell. **Activity 16.3 Q2**   ○ ○ ○

- give examples of the use of essential oils, such as: peppermint and spearmint essential oil in toothpaste; lemon, rosemary and orange essential oils in shampoos; lavender essential oil in hand creams and body lotions; eucalyptus essential oil in sinus sprays; and ginger essential oils in muscle massage oils. **Activity 16.2 Q2–6**   ○ ○ ○

- describe how essential oils are blended to make perfumes. **Activity 16.3 Q1**   ○ ○ ○

- explain why essential oils in perfumes are classified as top, middle and base notes. **Activity 16.3 Q6**   ○ ○ ○

17 Everyday consumer products 3: Plants for energy

This chapter includes coverage of:

N4 Everyday consumer products

You should already know:

- that plants are a source of carbohydrates and oils
- that carbohydrates and oils are used as fuels and foods
- that alcohol is made by fermentation of sugars in fruits and other plant materials
- that distillation is a method used to increase the alcohol concentration in alcoholic drinks
- that the alcohol content in drinks is measured in units.

Plants for food and energy

National 4

Learning intentions

In this section you will:

- learn that the foods we eat contain carbohydrates and fats and oils
- learn that our bodies burn carbohydrates and fats and oils for energy
- learn that fats are a more concentrated source of energy than carbohydrates.

Figure 17.1 *Fruit and vegetables are a source of carbohydrates and oils.*

Plants form an important part of our diet. The carbohydrates (sugars and starches) and oils in plants can provide us with the energy we need.

The energy released from burning sugars and fats can be measured using a **bomb calorimeter**.

Burning 1 g of oil produces more than twice the amount of energy that is produced from burning 1 g of sugar:

- burning 1 g of sugar releases 17 kilojoules (kJ) of energy
- burning 1 g of oil releases 37 kJ of energy.

📖 Word bank

- **Bomb calorimeter**
A device for measuring the heat energy given out by burning foodstuffs.

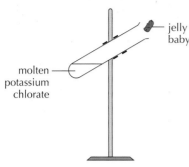

Figure 17.2 *Set-up for 'screaming jelly baby' experiment.*

'Screaming jelly baby' experiment

 The demonstration must only be carried out by a teacher.

This is a spectacular demonstration of the energy contained in sugars.

Some potassium chlorate is heated in a boiling tube until it melts. A jelly baby is then dropped into the tube. A violent exothermic reaction takes place, creating a lot of heat, light and noise.

This experiment can also be viewed on YouTube.

> ### 🟢 Activity 17.1
>
> ☻ **1.** Describe how you could show that carbohydrates and oils are energy foods and can be used as fuels.

National 4

Carbohydrates – What's in a name?

Learning intentions

In this section you will:

- learn that carbohydrates are compounds made only of carbon, hydrogen and oxygen atoms
- learn why starch and glucose can be described as carbohydrates
- learn about chemical tests that can be used to distinguish starch and glucose.

A simple demonstration sometimes called the 'carbon snake experiment' gives an idea of the elements in **carbohydrates**.

Your teacher may demonstrate this to you. Or you may see this experiment being carried out on YouTube.

When some concentrated sulfuric acid is mixed with sucrose – a carbohydrate – in a beaker and the mixture is stirred, after a short time the mixture begins to turn black, steamy fumes are given off and a 'black snake' rises out of the beaker. The black snake is carbon. Concentrated sulfuric acid removes water from the sucrose, producing carbon. The reaction is so exothermic that steam (water vapour) is produced, causing the carbon to expand into a cylinder shape that rises up out of the beaker.

★ You need to know

'Sugars' is a general name given to sweet carbohydrates. The sugar that some people put in their tea or coffee is actually sucrose.

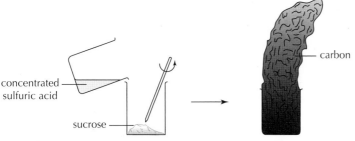

Figure 17.3 *When concentrated sulfuric acid is mixed with sucrose, a carbon snake is formed.*

Carbohydrates are made from only three elements: carbon, hydrogen and oxygen. The name for this group of compounds reflects this.

Table 17.1 shows the molecular formulae of some carbohydrate molecules.

Look at the formulae: the number of hydrogen atoms in the molecules is *always* twice the number of oxygen atoms. The hydrogen and oxygen atoms in carbohydrate molecules are in the ratio 2:1, just as in water, H_2O.

> **Word bank**
>
> • **Carbohydrate**
>
> A compound containing only C, H and O, with twice as many Hs as Os in the formula.

Table 17.1 *The molecular formulae of some carbohydrates.*

Carbohydrate	Molecular formula
glucose	$C_6H_{12}O_6$
fructose	$C_6H_{12}O_6$
maltose	$C_{12}H_{22}O_{11}$
sucrose	$C_{12}H_{22}O_{11}$
starch	$(C_6H_{10}O_5)_n$

n is a very large number, as high as 1000.

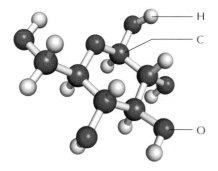

Figure 17.4 *A model of a glucose molecule, molecular formula $C_6H_{12}O_6$.*

Glucose and fructose are examples of single-unit sugar molecules. Both glucose and fructose have the molecular formula $C_6H_{12}O_6$ but the atoms within their molecules are arranged differently, as shown in Figure 17.5.

> **★ You need to know**
>
> Although carbohydrates have H : O in a 2 : 1 ratio, like water, there are no water molecules in carbohydrates.

(a)

H—O—⟨G⟩—O—H

(b)

H—O—⟨F⟩—O—H

Figure 17.5 *Representations of* **(a)** *a glucose molecule and* **(b)** *a fructose molecule.*

Maltose and sucrose are examples of sugar molecules with two units each.

Single-unit and two-unit sugars are described as **simple carbohydrates** as they are easy to digest.

> **Word bank**
>
> • **Simple carbohydrate**
>
> Single-unit sugars and two-unit sugars.

Converting glucose to starch in plants

During photosynthesis, plants change carbon dioxide and water into glucose. In order to store the Sun's energy, plants convert the soluble glucose into starch. Starch is a carbohydrate made of many glucose units joined together in a chain, as shown in Figure 17.6.

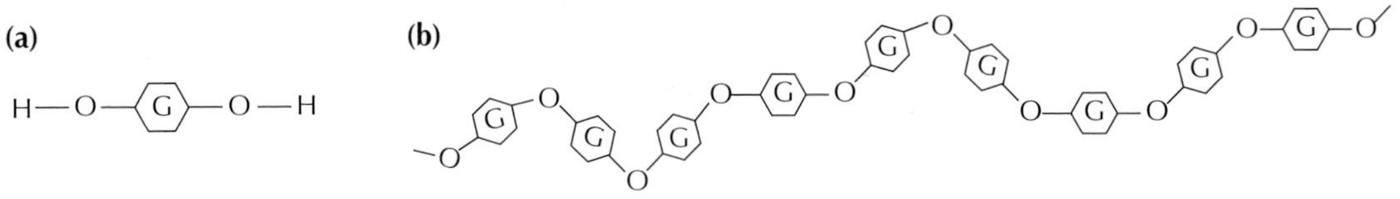

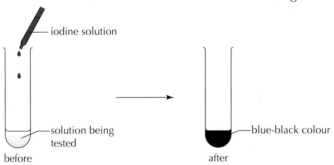

(a) **(b)**

Figure 17.6 (a) *A glucose molecule.* **(b)** *Starch molecules are long chains of glucose molecules strung together like a necklace.*

Because starch is made up of many glucose units joined together, it is described as a **complex carbohydrate**.

📖 Word bank

• **Complex carbohydrate**

A carbohydrate made of many sugar units in a chain.

Testing for glucose and starch

Benedict's solution and iodine solution are used to test whether a carbohydrate is glucose or starch. Starch gives a blue–black colour with iodine (Figure 17.7). Glucose gives a brick-red colour when heated with Benedict's solution (Figure 17.8).

iodine solution

solution being tested

before

blue-black colour

after

Figure 17.7 *Only starch solution gives a blue–black colour when iodine solution is added.*

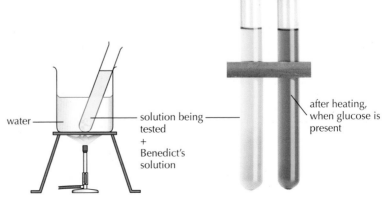

water

solution being tested + Benedict's solution

after heating, when glucose is present

🔆 Make the link

For more about testing foods, see Chapter 15.

Figure 17.8 *Benedict's solution (blue) turns brick-red when heated with glucose solution.*

Table 17.2 summarises the results of these tests.

Table 17.2 *Testing for glucose and starch.*

Carbohydrate	Reaction with Benedict's solution	Reaction with iodine solution
glucose	solution turns brick-red	no change
starch	no change	blue–black colour appears

GO! Activity 17.2

☻ **1.** Which of the following formulae does <u>not</u> represent a carbohydrate?

   A $C_{12}H_{22}O_{11}$ **B** $C_4H_6O_2$

   C $C_5H_{10}O_5$ **D** $C_6H_{12}O_6$

> 🔍 **Hint**
>
> The hydrogen : oxygen ratio in carbohydrates is 2 : 1

2. Glucose is described as a simple carbohydrate and starch is described as a complex carbohydrate.

   (a) Write the formula for glucose.

   (b) (i) Describe the structure of a starch molecule.

       (ii) Why is starch described as a complex carbohydrate?

3. Describe a chemical test that could be carried out to identify:

   (a) glucose   **(b)** starch.

Digestion

National 4

Learning intentions

In this section you will:

- learn what happens when starch is digested
- learn why glucose can be absorbed into the bloodstream but starch cannot.

> 📖 **Word bank**
>
> • **Digestion**
> The process of breaking down food in the body.

What happens to food during digestion?

Digestion is the process of breaking down food into forms that the body can absorb and use. The digestion process starts in the mouth and continues in the stomach. The molecules the body needs pass through the gut wall into the bloodstream and are carried around the body to be absorbed into cells.

Both glucose and starch are carbohydrates that provide the body with energy. Glucose molecules are small enough to pass through the gut wall and can be absorbed into the bloodstream. However, starch molecules are too large to pass through.

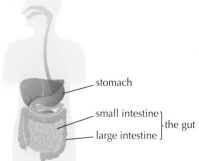

Figure 17.9 *Molecules that the body needs pass through the gut wall into the bloodstream.*

N3	L3	**N4**	L4

GO! Activity 17.3

1. Experiment

You will need: two beakers (400 cm³), two boiling tubes, short glass rods, filter funnel, paper clips, two lengths of visking tubing, glucose solution, starch solution, iodine solution, Benedict's solution.

The aim of this experiment is to show that starch molecules are larger than glucose molecules. Visking tubing has tiny holes in it. In this experiment the visking tubing represents the gut wall.

Method

(a) Place the lengths of visking tubing in warm water to allow them to soften.

(b) Tie a knot at one end of each length of tubing.

(c) Fill one length of tubing with starch solution. Use a clean filter funnel to make this easier and make sure no starch solution spills down the outside of the tubing.

(d) Wrap the open end of tubing around a glass rod and secure using a paper clip.

(e) Repeat using glucose solution.

(f) Place the filled visking tubing in boiling tubes containing water and place these in beakers of warm water.

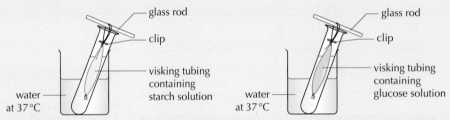

Set-up for the experiment.

(g) After 15 minutes, remove a sample of water from the boiling tube containing the visking tubing filled with starch solution and test with iodine solution.

(h) Remove a sample of water from the boiling tube containing the visking tubing filled with glucose solution and test with Benedict's solution.

(i) Record the results of your tests.

2. Use your results from the experiment to copy and complete the passage below.

When i_ _ _ _ _ is added to a sample of the water surrounding the visking tubing containing the starch it doesn't give a b_ _ _–b_ _ _ _ colour. The starch molecules are too l_ _ _ _ to pass through the visking tubing into the water.

When the water surrounding the visking tubing containing the glucose solution is tested with B_ _ _ _ _ _ _ _'_ solution, a b_ _ _ _ _-r_ _ _ colour is produced. Glucose molecules are s_ _ _ _ enough to pass through the visking tubing into the water.

N3 | L3 | **N4** | L4

Breaking down starch

Starch has to be broken down into smaller molecules so the body can use it. This is what happens during digestion. Special compounds called **enzymes** that are present in saliva begin the process. Enzymes are biological catalysts. They speed up biological processes.

> **📖 Word bank**
>
> • **Enzyme**
> A biological catalyst; that is, a compound that speeds up biological processes.

part of a large starch molecule

enzymes in saliva →

smaller glucose molecules

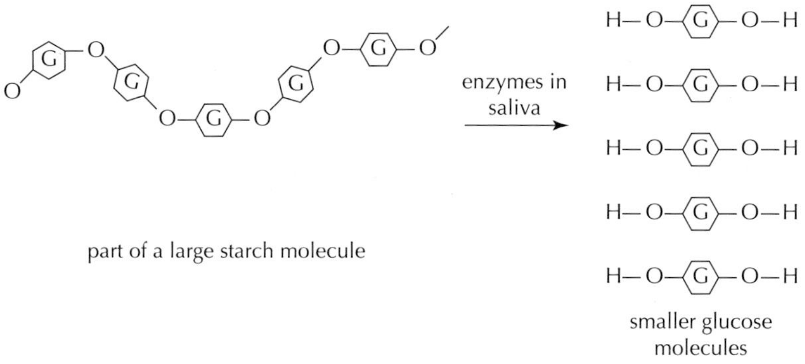

Figure 17.10 *Enzymes in saliva help to break down starch into smaller glucose molecules.*

Acids in the stomach also break down food. Digested food passes from the stomach into the small intestine, a part of the human gut (see Figure 17.9).

Glucose molecules are able to pass through the wall of the small intestine into the bloodstream. Blood carries glucose around the body, where it is absorbed into cells and used to provide energy. This is the process that we refer to as **respiration**. Respiration produces carbon dioxide and water, which we breathe out.

Figure 17.11 *You need enzymes to digest your pizza.*

> **📖 Word bank**
>
> • **Respiration**
> The reaction of glucose and oxygen in the cells of the body to produce energy.

Equations for respiration:

Word: glucose + oxygen → carbon dioxide + water

Formula: $C_6H_{12}O_6$ + O_2 → CO_2 + H_2O

GO! Activity 17.4

☺ **1.** During digestion, starch is broken down to glucose.

    (a) State the test for starch.

    (b) Amylase in saliva breaks down starch. What type of substance is amylase?

    (c) How could you test whether or not a starch solution has broken down to glucose?

(continued)

2. Experiment

In this experiment you will investigate the digestion of starch, using an enzyme and acid.

You will need: two beakers (400 cm³), two boiling tubes, starch solution, amylase solution, 2 mol/l hydrochloric acid, dimple tile, dropping pipette, thermometer, iodine solution, Benedict's solution.

Method

(a) Put 25 cm³ of starch solution into each of the two boiling tubes.

(b) To one tube add 5 cm³ of amylase solution; to the other add 5 cm³ of 2 mol/l hydrochloric acid.

(c) Place the boiling tube containing the starch + amylase in a beaker of water at 35–40 °C. (Amylase is a body enzyme and therefore works well at body temperature, that is, 37 °C.)

(d) Place the boiling tube containing starch + 2 mol/l hydrochloric acid in a beaker of boiling water.

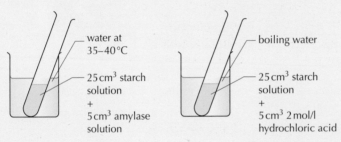

A — water at 35–40 °C — 25 cm³ starch solution + 5 cm³ amylase solution

B — boiling water — 25 cm³ starch solution + 5 cm³ 2 mol/l hydrochloric acid

Set-up for the experiment.

(e) Place a few drops of iodine solution in each of the dimples of a dimple tile.

(f) To the first dimple, add a few drops of the starch solution you are using in the experiment.

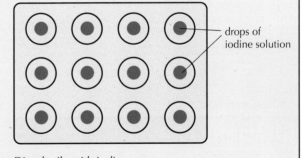

drops of iodine solution

Dimple tile with iodine.

(g) Every two minutes, use a dropping pipette to remove a small sample from the boiling tube containing the starch + amylase (A) and add it to a dimple.

(h) After 20 minutes test the solutions in each boiling tube using Benedict's solution.

Questions

(a) The blue–black colour, produced by the samples taken from boiling tube A, becomes less intense with each sample taken. Explain why.

(b) Explain why a brick-red colour is obtained when the starch + dilute hydrochloric acid mixture in B is tested with Benedict's solution.

Alcohol from plants

National 4

Learning intentions

In this section you will:
- learn about the conditions necessary for fermentation
- learn how distillation is used to increase the alcohol concentration.

Fermentation

Grapes are full of natural sugars. Naturally occurring yeasts live on the skins of the grapes. Yeasts contain enzymes that can change the sugars in grapes into alcohol (and carbon dioxide). As soon as the skin of the grape is broken, the sugars begin to ferment. Fermentation is the process of converting glucose into ethanol (alcohol).

In order to make wine, the grapes need to be collected and gently crushed – this releases the sugary juice and exposes it to the yeasts. Centuries ago, this was done by treading the grapes. Today, winemakers use wine presses.

Equations for fermentation:

enzymes

Word: glucose $\rightarrow$ ethanol   +   carbon dioxide

Formula: $C_6H_{12}O_6$ $\rightarrow$ C_2H_5OH + CO_2

(a)

(b)

Figure 17.12 (a) *Traditional method of using the feet to crush grapes.*
(b) *Modern method, using a wine press to crush the grapes.*

? Did you know ...?

Originally, grapes were pressed by foot to prevent the bitter seeds being crushed. Most of the port wine made in Portugal today is produced using modern methods, but some is still made by treading the grapes in stone tanks called *lagares*.

Figure 17.13 *Treading grapes to produce port wine.*

N3	L3	N4	L4

GO! Activity 17.5

1. Experiment

This experiment investigates fermentation of glucose.

You will need: glucose, 1 g fast-acting yeast, limewater, 100 cm³ conical flask, boiling tube, cotton wool.

Method

(a) Weigh 5 g of glucose and transfer it to a conical flask.

(b) Add 50 cm³ of warm water and swirl the flask to dissolve the glucose.

(c) Weigh 1 g of fast-acting yeast and add it to the flask.

(d) Place a cotton wool plug in the mouth of the flask and swirl the flask to mix the yeast with the glucose solution.

(e) Place the flask in a warm place, such as a window sill in sunlight or near a heater.

Set-up for the experiment.

(f) When you see a good froth on top of the liquid, remove the cotton wool plug and 'pour' the colourless gas on top of the liquid into a boiling tube containing limewater. Be careful not to pour any liquid in as well.

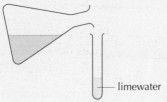

— limewater

Testing the gas produced by fermentation.

(g) Gently shake the tube of limewater and record any change.

(h) Replace the cotton wool plug and set the flask aside in a warm place until next lesson.

(i) Next lesson, remove the cotton wool plug and waft the vapours from the flask to your nose.

(j) Write a report about the experiment. Explain your results, making sure you answer these two questions:

- What evidence is there for a reaction taking place?

- What happened to the limewater? What caused this change?

Your teacher may ask you to collect the fermented glucose solution to distil it to produce alcohol.

Distillation

When natural yeasts on grape skins are used to ferment sugars, the alcohol concentration of the solution can increase to about 15%. At this stage, yeasts die naturally, the enzymes stop working, and fermentation stops. In beer making, a lower concentration of alcohol kills the yeasts that are used by the brewers.

Spirits are alcoholic drinks with an alcohol content of at least 20% alcohol by volume (ABV). Spirits such as whisky, gin, rum, brandy and vodka typically have an alcohol content of 40% ABV or greater. These drinks cannot be made only by fermenting sugars. The alcohol concentration is increased by **distilling** the alcoholic liquids obtained by fermentation.

Your teacher may demonstrate distillation using the liquids obtained by fermentation in Activity 17.5.

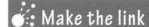

> ### Make the link
> You can read about distillation in Chapter 15.

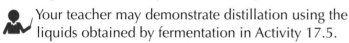
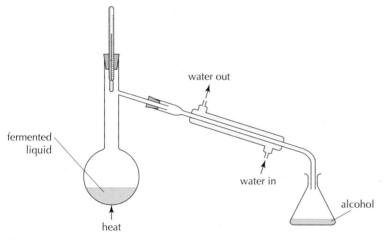

Figure 17.14 *Apparatus for distillation.*

> ### Word bank
> • **Distillation**
> The process of increasing the concentration of alcohol in order to make spirits.

Ethanol, the alcohol made by fermenting sugars, boils at 78 °C. When the fermented liquid is heated, the first vapours given off are mainly alcohol. These vapours can be condensed to produce an alcohol–water mixture with a high alcohol concentration.

To show that the liquid contains a high alcohol concentration, a sample can be burned. Alcohol evaporates easily and is very flammable. It burns with a clean blue flame.

Figure 17.15 *Alcohol burning.*

Plant sources of alcoholic drinks

It is not only barley and grapes that are used to make alcoholic drinks. Other alcoholic drinks are produced from different plant sources. Look back at Table 15.4 (page 221) which shows the plant sources for some alcoholic drinks.

✾ Chemistry in action: From barley to whisky

The flow chart in Figure 17.16 represents changes that take place when barley is used to make whisky.

Enzymes break starch in barley down to maltose (a two-unit sugar) and then glucose. This process is called malting.

Enzymes cause the malted barley to ferment, producing a 'beer' with an alcohol concentration of about 8%.

Distillation increases alcohol concentration.

Spirit is left to mature in barrels (to develop flavour and colour) for at least three years.

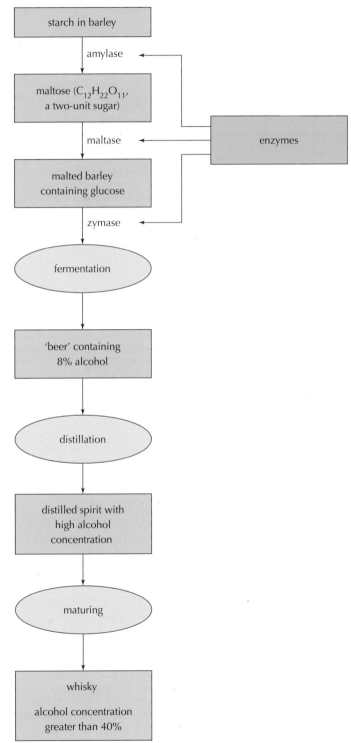

Figure 17.16   *The processes involved in making whisky.*

Units of alcohol

One unit of alcohol is equivalent to $10\,cm^3$ of alcohol. By law, the number of units of alcohol contained in an alcoholic drink must be shown on the label.

Example 17.1

A bottle of whisky states that it is 46% ABV. A $700\,cm^3$ bottle would contain:

$$700 \times \frac{46}{100} = 322 \; cm^3 \text{ of alcohol}$$

$$\text{Number of units} = \frac{322}{10} = 32.2$$

The label would, therefore, state that the bottle contains 32.2 units of alcohol.

Example 17.2

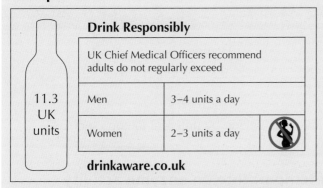

Drink Responsibly

UK Chief Medical Officers recommend adults do not regularly exceed		
Men	3–4 units a day	
Women	2–3 units a day	

drinkaware.co.uk

11.3 UK units

This label is taken from a $750\,cm^3$ bottle of wine.

The number of units in the bottle is 11.3.

Therefore, the bottle contains $10 \times 11.3 = 113\,cm^3$ of alcohol.

The percentage alcohol content of the wine is:

$$\frac{\text{volume of alcohol}}{\text{total volume of drink}} = 100 = \frac{113}{750} \times 100 = 15\%$$

↻ Keep up to date!

Recommended maximum limits for healthy drinking levels can change. Check online to see if official government or medical advice has changed.

🔵 Activity 17.6

😃 **1. (a)** Explain why yeast enzymes stop working when alcohol concentration reaches about 15%.

   (b) Explain how the alcohol concentration can be increased to 40%.

2. Write a paragraph describing how whisky is made from barley.

3. Alcoholic drinks can be made from plants other than grapes and barley.

   Give **two** examples and name their plant sources.

(continued)

4. (a) A 700 cm³ bottle of whisky shows that it contains 40% alcohol by volume.

Calculate how many units of alcohol are contained in the bottle of whisky.

(b) This is part of a label taken from a wine bottle.

Know your limits
This bottle contains 6 glasses
Units of alcohol per 125ml glass and 75cl bottle: **1.6** | **9.8 UK UNIT**

Hint

1 unit is equivalent to 10 cm³ of alcohol.

How many cubic centimetres of alcohol are contained in a 125 cm³ (125 ml) glass of the wine?

National 4

Effect of temperature and pH on enzyme activity

Learning intention

In this section you will:

- learn how temperature and pH affect enzyme activity.

Enzyme activity is affected by different factors, such as temperature and pH.

Effect of temperature on enzyme activity

Some yeast enzymes work best at particular temperatures.

Figure 17.17 shows the activity of a yeast enzyme that works best in the temperature range 25–30 °C. At low temperatures this yeast enzyme only works very slowly. As temperature increases above 30 °C, the enzyme activity decreases rapidly because higher temperatures kill the yeast cells.

Effect of pH on enzyme activity

pH also has an effect on enzyme activity.

Figure 17.18 shows how the activity of two enzymes that are involved in digestion, pepsin and amylase, are affected by pH. Pepsin is released in the stomach. It works best in acidic conditions. Amylase is a saliva enzyme and it works best in neutral conditions.

The conditions under which an enzyme works best are termed the **optimal conditions**.

Figure 17.17 *The yeast enzyme used in making wine works best at about 28 °C.*

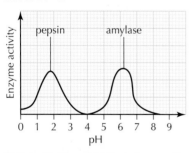

Figure 17.18 *pH affects how well an enzyme works.*

📖 Word bank

- **Optimal conditions**
The conditions under which an enzyme works best.

GO! Activity 17.7

☺ **1.** During fermentation, enzymes in yeast break down glucose to produce alcohol.

   (a) What are enzymes?

   (b) Name the gas given off during fermentation.

   (c) (i) The graph shows how pH affects the activity of an enzyme.

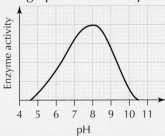

   State the optimal pH for this enzyme.

   (ii) State another condition that will affect enzyme activity.

☺☺ **2. Experiment**

The effect of temperature on the activity of amylase can be investigated using the digestion of starch.

You will need: 400 cm³ beaker, two boiling tubes, water bath, crushed ice, dimple tile, iodine, dropping pipettes, timer, starch solution, amylase solution.

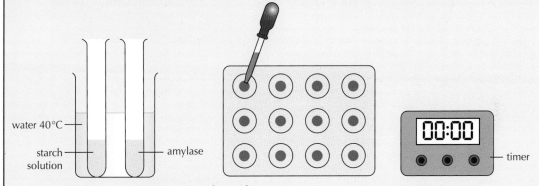

Experiment to investigate optimum conditions for enzymes.

   (a) Set up a water bath at 40 °C.

   (b) Set up a dimple tile with drops of iodine in each of the dimples.

   (c) Drop a sample of the starch solution you will use into the first dimple.

   (d) Put 20 cm³ of starch solution into a boiling tube. Put 20 cm³ of amylase into a second boiling tube. Place both boiling tubes in the water bath for 5 minutes.

   (e) After 5 minutes, mix the solutions together. Waggle the tube to ensure mixing and replace the tube in the water bath.

   (f) Start the timer.

   (g) Every minute, remove a few drops of mixture and place in a dimple of the dimple tile.

   (h) Time how long it takes before no blue–black colour is detected. This will be when all the starch has been digested. If it takes longer than 12 minutes – there are 12 dimples in your tile – record that you still detected starch after 12 minutes.

(continued)

(i) Repeat the experiment with the tubes in iced water (0 °C), room temperature (20 °C) and at 60 and 80 °C.

(j) Record your results in a table like the one below.

Temperature (°C)	Time taken for no blue–black colour to be detected (min)
0	
20	
40	
60	
80	

(k) Write up the experiment and explain your results. Describe the effect of temperature on the activity of amylase.

Variation

A similar experiment can be used to investigate the effect of pH on the activity of amylase.

You can use drops of 0.1 mol/l hydrochloric acid and 0.1 mol/l sodium hydroxide solution to vary the pH of the solutions. Try adding 2, 4 or 6 drops of the acid or alkali to the amylase solution before mixing.

Learning checklist

After reading this chapter and completing the activities, I can:

N3 L3 **N4** L4

- state that carbohydrates and oils are foods that can be burned to use as fuels. **Activity 17.1 Q1**   ◯ ◯ ◯

- state that carbohydrates are compounds which contain only carbon, hydrogen and oxygen atoms, with the hydrogen and oxygen in the ratio two to one (2 : 1). **Activity 17.2 Q1**   ◯ ◯ ◯

- state that glucose is a simple carbohydrate with formula $C_6H_{12}O_6$. **Activity 17.2 Q2(a)**   ◯ ◯ ◯

- state that starch is a complex carbohydrate made of many glucose units joined together. **Activity 17.2 Q2(b)**   ◯ ◯ ◯

- describe how Benedict's solution and iodine solution can be used to distinguish between glucose and starch. **Activity 17.2 Q3, Activity 17.3 Q1, Q2, Activity 17.4 Q1, Q2**   ◯ ◯ ◯

- describe how starch is broken down to glucose during digestion. **Activity 17.4 Q2**   ◯ ◯ ◯

N3 L3 N4 L4

- explain why glucose can pass through the gut wall and be absorbed into the bloodstream. **Activity 17.3 Q2, Activity 17.4 Q2**   ○ ○ ○

- explain why starch cannot pass through the gut wall. **Activity 17.3 Q2, Activity 17.4 Q2**   ○ ○ ○

- state that enzymes are biological catalysts. **Activity 17.4 Q1(b), Activity 17.7 Q1**   ○ ○ ○

- state that enzymes present in yeast can convert glucose into alcohol (ethanol) in a process called fermentation. **Activity 17.5 Q1**   ○ ○ ○

- explain why alcoholic drinks with alcohol concentrations greater than 15% cannot be produced only by fermentation. **Activity 17.6 Q1(a)**   ○ ○ ○

- state that distillation is used to increase the alcohol concentration of spirit drinks. **Activity 17.6 Q1(b)**   ○ ○ ○

- state that the alcohol content of drinks is measured in units and is noted on the labels of alcoholic drinks. **Activity 17.6 Q4**   ○ ○ ○

- state that different alcoholic drinks can be made from plant sources, for example, cider (apples), rum (sugarcane) and vodka (potatoes). **Activity 17.6 Q3**   ○ ○ ○

- explain that temperature and pH affect the activity of enzymes. **Activity 17.7 Q1(c), Q2**   ○ ○ ○

- state that the conditions under which enzymes work best are described as optimal conditions. **Activity 17.7 Q1(c)**   ○ ○ ○

- *calculate the alcohol content of an alcoholic drink.* **Activity 17.6 Q4**   ○ ○ ○

- *interpret a line graph.* **Activity 17.7 Q1(c)**   ○ ○ ○

- *work with others to carry out a practical activity.* **Activity 17.3 Q1, Activity 17.4 Q2, Activity 17.5 Q1, Activity 17.7 Q2**   ○ ○ ○

18 Plants to products

You should already know:

- that plants are a source of carbohydrates, fats and oils.

National 3

National 4

Products from plants

Learning intentions

In this section you will:

- learn that chemists have an important role in the manufacture of products from plants
- learn about some of the plant-based products that can be made.

Plants provide us with many of the things that we use in our everyday lives.

Chemists have an important role in the design and manufacture of products that can be obtained from plants.

This role includes:

- processing plants to produce food for us to eat

- extracting compounds from the plants such as dyes and compounds that can be used as medicines

- making new products such as soaps, shampoos and cosmetic products.

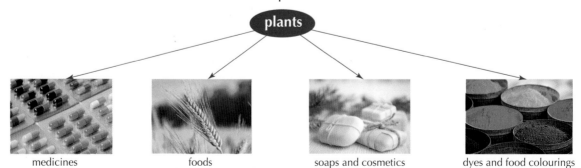

medicines   foods   soaps and cosmetics   dyes and food colourings

Figure 18.1 *Many of the products we obtain from plants enhance our daily lives.*

Foods

We eat a lot of plants as food. But we also process plants to produce other foods. The three most important plant foods are the **cereals** – maize (also called corn), wheat and rice.

(a)
(b)
(c)

Figure 18.2 *(a) Maize, (b) wheat and (c) rice are the three most important cereals we use for food.*

Originally, cereals were wild grasses. They were cultivated to produce the crops we know today. More food energy is obtained from cereals than from any other food source. They are, therefore, often referred to as **staple foods**.

Cereals are often ground down to produce flours. These can then be processed into breakfast cereals, bread, pasta, biscuits, cake and many more foods.

Word bank

- **Cereal**
A grain used for food, for example maize, wheat and rice.

- **Staple food**
A food that is eaten regularly and from which a large proportion of our energy is obtained.

? Did you know …?

Over 800 million tonnes of maize is grown in the world each year. Some of it is used to make the cornflakes we enjoy for our breakfast.

In other countries, it is used to make other foods. For instance, in Mexico it is used to make tortillas and in Sub-Saharan Africa mealie pap, made from ground maize, is a traditional staple for many people.

Cornflakes were invented by Dr John Harvey Kellogg who was trying to develop a breakfast food that was more wholesome than meat for the patients of the hospital where he was in charge. His brother, Will, also worked with him in the hospital, but obviously had an eye for business. In 1906 he bought the rights to produce and market the cornflakes. He added sugar to the recipe to make the flakes taste better. His brother had resisted doing this, fearing it would affect his reputation as a doctor. This was the beginning of the Kellogg company, now worth in excess of 25 billion US dollars.

Figure 18.3 *Tortillas are made using maize flour.*

Figure 18.4 *Cornflakes were developed as a wholesome alternative to meat for breakfast.*

Figure 18.5 *In Greece traditional handmade soaps are made using olive oil.*

Make the link

You can learn more about how edible fats and oils, and essential oils are used in cosmetic preparations in Chapter 16.

Figure 18.7 *Naturally dyed threads.*

Figure 18.8 *Mel Gibson portrayed William Wallace in the movie* Braveheart. *His face was painted blue for the battle scene.*

Soaps and shampoos

The edible oils from plants can also be processed to produce soaps. This involves reacting the oils from the plants with a strongly alkaline material. This was originally a substance called lye, which was produced from wood. Nowadays, sodium or potassium hydroxide is used. Oils such as coconut oil, palm oil and olive oil can be used.

Dyes and food colourings

Plants also contain compounds that can be used as dyes.

(a)    (b)

Figure 18.6 **(a)** *Saffron rice is dyed yellow using* **(b)** *saffron strands.*

We are probably all familiar with rice that has been dyed yellow. This is often done using saffron strands, which are the dried stamens of the saffron crocus. As well as having a powerful taste, they contain a yellow-coloured dye.

Other dyes that can be used to colour textiles can be extracted from plants.

A red dye that is extracted from plants is called alizarin. Natural alizarin is extracted from the roots of the madder plant, which are crushed and boiled to release the dye.

Woad is a blue dye that was used by ancient tribes in Britain to paint their bodies before going into battle. It was extracted from the leaves of a plant called woad. In other parts of the world, this dye was obtained from the leaves of the indigofera plant and was called indigo.

Chemistry in action: Indigo blue dye

Denim jeans are coloured using indigo dye. To meet market demand, indigo dye must be manufactured rather than obtained from natural sources. About 80 000 tonnes are made each year. Indigo is not very water soluble, so denim jeans retain their colour even after many washes.

Figure 18.9 *Indigo is used to dye denim blue.*

GO! Activity 18.1

1. Experiment:

Extracting dyes from plants

Dye can be extracted from different plants, including onion skins (red or yellow), beetroot, red cabbage, frozen spinach (don't use fresh spinach – it's not as good) and blueberries.

You will need: plant material to extract dye from, 400 cm³ beaker, pieces of white wool or cotton fabric, potassium aluminium sulfate, potassium hydrogen tartrate.

(a) Chop your plant material into small pieces.

(b) Place the plant material you are using in a beaker and add 100 cm³ of water.

(c) Heat the beaker and simmer the mixture until the water has become coloured. The longer you heat the mixture the more coloured the water should become.

(d) When the water has become coloured, allow to cool and filter to obtain the dye solution.

(e) Use your dye to colour lengths of white wool or pieces of white cloth. Place the wool or piece of cloth in the dye bath and heat the mixture together for an hour.

Notes

- Your dye will adhere (stick) better to the fabric, especially if you are using cotton, if you soak the cloth in a **mordant** (a chemical mixture that helps the dye molecules to stick to the fabric) for 24 hours before you dye your fabric.

  You can make a mordant by dissolving 8 g of potassium aluminium sulfate and 7 g of potassium hydrogen tartrate in 100 g of water.

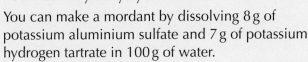

> 📖 **Word bank**
>
> - **Mordant**
>   A chemical that helps dyes to stick to fabrics.

- Some dyes give different colours in acid and alkaline conditions. If you are using red cabbage or blueberries, split your dye solution into three portions. Add some vinegar to one portion to make the conditions acidic and add baking soda to another to make the conditions alkaline. Leave one portion neutral.

2. (a) Name the dye that is used to dye denim blue.

(b) Madder produces a red dye. What part of the plant is used to make the red dye?

(c) Name the type of chemical that is used to help dyes stick to fabrics.

National 3

National 4

Medicines from plants

Learning intentions

In this section you will:

- learn about the active compounds in plants that are used as medicines
- learn that many of the world's medicines are obtained from plants.

Figure 18.10 *Coca leaves are available at Machu Picchu.*

📖 Word bank

- **Synthetic compound**

A compound that is made from chemicals, rather than obtained from natural sources.

Since earliest times, people have gathered plant materials and used these to create herbal medicines to treat diseases.

Even today in the developed world, about a quarter of all prescriptions contain material derived from plants. It is also estimated that more than 70% of the world's population, over 5 billion people, still rely on plant-based medicines.

Many of the tribes living in remote situations have valuable knowledge of the medicinal benefits of plants.

Of all the known plants, only about 20% have been investigated to see if they contain compounds that could be used as medicines.

Plants that are used as medicines can be harvested from the wild. The demand can put them under great pressure. In Europe alone, an estimated 150 medicinal plant species are at risk from over-harvesting.

It is known that in the world's rainforests there are many plants that still have to be identified. These could provide a source of future medicines. Unfortunately, rainforest destruction means that many of these plant species might be lost.

In 1962, bark from the Pacific yew tree was collected as part of a programme to test plants for potential use in medicine. It was discovered that the bark contains a compound, named paclitaxel, that kills cancer cells. Research into its use is still going on.

Chemists have been responsible for the extraction and identification of the compounds in plants that can be used in medicine. They have also been responsible for developing **synthetic compounds** with structures and properties similar to those found in plants.

The role of chemists is to isolate and identify the active compounds that give plants their medicinal properties. They then research how the compounds can be made in the lab, and how they can be changed to produce other compounds with similar properties.

✳ Chemistry in action: The story of aspirin

The story of the development of aspirin, one of the most commonly used medicines in the world, helps us understand the importance of the work of chemists in the production of new medicines.

The use of willow bark to treat ailments has been recorded for thousands of years. Willow bark was used in ancient Egypt to relieve inflammation and to ease general aches and pains. In ancient Greece, a tea brewed from willow leaves was given to women to ease pain during childbirth.

Figure 18.11 *Aspirin is commonly used for pain relief. It is also probably the world's most researched medicine.*

In more modern times, research was carried out by chemists to isolate and identify the active ingredient in willow bark. Willow bark contains a chemical called salicin. This is the active compound that chemists isolated and made into aspirin. Here is a timeline for the development of aspirin from willow bark.

1763   A report is published by the Royal Society on the use of dried powdered willow bark in curing fevers.

1828   Johann Buchner, a German chemist, isolates salicin, the active ingredient in willow bark.

1830   Meadowsweet flowers are found to contain salicin.

1853   The structure of salicin is worked out and a new compound, acetyl salicylate, is produced from salicin.

Figure 18.12 *Salicin, the chemical used to make aspirin, can be extracted from willow bark.*

1876   Studies by a Scottish doctor, Thomas MacLagan, in Dundee show salicin reduces fever and joint pain in patients with rheumatism.

1897   Studies show converting salicin to acetyl salicylate reduces the irritant effects of the medication.

1897   Acetyl salicylate is named aspirin.

Recent studies show aspirin has much wider use than simply pain and inflammation relief. Many studies are now focusing on its preventative role in medicine. Here are some of the benefits that have been shown in recent studies.

1991–92   Studies show benefits of using aspirin in cancer treatment.

1998   Studies show reduced risk of heart attack in patients with high blood pressure who take aspirin.

2005   Studies indicate that use of aspirin to thin the blood reduces the risk of stroke.

❓ Did you know ...?

It is estimated that the average cost to research and develop each successful drug is between $800 million and $1 billion.

Plants species used in medicine

Table 18.1 lists some examples of the plants that are used in medicine and their **active compounds**.

Table 18.1 *Some of the plants that are used in medicine.*

Plant	Willow	Meadowsweet	Pacific yew
Grown	worldwide	wild in damp meadows in Europe and Western Asia	North Western America coast from Alaska to California
Active compound	salicin (salicylic acid)	salicin (salicylic acid)	paclitaxel
Use	production of aspirin	production of aspirin	in chemotherapy to treat breast and ovarian cancer

Plant	Foxglove	Opium poppy	Cinchona tree
Grown	native to Europe; treated as an invasive weed in USA	extensively throughout the world	South America, Congo
Active compound	digitoxin (digitalin; digoxin)	morphine, codeine	quinine
Uses	treatment of heart conditions such as atrial fibrillation and heart failure	pain relief	anti-malaria drug

📖 Word bank

- **Active compound**

The compound in a medicine that is being used to treat a condition or illness.

❓ Did you know ...?

Traditional Chinese medicine uses over 5000 plant species.

Illegal drugs

Plants are a source of many compounds that are beneficial as medicines, but they are also the source of compounds that can be harmful and may be abused. These compounds tend to be addictive, meaning people who use them become dependent on them. Governments often make such drugs illegal if they can cause harm to individuals or to society as a whole.

Misuse of some medicines is also illegal. Codeine is obtained from poppies and is widely used as a medicine for pain relief. It is also classified as a Class B illegal drug, due to its potential to be misused.

The opium poppy is a source of medical drugs such as morphine and codeine, but is also the source of the illegal drug heroin, which is highly addictive and harmful.

Marijuana, or cannabis, is an illegal drug in the UK. It is known to produce harmful side effects such as anxiety, hallucinations and personality disorders. Recent research, however, has shown that it can be of benefit in controlling pain for sufferers from multiple sclerosis. And there is also research showing that compounds obtained from cannabis may be of help in treating certain types of cancer.

Some drugs which are legal, such as alcohol and nicotine, are known to be addictive and can be harmful. The dangers of smoking are so great that many governments insist that cigarettes are sold in plain packaging with prominent health warnings.

'Legal highs' are drugs that have been made to have the same effect as other drugs that are illegal. However, just because a drug is legal doesn't mean it is harmless. From 2011 to 2013, the number of deaths in England and Wales due to taking legal highs more than tripled from 7 to 23.

Figure 18.13 *The sap from the seedpod of the opium poppy is used to make opium, from which heroin can be made. Both opium and heroin are illegal drugs.*

Figure 18.14 *Cannabis plants are very distinctive.*

🔵 Activity 18.2

☺ **1. (a)** Name a drug that can be obtained from poppies and which is used in medicines.

   (b) Name a drug that can be obtained from poppies and which is illegal.

   (c) Give a reason why drugs are made illegal.

☺☺ **2.** Research how plants are used to make products such as medicines.

   For each plant you should try to find out:

   - where the plant is grown
   - the active compound that gives the plant its medicinal properties
   - how chemists extract and use the active compound
   - how products made from the plant are used
   - benefits of using the products.

   Produce a poster or presentation showing the importance of the plant you have researched.

Labelling of medicines

Learning intentions

In this section you will:

- learn that labels on medicines describe the contents, what the medicine should be used for and how it should be used
- learn that medicines contain an active ingredient.

The labels on medicines give a lot of important information. This is true whether we buy them over the counter in pharmacies and supermarkets or are prescribed them by a doctor. Labels describe the contents of the medicine, including the active compound, storage information, what the drug can be used for and how it should be used. You should always read the label carefully.

paracetamol —

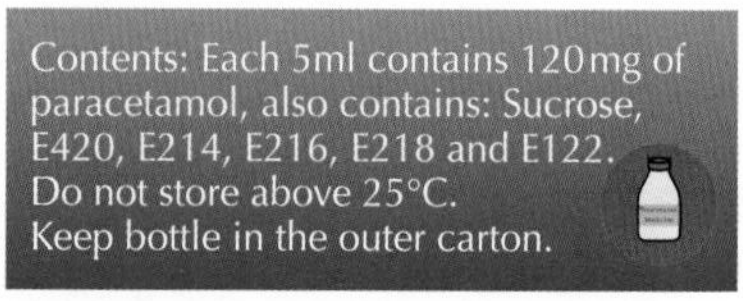

Contents: Each 5ml contains 120mg of paracetamol, also contains: Sucrose, E420, E214, E216, E218 and E122. Do not store above 25°C. Keep bottle in the outer carton.

Figure 18.15 *Medicine labels clearly identify the active compound, in this case paracetamol.*

Figure 18.16 *A label showing contents and storage conditions for a medicine.*

The label in Figure 18.16 shows that the active ingredient (compound) is paracetamol. The sucrose and compounds indicated by E numbers are added to make the medicine look and taste pleasant, and to increase its shelf-life.

Table 18.2 explains the purpose of some common E numbers.

Table 18.2 *Some common E numbers and their purposes.*

E number	Purpose
E420	Sweetener
E214	Preservative
E216	Preservative
E218	Preservative
E122	Red dye

The labels on medicines will indicate how much and how often a medicine should be given, as shown in Figure 18.18. They will also mention any restrictions and warnings such as 'Don't take if you are pregnant' or 'May cause drowsiness' or 'Don't take if you are taking another type of medicine'. A label is also likely to give advice about what to do if the medicine doesn't seem to be working.

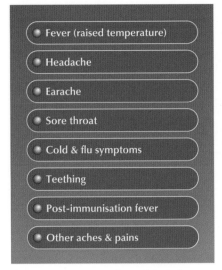

Figure 18.17 *A medicine label also indicates what conditions and symptoms the medicine can be used to treat.*

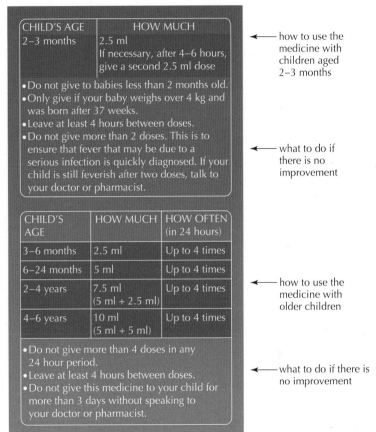

Figure 18.18 *The labels on medicines indicate the dose (how much) and how often a medicine should be given. They also mention any restrictions and warnings, and may give advice about what to do if the medicine doesn't appear to be working, or what to do if there are unwanted side effects.*

GO! Activity 18.3

☺ 1. (a) What term is used to describe the compound in a medicine that is used to treat an illness?

  (b) Why are other compounds like sucrose added to medicine?

2. Look at the label in Figure 18.18.

  (a) (i) What is the dose for a 2–3 month old baby?

    (ii) Why is the number of doses that a 2–3 month old baby can be given restricted to two?

  (b) What is the maximum number of doses that a 4–6 year old child can be given in 24 hours?

☺☺3. Collect the packaging from different medicines.

  Create a poster highlighting what the medicines contain, how they should be stored, what they can be used for and how they should be used.

Learning checklist

After reading this chapter and completing the activities, I can:

| N3 | L3 | N4 | L4 |

- describe how material such as dyes can be extracted from plants and used. **Activity 18.1 Q1, Q2** ○ ○ ○

- state that a large proportion of medicines are made from plants. **Activity 18.2 Q1, Q2** ○ ○ ○

- state that plants contain active compounds that have medical benefits. **Activity 18.2, Activity 18.3 Q1** ○ ○ ○

- state that labels on medicines give information about the contents, what the product can be used for and how the product should be used. **Activity 18.3 Q3** ○ ○ ○

- extract information from a label. **Activity 18.3 Q2** ○ ○ ○

Unit 2 practice assessment

Section 1 National 3 Outcomes

Total: 24 marks

N3 Fuels and energy

1. Most of the energy used in Scotland is produced by burning natural gas and products obtained from crude oil.

   (a) State the term used to describe substances that can be burned to produce energy. **1**

   (b) Natural gas consists of hydrocarbons.

   State what is meant by a hydrocarbon. **1**

   (c) The apparatus below can be used to identify the products when the hydrocarbons in candle wax burn.

   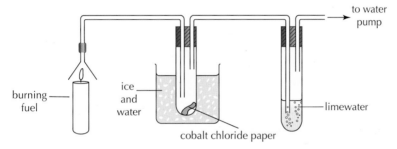

   (i) The cobalt chloride paper turns from blue to pink.

   Name the product of burning hydrocarbons that turns blue cobalt chloride pink. **1**

   (ii) The limewater shows that burning hydrocarbons produce carbon dioxide.

   Describe how limewater changes when carbon dioxide is bubbled through it. **1**

2. The fire triangle indicates three things are required for a fire.

   (a) Name the **three** things that are required for a fire. **1**

   (b) Information on a fire extinguisher indicates the types of fire that it can be used to extinguish.

   Explain why water extinguishers should never be used on electrical fires. **1**

269

3. Scotland is replacing fossil fuels with sustainable energy sources for electricity generation.

   (a) State **one** reason why fossil fuels are being replaced by sustainable energy sources. **1**

   (b) Scotland has many hydroelectric schemes that generate electricity.

   The Ben Cruachan hydroelectric plant uses a unique scheme described as a pumped storage system.

   Explain how pumped storage hydroelectric schemes help meet electricity demand. **2**

   (c) The main sustainable energy source being developed in Scotland is wind power.

   State **one** disadvantage of relying on wind power for electricity generation. **1**

   (d) Biomass is also being used in electricity generation.

   Give an example of a biomass material. **1**

   (e) Sugar beet was once an important crop grown in Scotland.

   Explain how sugar beet could be used to help meet Scotland's energy demands. **1**

N3 Everyday consumer products

4. Plants are a source of different food groups that we need in our diets.

   (a) Fertilisers are added to soils to help plants grow better.

   Explain why fertilisers help plants grow better. **1**

   (b) State the purpose of carbohydrates in our diets. **1**

   (c) Carbohydrates are used to produce alcohol.

   Name the process used to produce alcohol from carbohydrates. **1**

   (d) Palm oil is a source of fatty acids.

   The chart shows how world production of palm oil has increased since 2010.

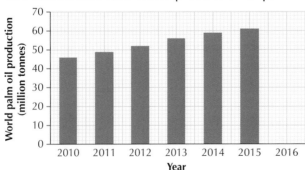

   Predict the level of palm oil production in 2016. **1**

5. The apparatus below can be used to extract essential oils from plant material.

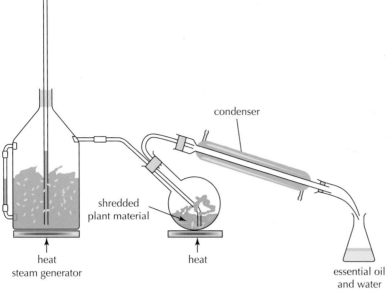

(a) Name the process that is being used to extract the essential oils.   **1**

(b) State **one** way of knowing that an essential oil has been extracted.   **1**

(c) Name an essential oil that is used to flavour toothpaste.   **1**

(d) Perfumes are blends of different essential oils. The essential oils evaporate from the skin at different rates.

State the term used to describe essential oils whose scents remain on the skin for the longest time.   **1**

N3 Plants to products

6. Many natural dyes are extracted from plants.

  (a) Describe **two** things you would do to extract a dye from a plant.   **2**

  (b) State a use for a plant dye.   **1**

7. Aspirin is commonly used to give relief from pain.

Name a plant from which aspirin can be obtained.   **1**

Total 24

A score of 12 or more is a pass!

Section 2 National 4 Outcomes

Total: 26 marks

N4 Fuels

1. Crude oil is a fossil fuel.

   Describe how crude oil was formed.   2

2. In some power stations fossil fuels are burned to produce heat energy.

   (a) Give another name for burning.   1

   (b) State the term used to describe chemical reactions that give out heat energy.   1

   (c) Fossil fuels are mainly hydrocarbons. Burning hydrocarbons in a plentiful supply of air produces carbon dioxide and water.

   Name another substance that is formed when hydrocarbons burn in a limited supply of air.   1

   (d) In some countries that burn fossil fuels to produce electricity, carbon capture units are being built next to power stations.

   Describe how carbon can be captured and prevented from being released to the atmosphere.   2

3. Petrol is very flammable.

   The fire triangle can be used to explain how fires are extinguished.

   Explain how using a foam extinguisher on a burning petrol spillage will put out the fire.   1

4. The bar chart shows how electricity generation capacity from renewables has increased in Scotland from 2007 until 2015.

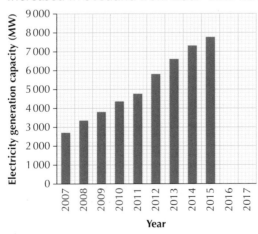

   (a) Predict what the electricity generation capacity, in megawatts, would have been in 2016.   1

   (b) Burning biomass is one source of renewable energy. Biomass can be converted into biofuels.

   Name a biofuel that can be produced from biomass.   1

N4 Hydrocarbons

5. In oil refineries crude oil is separated into fractions.

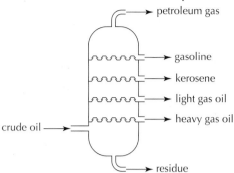

(a) State the property of compounds in crude oil that allows them to be separated into fractions. 1

(b) Which of the following statements is true for kerosene and light gas oil? 1

    A Kerosene is less flammable and less viscous than light gas oil.

    B Kerosene is less flammable and more viscous than light gas oil.

    C Kerosene is more flammable and less viscous than light gas oil.

    D Kerosene is more flammable and more viscous than light gas oil.

(c) Petroleum gas is mainly a mixture of propane and butane.

    (i) Write the molecular formula for propane. 1

    (ii) Draw a structural formula for butane. 1

6. The diagram shows how octane can be cracked in the lab to produce shorter-chain hydrocarbon molecules.

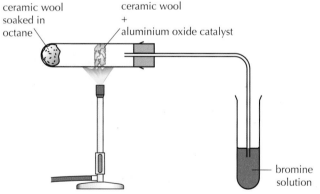

(a) Name the family of hydrocarbons to which octane belongs. 1

(b) Explain why the bromine solution changes from red/brown to colourless when the gas produced by cracking octane is bubbled through it. 1

(c) The equation below represents one of the reactions that take place when molecules of octane are cracked.

$$C_8H_{18} \rightarrow C_5H_{12} + \text{product X}$$

Write the molecular formula for product X. 1

N4 Everyday consumer products

7. The flow chart represents changes that take place when barley is used to make whisky.

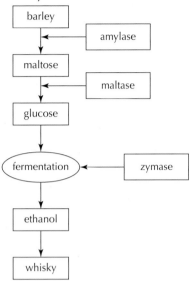

(a) Barley grains contain starch.

Describe a test that will show barley grains contain starch. **1**

(b) Amylase, maltase and zymase are enzymes.

State what is meant by the term 'enzyme'. **1**

(c) Maltose has the formula $C_{12}H_{22}O_{11}$.

Explain why the formula indicates that maltose is a carbohydrate. **1**

(d) The graph shows how the percentage of alcohol obtained by fermentation varies with temperature.

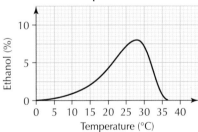

State the temperature that is best for fermentation. **1**

(e) The percentage of alcohol in the liquor obtained by fermentation needs to be increased to make whisky.

Name the process used to increase the alcohol concentration to produce whisky. **1**

N4 Plants to products

8. The graph shows how brewing time affects the mass of caffeine that is extracted from a tea bag when making tea.

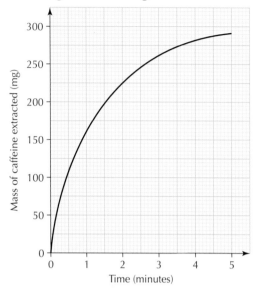

(a) The rate at which caffeine is extracted changes with time.

Complete the statement by selecting the correct term.

As time passes the rate at which caffeine is extracted from the tea bag **increases / stays the same / decreases**.   **1**

(b) What mass of caffeine, in mg, is extracted from the tea bag in 2 minutes?   **1**

9. A medicine has codeine listed in its ingredients. Codeine has pain-relieving properties. The other ingredients in the medicine give it colour and a pleasant taste.

(a) State the term that would be used to describe the role of codeine in the medicine.   **1**

(b) Codeine is obtained from the opium poppy.

Name another drug that is obtained from the opium poppy.   **1**

Total 26

A score of 13 or more is a pass!

19   Properties of materials 1

Materials   •   Fibres and fabrics   •   Ceramics   •   Novel materials

20   Properties of materials 2

Polymers and polymerisation   •   Ceramics   •   Novel materials

21   Properties of metals 1

Metals   •   Metals and batteries

22   Properties of metals 2

Reactivity of metals   •   Alloys   •   Corrosion
•   Chemical cells and the electrochemical series

23   Properties of solutions

Solubility of substances

24   Fertilisers

Growing healthy plants   •   Natural and synthetic fertilisers
•   Environmental impact of fertilisers

25   Nuclear chemistry

Formation of elements   •   Background radiation

26   Chemical analysis 1

Chemical hazards

27   Chemical analysis 2

Analytical chemists   •   Simple analytical techniques

Unit 3 practice assessment

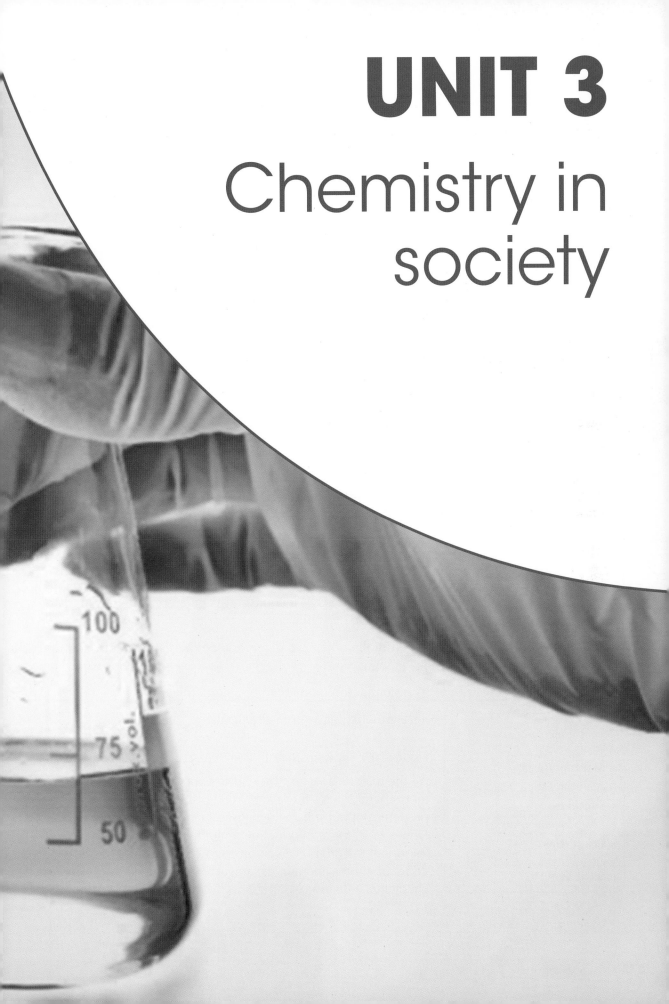

UNIT 3
Chemistry in society

19 Properties of materials 1

This chapter includes coverage of:

N3 Properties of materials

You should already know:

- how to sort materials depending on their properties such as hard/soft and strong/weak
- which material to choose for a job because of its properties.

National 3

Materials

Learning intentions

In this section you will:

- learn about the properties of natural, manufactured and synthetic materials
- learn about how we make use of the properties of materials.

A **material** is a substance that can be used to make something that is useful to us. We come across different materials in our everyday life. Materials are:

- **natural**: come from plants and animals and are also found in the ground

- **manufactured**: made from natural materials

- **synthetic**: made from chemicals.

Figure 19.1 shows some different types of materials.

All the materials we use have things about them that make them useful to us – these are called **properties**. Some of the properties of our most widely used materials are shown in Figure 19.2. More information about the properties of our most important materials is given on page 282 (plastics), pages 283–287 (fibres and fabrics), pages 289–291 (ceramics), and in chapter 21 (metals).

📖 Word bank

- **Material**

A substance that can be used to make something useful to us.

- **Natural**

Comes from plants and animals or in the ground.

- **Manufactured**

Made from natural materials.

- **Synthetic**

Made from chemicals.

- **Properties**

Things about a material that make it useful to us.

Materials

Aluminium

Aluminium, a metal used to make cooking foil, is a natural material found as compounds in the earth.

Plastics

Plastics are synthetic and have replaced many items traditionally made from natural materials.

Sandstone

Many of the buildings in Edinburgh are made from sandstone, a natural material.

Porcelain

Porcelain, a type of ceramic, is manufactured. It is scratch resistant and used to make bathroom tiles, toilet bowls and basins.

Leather and cotton

Leather, used to make shoes, comes from animals. Jeans are made from cotton, which comes from a plant. Both are natural materials.

Wood

Wood, a natural material, is still widely used in the building trade.

Figure 19.1 *Different types of materials.*

Useful properties of materials

Metal radiators

Central heating radiators are made of metal because metals are good **conductors of heat**. This means heat can travel through them easily. Most metals are also strong and can be moulded into shapes.

Copper wires

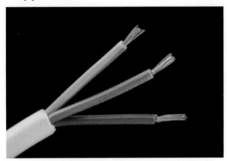

Electrical cables used for household electrical appliances usually have copper (metal) wires covered by plastic. The copper wires allow electricity to pass through easily – they **conduct the electricity**. The plastic coating does not conduct electricity – it is an **electrical insulator**. Both the wires and the plastic coating are **flexible** – they bend easily. An electrical cable is sometimes known as a flex – it comes from the word flexible.

Wool jumpers

Wool is a good **insulator of heat**. This means it does not allow heat to travel through it very easily. Wool can therefore trap your body heat next to your skin and keep you warm. Wool forms **flexible** threads that can be knitted into many patterns.

Windows

Transparent materials let light through (you can see through them). Glass is transparent and **scratch resistant**, which makes it useful for windows. Many windows have plastic frames that are **strong and waterproof** – they don't let water through.

Plastic cups

Plastic is a good **heat insulator** – it doesn't let heat out. Plastic cups keep drinks hot for longer than a ceramic cup.

Figure 19.2 *Examples of objects and the properties of the materials they are made of.*

N3 | L3 | N4 | L4

GO! Activity 19.1

Answers to all activity questions and assessments in this Unit are available online at: www.leckieandleckie.co.uk/page/Resources

☺ Use the information in Figures 19.1 and 19.2 to help you answer these questions.

1. **(a)** Here are some useful properties of materials:

   transparent   scratch resistant

   heat conductor   heat insulator

   electrical conductor   electrical insulator

   flexible   strong

   Draw up the table shown.

   Select a property that matches each of the materials in the list.

   The first one has been done for you. For some materials there may be more than one property.

Material	Property
ceramic	scratch resistant
glass	
metal	
plastic	
wool	

   (b) (i) Draw up the table shown and complete it.

   (ii) Give an example of **one** item made from each material listed in the table.

Material	Natural/manufactured/synthetic
ceramic	
glass	
metal	
plastic	
wool	

2. **Experiment**

   ⚠ You must follow the normal safety rules and take care when using hot water!

   You will need: a cup, a metal teaspoon, a plastic teaspoon, electric kettle, stop clock.

   Method

   (a) Boil some water in the kettle.

   (b) Once it has boiled, leave the water in the kettle for 1 minute.

   (c) Pour the water into the cup until it is about three quarters full.

   (d) Put the spoons in the cup.

   (e) Start the stop clock.

   (f) Stop the clock after 5 minutes.

   (g) Carefully touch the spoons.

   (h) Note which one feels warmer.

   Questions

   (a) Which spoon felt warmer after being in the hot water for 5 minutes?

   (b) Which material was the better conductor of heat?

   (c) State one thing you could change about the experiment to make it a fairer comparison.

N3	L3	N4	L4

Plastics

Plastics are synthetic. They are materials made from chemicals obtained from oil. They have many useful properties (see Figures 19.1 and 19.2) and in the past 60 years they have replaced materials such as wood and metal in items such as window frames and furniture.

Plastics can be classified as thermoplastic or thermosetting. **Thermoplastics** can be melted and shaped, then melted again and reshaped. **Thermosetting** plastics cannot be melted and reshaped once they are formed. Many have complicated names that are often shortened or trade names used rather than chemical names, for example PET (or PETE), PVC and Teflon™.

📖 Word bank

• **Thermoplastic**

A plastic that can be melted and shaped then melted again and reshaped. Thermoplastics are also known as thermosoftening plastics.

• **Thermosetting**

Plastics that cannot be melted and reshaped once they are formed.

Figure 19.3 *PET is a thermoplastic used to make drinks bottles because it is strong, light and doesn't react with chemicals.*

Figure 19.4 *Urea-formaldehyde is a thermosetting plastic used for plugs and sockets because it is hard and is a heat and electrical insulator.*

❓ Did you know ...?

A number of thermoplastic items can be recycled – melted and used to make other items like bin bags and carrier bags. Many plastic items have a symbol on them to show what they are made from and whether they can be recycled. This is the symbol on bottles made from PET.

The number and initials tell us which plastic the item is made from and the triangle indicates it can be recycled.

♳ **PET**

GO! Activity 19.2

☺ **1.** Many household items are made of plastic. Some can be described as thermoplastic and others thermosetting.

(a) State what is meant by 'thermoplastic'.

(b) State what is meant by 'thermosetting'.

(c) Suggest why many thermoplastics can be recycled but thermosetting plastics cannot.

2. (a) Draw up the table below.

Item	Code	Type of plastic
cling film film for food packaging	3	LDPE

Look at plastic items in the laboratory (or at home) and find out what plastic they are made from. The symbol is usually stamped on the bottom. Use Table 20.4 (page 301) to find out what the plastic is.

Make sure you include different types of bottles and food containers, a washing-up bowl and plastic plates and cups. If an item has a recycle symbol but no code, add it to the table and write 'no code'.

Add the name and details of the plastic items you looked at to the table.

(b) Look at a selection of supermarket plastic carrier bags. Note the names of the ones that say they are made from recycled plastic.

Fibres and fabrics

National 3

Learning intentions

In this section you will:

- learn about different types of fibres
- learn about the properties of fibres and fabrics
- learn about the uses of fibres and fabrics.

Fibres are usually long and thin and can be twisted together to make threads. The threads made from fibres can be woven or knitted together to make fabrics.

Fabrics are used to make the likes of furniture coverings and clothes for everyday use, as well as for specialised use in sports and even space suits.

Different fibres have different properties and so have to be carefully selected for the use they are put to.

 $\rightarrow$  $\rightarrow$

raw cotton fibres $\rightarrow$ cotton threads $\rightarrow$ cotton fabric

Figure 19.5 *Cotton is a natural fibre that comes from plants.*

Natural fibres

Natural fibres can be obtained directly from a plant or an animal. Plant fibres are usually made of cellulose. Cotton is the purest form of cellulose and one of the most commonly used natural fibres. Linen, obtained from the flax plant, is used to make clothes that keep you cool in hot weather.

Figure 19.6 *Flax ready for harvesting in Normandy, France.*

Animal fibres are mainly made of proteins and are usually obtained from the hair of animals, mainly wool from sheep. Another source of animal fibres is the silkworm. Silkworms are like caterpillars. They spin a cocoon made of silk fibres. Over time, the caterpillar changes into a moth inside the cocoon. The long silk fibres can be unwound from the cocoon.

Figure 19.7 *A silkworm cocoon in the branch of a tree. The silk fibres can be clearly seen.*

Manufactured fibres

Manufactured fibres come from a natural source but are chemically changed to make their properties more useful to us. Rayon was the first manufactured fibre made from wood fibre, in the 1880s. It was used as a substitute for silk. Nowadays, bamboo is a popular source of plant fibre for chemical processing. Almost all the bamboo used is grown in China. One of the reasons for choosing bamboo is it is the fastest growing grass in the world. The fibres manufactured from bamboo are as soft as cotton and are good at absorbing water. They are also biodegradable – they break down naturally in the soil (see page 308).

Figure 19.8 *Bamboo, the fastest growing grass, is used to make manufactured fibres.*

Eucalyptus trees are also used to make manufactured fibres. They are grown on land not fit for growing plants for food and don't need much water. The fibres produced are very smooth and soft, which makes them good for making clothing. They are also biodegradable.

Figure 19.9 *Fibres manufactured from bamboo and eucalyptus are very soft and comfortable next to the skin.*

Synthetic fibres

Synthetic fibres are made by reacting chemicals that come mainly from oil. Nylon was the first truly synthetic fibre. It was developed by Wallace Carothers, an American researcher at the chemical firm DuPont in the 1930s. Nylon fibres have great strength. Nylon was first used as bristles in a toothbrush. On a large scale nylon was a replacement for silk, which was used to make parachutes, and for other military items like rope. Nylon also replaced silk in women's stockings.

The first polyester fibre was made by British chemists working at the Calico Printers' Association, in 1941. They called it Terylene, and it was even tougher and more hard wearing than nylon.

Polyester is still one of the most widely used synthetic fibres today. It is strong, resistant to shrinking, stretching and creasing, and is easily dyed. It is so strong it is used to make car seat belts. Polyester is lightweight and is a good heat insulator.

Figure 19.10 *The first use of synthetic nylon fibres was for the bristles in a toothbrush. Nylon is still used today.*

Figure 19.11 *Polyester has good heat-insulating properties and is used as the filling in duvets and winter jackets.*

> **❓ Did you know ...?**
> The world production of synthetic fibres is over 55 million tonnes a year. More than half of this is polyester.

Properties of some fabrics

The properties of a fabric may be different to the properties of the fibres or the threads from which it is made. For example, nylon thread is waterproof, but when woven, the fabric often has tiny holes, which means it can absorb water. However, compared with cotton, nylon is very poor at absorbing water. Cotton can absorb up to 25% of its weight of water, whereas nylon can only absorb 10%. This makes cotton useful for making towels. The cotton threads are also looped during the manufacturing process, which makes the fabric even more absorbent.

Nylon's resistance to absorbing water makes it useful for making products that can keep water out. They dry quickly and usually require little or no ironing.

Fabrics like cotton can be treated with chemicals such as wax to fill the holes in the fabric and so make them more waterproof. However, waxed cotton has mostly been replaced by modern materials like Trestex, which combines the properties of different fibres. Trestex is made up of three layers of fabric each doing different jobs. Figure 19.14 shows how layering materials can make fabrics waterproof and windproof on the outside, and also trap heat to keep you warm. It also allows moisture to escape from your body and so keep you dry.

Figure 19.12 *Towels are mostly made of cotton because it is good at absorbing water and is soft against the skin.*

Figure 19.13 *Nylon's strength and water resistance makes it one of the most widely used synthetic fibres.*

| N3 | L3 | N4 | L4 |

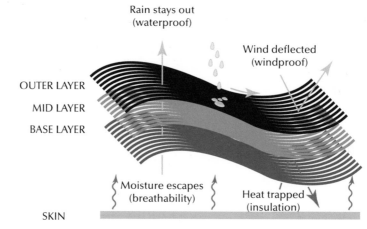

Figure 19.14 *Layering different fabrics gives a combination of useful properties.*

The way a fabric is formed from threads can lead to different properties. For example, a fabric made from woven threads has a higher sun protection factor (SPF) than the same threads when knitted together. This is because the holes in a woven fabric are smaller than in a knitted fabric.

Blending (mixing) fibres

Mixing natural and synthetic fibres produces fabrics that combine their properties.

Synthetic fibres like polyester are stronger than natural fibres like cotton. On the one hand, cotton is more breathable than synthetic fibres and so feels better next to the skin. On the other hand, pure cotton creases easily and can shrink when washed. Polyester fibres can be mixed with natural fibres like cotton to produce a cloth with blended properties. Polyester–cotton blends (polycotton) are strong, wrinkle and tear-resistant, and resist shrinking. Synthetic fibres tend to be more fire resistant than natural fibres. This is helpful when in, say, a polycotton blend, but the polyester fibres can melt, causing severe burns to the skin. Fire-retardant chemicals are used on fabrics chosen to make furniture covers and curtains. It is not practical to add fire-retardant chemicals to clothes as they will quickly wash out.

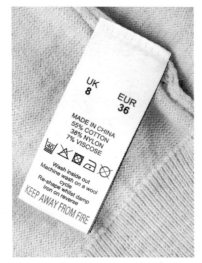

Figure 19.16 *Clothes care label showing the mixed fibre content of a fabric and fire risk warning.*

☀ Chemistry in action: Forensic science

Different fibres and threads look different when viewed under a microscope. This is of great importance in forensics – tests used in connection with crimes. Fibres, threads and pieces of fabric are often left at a crime scene. Scenes of Crime Officers (SCOs) carefully collect evidence from a crime scene.

Specially trained officers may be able to identify the fibres in a fabric and match the fabric to a suspect's clothing. Sometimes, there may be evidence trapped in the fabric, such as soil or pollen, that could be used to tell where the fabric came from.

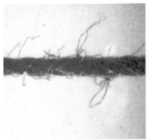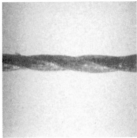

Figure 19.17 *Under a microscope threads made from natural fibres, like cotton (left), look different from threads made from synthetic fibres, like nylon (right).*

Figure 19.18 *Forensic scientists analyse evidence, such as fibres from clothes found at a crime scene.*

☀ Chemistry in action: Space suits

Space suits are made from 14 different layers of material. The layers all do different jobs so have different properties. The inner layers need to fit close to the body. They are made of Lycra®, a synthetic fibre, which has excellent elastic properties and so forms a tight fit. The next layers have to be tough enough to resist accidental tearing. There are seven layers of a plastic sheet called Mylar®, which keeps the body temperature at the correct level. One of the three outer layers is a protective layer made from Kevlar® – the same material as is used in some bulletproof vests. The outer two layers are made from waterproof and fire-resistant materials.

Figure 19.19 *Space suits are made from 14 layers of fabric each made from fibres with different properties.*

The space suits of the future are likely to have shape-memory materials in them to support the astronaut's body. The heat of the body causes the material to form the shape of the body and support it. When cooled the material relaxes and the suit can be easily removed. Materials like this are called **smart materials** (see page 292). Space suits of the future are likely to be closer fitting and allow easier movement because of smart materials.

📖 Word bank

- **Smart materials**

Materials that are specially designed so that their properties change when conditions such as temperature or pH change.

GO! Activity 19.3

☺ **1.** Wool, nylon, polyester, cotton, rayon and silk are all examples of fibres used to make fabrics.

(a) (i) Identify which fibres are: natural; manufactured; synthetic.

(ii) For the natural fibres you identified in part **(a)(i)**, state their source.

(b) Describe one way fibres can be made into fabrics.

(c) (i) State one property of cotton that makes it useful for making clothes.

(ii) State one property of cotton that can be a disadvantage when using it for making clothing.

(iii) State one property of polyester that makes it useful for making car seat belts.

(d) Polyester and cotton can be blended together to make polycotton.

(i) State one advantage of using polycotton to make a shirt instead of using pure cotton.

(ii) State one disadvantage of using polycotton to make a shirt.

☺☺ **2. Experiment**

Fibre strength and elasticity are two important properties when it comes to deciding which ones to use to make fabrics for a particular use. Threads made from particular fibres can be tested using the arrangement shown in the diagram.

You can work in a group to test the strength of some threads using the arrangement shown. However, you must collect your own results and make your own observations and conclusions.

You will need: threads to be tested, scissors, ruler, clamp stand with two bossheads and two clamps, slotted masses, padding.

Method

(a) Use a ruler to measure 25 cm lengths of each thread you want to investigate.

(b) Tie a loop at each end of the threads.

(c) Set up a stand, bossheads and clamps, with a ruler as shown in the diagram.

(d) Draw a suitable table for your results.

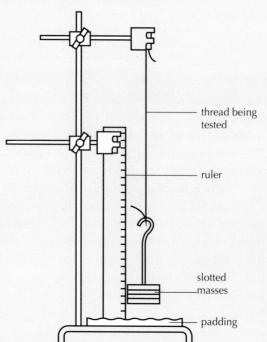

thread being tested

ruler

slotted masses

padding

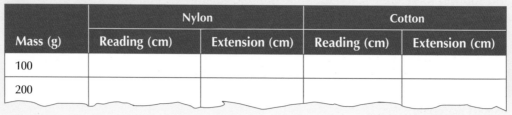

| Mass (g) | Nylon | | Cotton | |
	Reading (cm)	Extension (cm)	Reading (cm)	Extension (cm)
100				
200				

(e) Attach a thread (e.g. nylon) to the clamp, and hang the 100 g holder for the slotted masses on it.

(f) Using the ruler, measure the starting position of the bottom of the slotted masses. Record this reading.

(g) Put on another 100 g mass, and measure the new position of the masses. Record this reading and calculate the extension (how much the thread has stretched).

(h) Continue to add 100 g masses until the thread breaks or until you reach 1000 g.

(i) Repeat the experiment for all the threads you want to investigate.

(j) Write a short report. Include:
- the aim of the experiment
- what you did
- your results and any observations you made
- your conclusions (Which thread is the strongest? Which type of thread is stronger, natural or synthetic?)
- how you could change the experiment to get more accurate results (suggest two things).

(k) Discuss your report with a partner then your teacher.

Ceramics

National 3

Learning intentions

In this section you will:
- learn about the properties of ceramics that make them one of our most useful materials
- learn how ceramics are used in everyday life.

The earliest ceramics made by humans were pottery objects, made from clay, a natural material dug from the earth, usually mixed with sand. The clay mixture was hardened by heating in a fire to remove all the water. Some pottery objects as old as 27 000 years have been found. Later ceramics were glazed and fired to make smooth, coloured surfaces. Ceramics now include household, industrial and building products. In the twentieth century, new ceramic materials were developed for use in advanced ceramic engineering, such as in microchips used in the electronics industry.

Figure 19.20 *An eighteenth-century Chinese vase, made from a type of ceramic called porcelain, became the world's most expensive, when it sold for £53 million in 2010.*

| N3 | L3 | N4 | L4 |

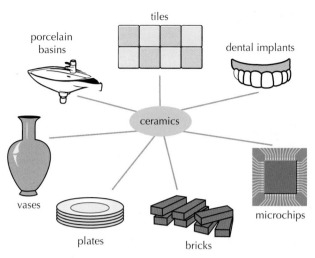

Figure 19.21 *Ceramics have a range of different uses at home, in the building trade and in the electronics industry.*

Properties of ceramics

Ceramics are probably the most widely used materials in the world because of their properties. Most are:

- hard, resistant to wear and easy to clean
- heat insulators – they don't let heat travel through them very well
- electrical insulators – they don't let electricity travel through them very well
- resistant to chemical attack, including from oxygen in the air.

There are disadvantages to some ceramic materials:

- they are brittle – they shatter (break) easily
- they cannot be made into wires.

Figure 19.22 *Advanced ceramics in body armour are lightweight but can absorb the shock of a bullet.*

Hardly any naturally occurring materials are used to make modern advanced ceramics. Many are made from synthetic silicon carbide and tungsten carbide. Both are very hard wearing and so are used in heavy industry such as in crushing equipment in mining operations. Advanced ceramics are also used in body armour, in medicine, and in the electronics and engineering industries. Some advanced ceramics can conduct electricity.

Make the link

You may have worked with clay in the art department to make figures or other objects. To dry out the clay completely the clay is left in the air for a while then heated in a type of oven known as a kiln. The kiln itself is likely to be lined with heat-resistant ceramic bricks or tiles.

Figure 19.23 *A tile being fired in a small pottery kiln.*

? Did you know ...?

When NASA's Space Shuttle returned from space, thousands of heat-resistant tiles protected it from overheating due to friction caused by Earth's atmosphere. The Shuttle used different tiles made from carbon, ceramics and silica. Future reusable space planes are expected to use a new, lightweight ceramic material made from hafnium and zirconium carbides that will resist temperatures of up to 2400 °C.

Figure 19.24 *The outside of the space shuttle was covered with ceramic tiles to resist temperatures of over 1500 °C.*

GO! Activity 19.4

☻ **1.** Copy and complete the paragraph below, which is about ceramic material.

Use the following words to help you.

ceramics   clay   conductivity   dishes   insulators
material   natural   properties   tiles

Traditional ceramics are generally made from
(a) _____ , a natural **(b)** _____ .
(c) _____ have a number of useful properties
such as hard wearing and heat and electrical
(d) _____ . Ceramics are used to make
(e) _____ , wash hand basins for the bathroom
and **(f)** _____ for the kitchen. Hardly any
modern advanced ceramics use **(g)** _____
materials. They can have **(h)** _____ that
traditional ceramics don't have, including toughness
and electrical **(i)** _____ .

2. (a) An archaeologist found pieces of a ceramic pot,
thought to be over 2000 years old, buried in the
ground. When the pot was put back together it was
found that all the pieces of the pot were there.

State **one** of the properties of ceramics illustrated
by the information in the paragraph above.

(b) When making pottery the clay has to be heated to
a very high temperature.

State the reason for this.

National 3

Novel materials

Learning intention

In this section you will:
- learn about materials with unusual properties that are useful to us.

Novel materials have unusual or interesting properties. They are often designed to do a particular job and can be made by changing the properties of existing materials like plastics, metals and ceramics. Many novel materials are referred to as smart materials – they can change their properties depending on their surroundings. This includes changes in temperature and pH. The change is reversible and can be repeated many times.

- **Hydrogels** can absorb many times their own mass of water. They are used in babies' nappies because they absorb more liquid than, say, cotton wool on its own. They can be used in hospitals to dress wounds because they help the wound to heal. The high water content of a hydrogel makes it more similar to living body tissues than any other material. There are even types of hydrogel in hair gel.

- **Smart colours** change colour as the temperature changes. These are used on contact thermometers made from plastic strips. When the strip is held on the forehead a colour will appear at the body's temperature. They are also used in food packaging materials that show you when the product they contain is cooked to the right temperature.

- For most materials, if they are bent out of shape, they stay that way. However, if a part made from a **shape-memory metal** is bent out of shape, when it is heated above a certain temperature it will return to its original shape. This property makes it useful for making spectacle frames – they return to their original shape if they are put in hot water after bending them.

- **Electrically conductive fabrics** such as electroLycra change their electrical conductivity as they are stretched. This is because it is silver plated so it conducts and the Lycra allows the material to stretch. It can be used to make 'intelligent clothing', such as sportswear that can monitor heart rate.

Figure 19.25 *A disposable nappy has layers of cotton wool with hydrogel spread through it.*

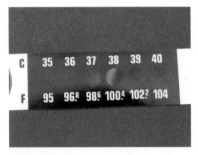

Figure 19.26 *Smart colours are used in strip thermometers.*

Figure 19.27 *Shape-memory metals will go back to their original shape after being bent out of shape.*

Activity 19.5

☺ **1.** Write a report on **two** novel materials. Include:

- the name of the material
- what its unusual property is
- an example of how it is used.

Discuss your report with a partner then your teacher.

☺☺ **2. Experiment**

You are going to investigate how much water a hydrogel can absorb. You can work in a group but you should record your own results and observations and write your own report.

 You should follow the normal safety rules and avoid breathing in the hydrogel.

You will need: a disposable nappy, scissors, a 250 cm³ beaker, tap water, a dropper.

Method

(a) Cut open the disposable nappy.

(b) Shake the hydrogel out of the nappy into the beaker.

(c) Stop adding the hydrogel when the bottom of the beaker is covered in a thin layer of hydrogel.

(d) Add water from the dropper to the hydrogel in the beaker.

(e) Stir the mixture with a glass rod until the water is absorbed.

(f) Keep adding water until no more is absorbed.

(g) Note the number of dropperfuls of water that are added before the gel stops absorbing the water.

(h) Note how the gel looks at the end of the experiment compared to the start.

Report

(i) Write a short report. Include:

- the aim of the experiment
- what you did
- your results and any observations you made
- your conclusion (think about how much water is absorbed compared to the amount of gel in the beaker)
- how you could change the experiment to get more accurate results.

Learning checklist

After reading this chapter and completing the activities, I can:

N3 L3 N4 L4

- state that materials have properties which are useful to us, such as: heat and electrical conduction; heat and electrical insulation; scratch resistance; transparency; flexibility. **Activity 19.1 Q1(a)** ○ ○ ○

- give one example of each of the following materials: natural; manufactured; synthetic. **Activity 19.1 Q1(b)** ○ ○ ○

- give an example of one item made from each of the following materials: natural; manufactured; synthetic. **Activity 19.1 Q1(c)** ○ ○ ○

- state that metals are good conductors of heat and plastics are poor conductors of heat (insulators). **Activity 19.1 Q2** ○ ○ ○

- state that thermoplastics can be melted, shaped, melted again and reshaped. **Activity 19.2 Q1(a)** ○ ○ ○

- state that thermosetting plastics cannot be melted once they are formed. **Activity 19.2 Q1(b)** ○ ○ ○

- state that many thermoplastics can be recycled. **Activity 19.2 Q1(c), 2(b)** ○ ○ ○

- state that fibres are natural, manufactured or synthetic. **Activity 19.3 Q1(a)** ○ ○ ○

- state that synthetic fibres are generally stronger than natural fibres. **Activity 19.3 Q1(d)(i)** ○ ○ ○

- state that fibres can be woven or knitted together to make fabrics. **Activity 19.3 Q1(b)** ○ ○ ○

- state that cotton is used to make clothes because it is a breathable fabric which feels comfortable next to the skin. **Activity 19.3 Q1(c)(i)** ○ ○ ○

- state that cotton creases easily and shrinks when washed, which is a disadvantage when using it to make clothes. **Activity 19.3 Q1(c)(ii)** ○ ○ ○

N3 **L3** N4 L4

- state that polyester fibre is so strong that it is used to make seat belts. **Activity 19.3 Q1(c)(iii)**   ○ ○ ○

- state that blending (mixing) natural and synthetic fibres changes the properties of a fabric. **Activity 19.3 Q1(d)**   ○ ○ ○

- state that traditional ceramics used in the home are generally made from clay, a natural material. **Activity 19.4 Q1**   ○ ○ ○

- state that when clay is used to make ceramics it has to be heated to a high temperature to dry it out completely. **Activity 19.4 Q2(b)**   ○ ○ ○

- state that ceramics have useful properties, including: hard and resistant to wear and attack from chemicals; easily cleaned; heat and electrical insulators, but is brittle so can break easily. **Activity 19.4 Q1, Q2(a)**   ○ ○ ○

- give two uses of ceramics in the home. **Activity 19.4 Q1**   ○ ○ ○

- state that hardly any modern advanced ceramics use natural materials. **Activity 19.4 Q1**   ○ ○ ○

- state that modern advanced ceramics have properties that traditional ceramics don't have, including one of the following: toughness and electrical conductivity. **Activity 19.4 Q1**   ○ ○ ○

- give an example of a novel material, say what its unusual property is and give an example of how it is used. **Activity 19.5 Q1**   ○ ○ ○

- *present information in the form of a table.* **Activity 19.2 Q2(a)**   ○ ○ ○

- *apply my scientific knowledge in an unfamiliar situation.* **Activity 19.4 Q2(a)**   ○ ○ ○

- *work with others to plan and carry out a practical activity.* **Activity 19.3 Q2, Activity 19.5 Q2**   ○ ○ ○

20 Properties of materials 2

You should already know:

- that the materials we use are natural, manufactured or synthetic
- that plastics are one of our most widely used synthetic materials
- that the useful properties of plastics include heat and electrical insulators; lightweight; waterproof; can be moulded into the required shape
- that plastics are either thermoplastic or thermosetting
- that traditional ceramics used in the home are generally made from clay, a natural material, and include tiles and wash hand basins
- that ceramics have useful properties, including: hard and resistant to wear and attack from chemicals; easily cleaned; heat and electrical insulators, but are brittle so can break easily
- the name of a novel material, what its unusual property is and an example of how it is used.

Polymers and polymerisation

National 4

Curriculum level 4

Materials: Properties and uses of substances SCN 4-16a; Topical science SCN 4-20a, SCN 4-20b

Learning intentions

In this section you will:
- learn that plastics are polymers
- learn that polymers are long-chain molecules
- learn how polymers are formed
- learn how to name polymers
- learn about different plastics and their properties
- learn that most plastics burn and some produce toxic gases

(continued)

- learn about ways of disposing of plastics
- learn that bioplastics and biodegradable plastics have been developed
- learn about environmental issues surrounding plastics
- learn which types of polymer can be recycled.

Polymers are long-chain molecules formed when lots of small molecules join together. The small molecules are known as **monomers** and the process by which they join with each other is called **polymerisation**.

polymerisation

monomers ⟶ polymers

If you think of a paper clip as a monomer, then lots of paper clips joined together is a polymer.

polymerisation

monomers → part of a polymer structure

Figure 20.1 *Using paper clips as monomers to show polymerisation.*

Polymers can be natural, as in wool and cotton, or synthetic, like polythene and polyesters.

Cotton is made of the natural polymer cellulose. Cellulose is made in nature when lots of glucose molecules (the monomer) join together (polymerise).

polymerisation

glucose ⟶ cellulose

(monomer) (polymer)

Figure 20.2 is a simplified diagram showing how glucose molecules are joined to form long-chains in cellulose. The glucose molecules are represented as hexagons.

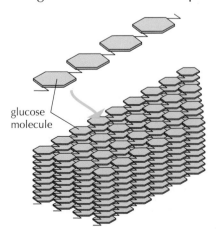

glucose molecule

Figure 20.2 *Diagram of glucose monomers joined to make the natural polymer cellulose.*

Word bank

- **Plastics**

Synthetic polymers made from chemicals that come from oil.

- **Polythene**

The common name of a plastic made from ethene. Its scientific name is poly(ethene).

Make the link

For more about cracking long-chain hydrocarbons see Chapter 14.

Plastics are synthetic polymers and are usually made from chemicals that come from oil. **Polythene**, for example, is made from ethene (C_2H_4), which is obtained by cracking long-chain hydrocarbons that come from oil. The scientific name for polythene is poly(ethene). Its common name is polythene.

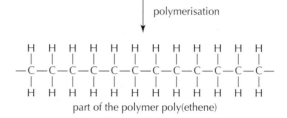

ethene monomers

polymerisation

part of the polymer poly(ethene)

Figure 20.3 *The formation of the polymer poly(ethene) from ethene monomers.*

Notice that at each end of the polymer chain there is a bond. This indicates that only part of the polymer structure is shown.

The scientific name of a polymer comes from the name of the monomer. A bracket is put around the name of the monomer and the word 'poly' put in front of that. Some examples of how to work out the scientific name of polymers from their monomer are shown in Table 20.1.

Table 20.1 *The scientific names of polymers are based on the monomers they are made from.*

Monomer	Scientific name of polymer	Common name of polymer
ethene	poly(ethene)	polythene
propene	poly(propene)	polypropylene
chloroethene	poly(chloroethene)	polyvinyl chloride (PVC)
phenylethene	poly(phenylethene)	polystyrene

Scientists are continually changing the structure of polymers to change their properties to make them useful for a specific purpose.

Poly(ethene) is one of the most widely used synthetic polymers. Two of its common forms are LDPE (low-density poly(ethene)) and HDPE (high-density poly(ethene)). Chemically they are the same, but differences in the polymer chains gives them different properties. LDPE has a lot of branches coming off the main polymer chain. This means the long chains do not pack closely together. This results in LDPE being easy to stretch but not very strong. It is used widely as a plastic film for making bags and film wrap.

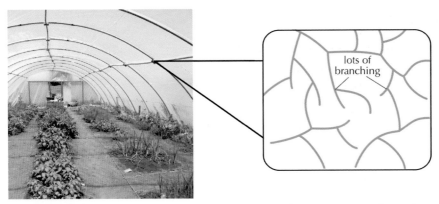

Figure 20.4 *The polymer chains in low-density poly(ethene) (LDPE) used in polytunnels have lots of branching.*

HDPE has few branches coming off the main polymer chain. This means the long polymer chains can pack closely together. This results in HDPE being much stronger than LDPE, which has more branches. It is used to make plastic milk and detergent bottles, bulk refuse bins and mains water pipes.

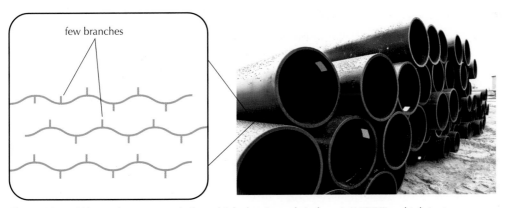

Figure 20.5 *Water pipes are made from high-density poly(ethene) (HDPE), which is stronger than low-density poly(ethene) (LDPE) because it has fewer branches on the polymer chains.*

Rubber is a natural polymer. Natural rubber is sticky and will rot over time. Scientists have made a variety of synthetic rubbers made from monomers that come from oil. The monomers used and other chemicals added control the type of rubber made. Synthetic rubbers are much easier to handle than natural rubber and are more resistant to attack from chemicals in the air. Much of the world's rubber production is used to make tyres for vehicles. To make the rubber tougher and more resistant to rot, it has sulfur added. This is called vulcanisation. Vulcanisation results in the long polymer chains linking with each other through the sulfur atoms. This is known as cross-linking. Vulcanisation is an irreversible process, like baking a cake. The normally soft and springy rubber molecules become locked together, resulting in a harder material.

> ### 📖 Word bank
>
> - **Rubber**
> A natural polymer obtained from trees.

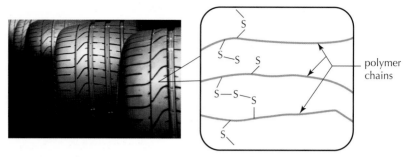

Figure 20.6 *Vulcanised rubber used to make tyres has sulfur added, which causes the polymer chains to cross-link.*

Thermosoftening and thermosetting polymers

Poly(ethene) is an example of a thermosoftening polymer. Thermosoftening polymers can be melted and shaped, then melted again and reshaped. This is because the polymer chains are not linked with each other (see Figures 20.4 and 20.5) so can move around when heated.

> ★ **You need to know**
>
> Thermosoftening plastics are also known as thermoplastics.

Vulcanised rubber is an example of a thermosetting polymer. Thermosetting polymers can be shaped when molten but can't be melted again or reshaped. This is because the polymer chains are cross-linked (see Figure 20.6) so the polymer chains are fixed and can't move when heated.

Both thermosoftening and thermosetting polymers have properties that are useful to us. These are summarised in Tables 20.2 and 20.3.

Table 20.2 *Properties and uses of some thermosoftening plastics.*

Common name	Properties	Main uses
Nylon	Tough, resists wear, slippery, good resistance to chemicals	Curtain rail fittings, clothing, food packaging. bristles in tooth and hair brushes
Polyvinyl chloride (PVC)	Can be made into a thin film, doesn't let smells through	Film for wrapping food, pipes, window frames, vinyl records
Polypropylene	Light, hard but scratches easily, tough, good resistance to chemicals	Medical equipment, banknotes, carpets
Polystyrene	Light, hard, transparent, good water resistance, good heat insulator when expanded (expanded polystyrene (EPS))	Toys, packaging, 'plastic' boxes, heat insulation in buildings
Low-density polythene (LDPE)	Tough, good resistance to chemicals, flexible, fairly soft, good electrical insulator	Packaging, toys, packaging film, disposable gloves and bags
High-density polythene (HDPE)	Hard, stiff, able to be sterilised	Detergent and milk bottles, water pipes, medical items
PET	Good resistance to chemicals, wear resistant	Food storage, bottles for water and fizzy drinks

Table 20.3 *Properties and uses of some thermosetting plastics.*

Name	Properties	Main uses
Epoxy resin	Good electrical insulator, hard, resists chemicals well	Adhesives, bonding of other materials
Melamine formaldehyde	Hard wearing, strong, resists some chemicals and stains	Laminates for work surfaces, electrical insulation, tableware
Polyester resin	Hard wearing, good electrical insulator, resists chemicals well	Bonding of other materials, reinforced plastic wall panels
Urea-formaldehyde	Hard, strong, good electrical insulator, flame resistant	Electrical fittings, handles and control knobs, adhesives

Because they can be melted and reshaped many thermosoftening polymers can be easily recycled. They are melted down and made into items like refuse sacks and packaging. Many items made of thermosoftening polymers have their abbreviation stamped alongside a recycling symbol. There are more than 50 different types of plastics, but six common types have been given codes that help identify them for recycling. There is also a seventh category for 'other' plastics, including perspex, nylon and mixed plastic. Most local authorities now collect recyclable polymers from houses and businesses. Examples are given in Table 20.4.

Table 20.4 *Symbols of some common recyclable plastics.*

Code	1 PET	2 HDPE	3 PVC	4 LDPE	5 PP	6 PS
Abbreviation	PET (or PETE)	HDPE	PVC	LDPE	PP	PS
Common name	Polyethylene terephthalate	High-density polythene	Polyvinyl chloride	Low-density polythene	Polypropylene	Polystyrene

? Did you know ...?

Supermarket carrier bags used to be made from LDPE (low-density polythene), a thermoplastic, because it is very light and can be made into a very thin but strong film. Most supermarkets now use plastic bags made from recycled plastics. Since October 2014 shops in Scotland have charged customers 5 p for a carrier bag. This is to encourage us to reuse bags. It has been estimated that in the first year of the introduction of the charge, 650 million fewer carrier bags, mostly plastic, were used.

Figure 20.7 *Most plastic carrier bags are now made from recycled plastics.*

? Did you know …?

Many banks, including Scottish banks, make banknotes out of a type of polypropylene instead of paper. There are a number of reasons for changing to plastic notes – they last nearly three times longer than paper notes and they are harder to forge. They are harder to tear, are resistant to folding and are waterproof – they can go through the washing machine without being damaged. However, being thermoplastic they can melt so can't be ironed! At the end of their useful life they can be shredded and recycled. The Royal Bank of Scotland's £10 note will feature a portrait of Scottish scientist Mary Somerville, considered to be Scotland's first female scientist.

Figure 20.8 *The Clydesdale Bank plastic £5 note features the Forth Bridge.*

☀ Chemistry in action: The Airlander 10 and Vectran™

Vectran™ is a type of polyester that has exceptional strength and ability to keep its shape (rigidity). Its manufacturers claim it is five times stronger than steel and 10 times stronger than aluminium. It is lightweight, resistant to chemical attack and can withstand heat over a range of temperatures. It is also highly flame resistant. Vectran is the material used for the body of the Airlander 10 – the world's largest aircraft. It is called a hybrid aircraft because it combines some of the features of an airplane and a balloon.

There is no internal structure in the Airlander – it keeps its shape as a result of the pressure of the helium inside it, and the strong, lightweight Vectran material it is made of. Carbon composites (see page 312) are used throughout the aircraft for strength and weight savings.

The Airlander is designed to lift and transport heavy goods, and can stay in the air for days when manned and possibly weeks when unmanned.

Figure 20.9 *The body of the Airlander 10 is made of Vectran™, a type of polyester.*

GO! Activity 20.1

1. You can work in a small group for this activity.

   You will need: a molecular model building kit (per person).

   The structure of chloroethene is shown.
   Each person should build a model of a chloroethene molecule.
   (Use black for carbon; white for hydrogen; green for chlorine.)

   $$\begin{matrix} Cl & H \\ | & | \\ C & = & C \\ | & | \\ H & H \end{matrix}$$

   Break one of the bonds between the carbons in each of the chloroethene molecules.

   Join your molecule with a molecule from each person in your group to make a long-chain molecule.

   (a) Draw part of the long-chain molecule formed from chloroethene molecules, showing three of the molecules joined. Use symbols for the elements in the structure in the same way as shown for poly(ethene) on page 298.

   (b) What name is given to the type of long-chain molecules that are formed in this way?

   (c) What name is given to this type of reaction?

   (d) Name the long-chain molecule that is formed from chloroethene.

   (e) What name is given to small molecules like chloroethene that join in this way?

2. Poly(propene) is a thermoplastic polymer.

   (a) Explain what is meant by thermoplastic.

   (b) Name the small molecule from which poly(propene) is made.

3. Urea-formaldehyde is a plastic used to make electrical plugs and sockets.

   (a) State **one** property of urea-formaldehyde that makes it suitable for this use.

   (b) Urea-formaldehyde is a thermosetting plastic. Explain the meaning of thermosetting.

4. (a) From the list of plastics below, select which can and which can't be recycled, and present the results in a table. (Use Table 20.4 to help you.)

   | polystyrene | urea-formaldehyde | PVC | polythene |
   | epoxy resin | polypropylene | melamine formaldehyde | |

   (b) What type of plastic are those in the list that can be recycled?

   (c) What type of plastic are those in the list that can't be recycled?

   (d) With reference to their structure, explain why the plastics in part (c) can't be melted and recycled.

5. Read the Chemistry in action section 'The Airlander 10 and Vectran™'.

   (a) What evidence is there to suggest that Vectran™ is a thermosetting polymer?

   (b) State **two** properties of Vectran™ that would make it useful in an aircraft.

   (c) Suggest why helium gas is used inside the aircraft.

6. In London in October 2016, cars were crushed against the ceiling of an underground car park. Water had flooded into the basement and caused expanded polystyrene (EPS) blocks under the floor to swell, which, in turn, caused the floor to rise over 1 metre.

   From your knowledge of the general properties of plastics, suggest:

   (a) why the polystyrene blocks were used under the floor of the building

   (b) why some engineers didn't think water would cause the polystyrene to swell.

Burning plastics

Many plastics burn. This is a danger in itself, especially when so many household items are made of plastic. An added danger comes from the gases produced when plastics burn. Plastics are made up of polymer chains that contain carbon. When they burn in plenty of air the carbon reacts with oxygen to form carbon dioxide.

Word equation: carbon + oxygen → carbon dioxide
(lots of)

Formula equation: $C + O_2 → CO_2$

Although carbon dioxide contributes to environmental problems (see Chapter 13) it is not toxic to animals.

However, when a plastic burns in a limited amount of air, the poisonous gas carbon monoxide is also formed.

Word equation: carbon + oxygen → carbon monoxide
(not very much)

Formula equation: $C + O_2 → CO$

Many plastics contain other elements that form harmful gases when burned. Polyurethane contains nitrogen so when it burns the poisonous gas hydrogen cyanide (HCN) is produced. PVC contains chlorine so the harmful gas hydrogen chloride (HCl) is formed. By looking at the formula or structure of a monomer or part of a polymer structure you can identify elements that can result in harmful gases being produced when a plastic burns. Table 20.5 gives some examples of plastics and the gases they produce.

Table 20.5 *The harmful gases produced by some plastics.*

Plastic (common name)	Structure of monomer	Harmful gas
polythene	$\begin{array}{cc} H & H \\ \mid & \mid \\ C &= C \\ \mid & \mid \\ H & H \end{array}$	carbon monoxide (CO)
PVC	$\begin{array}{cc} Cl & H \\ \mid & \mid \\ C &= C \\ \mid & \mid \\ H & H \end{array}$	carbon monoxide (CO) hydrogen chloride (HCl)
nylon	$\begin{array}{ccc} H & & H \\ \mid & & \mid \\ N &- (CH_2)_4 - & N \\ \mid & & \mid \\ H & & H \end{array}$	carbon monoxide (CO) hydrogen cyanide (HCN)

N3 | L3 | N4 | L4

Polyurethane is used in soft furnishings and cushions. Because it gives off poisonous carbon monoxide and hydrogen cyanide when it burns, polyurethane is treated to make it very flame resistant. Furnishings that use materials like polyurethane have to undergo the cigarette and match test. These tests simulate what happens when a smouldering cigarette or burning match are dropped onto a material. They must also have a label to indicate they have passed the fire-resistance tests. Dropping lit cigarettes and matches is the most common cause of furniture catching fire.

> ★ **You need to know**
>
> You need to be able to identify which harmful gases could be produced given the structure of a monomer or part of a polymer structure. You do not need to be able to name the monomer or polymer from its structure.

Filling material(s) and covering fabric(s) meet the requirements for resistance to cigarette and match ignition in the 1988 safety regulations
CARELESSNESS CAUSES FIRE

Figure 20.10 *Soft furnishings must have labels indicating they meet the fire-resistance regulations.*

Burning plastics to identify them

What happens to a plastic when a flame is brought near it can help identify it. Some thermosetting plastics don't catch fire. However, they may smoulder and give off smoke and fumes. Most thermosoftening plastics do burn, and the way they burn can help identify them. The flame colour can vary. Some produce smoke, while others burn with a distinctive smell. Table 20.6 summarises the features shown by some common plastics when they are exposed to a flame.

Table 20.6 *The features of some plastics when exposed to a flame.*

Type of plastic	Still burns when taken out of flame?	Flame colour	Drips?	Smell	Smoke?	Speed of burning
Nylon	✓	Blue, yellow tip	✓	Burning wool	Grey	Slow
Polystyrene	✓	Yellow	✓	Choking	Thick black, with soot in the air	Fast
Polythene	✓	Blue, yellow tip	✓	Burning wax	Black	Slow
PVC	X	Yellow, blue edges	X	Choking	Grey	Doesn't burn out of flame

★ You need to know

You don't need to memorise the details of how different plastics burn. You do need to be able to use the data to help identify a plastic (see Activity 20.2 Q1). Other tests have to be carried out to be sure of the identity of a plastic. Identification of plastics has been made easier by the introduction of codes on plastics that can be recycled.

GO! Activity 20.2

1. The structure of the monomer used to make polystyrene is shown.

$$\begin{array}{ccc} H & & C_6H_5 \\ | & & | \\ C & = & C \\ | & & | \\ H & & H \end{array}$$

(a) State which toxic gas would be produced when polystyrene burns.

(b) Use the information in Table 20.6 to state what you would see when polystyrene burns that would help identify it.

(c) Tiles made from expanded polystyrene used to be stuck to the ceilings in houses.

(i) Suggest why this was done.

(ii) Give a reason why this was not a good idea, especially in the kitchen.

2. Polyurethane is used to make the insides of cushions used in sofas.

Explain why materials like polyurethane have to undergo the cigarette and match test.

3. (a) Part of the structure of a nylon polymer is shown.

$$\begin{array}{cccccc} H & O & & & O & H \\ | & \| & & & \| & | \\ -N & -C & -(CH_2)_6 & -C & -N- \end{array}$$

Name **two** toxic gases that might be produced when nylon is held in a flame.

(b) Hydrogen chloride is a toxic gas formed when PVC burns.

Which elements does this suggest are present in PVC?

Getting rid of plastics

Disposing of plastic items is an increasing problem as the number of plastic items we use increases. In Britain alone, people use 8 billion disposable plastic bags each year. About 80% of the waste that washes up on our beaches is plastic, including bottles, bottle tops, and tiny plastic pellets known as 'mermaids' tears', which look a bit like fish eggs. The plastic pellets are washed down the drains in factories producing plastic items. The campaign group Surfers Against Sewage (SAS) are campaigning to have a 50% reduction in marine litter by 2020.

↻ Keep up to date!

You can find out if Surfers Against Sewage are successful in their campaign to reduce the amount of plastic waste in the seas by searching online.

Figure 20.11 *Plastic waste washed up on a beach.*

Microbeads are tiny plastic beads used in cosmetics, shower gels and toothpastes. They act as an abrasive – they help remove layers of dead skin and stains from teeth. They are washed away into the sewage system and end up in rivers, lakes and the oceans where they can cause water pollution. The beads can absorb pollutants like pesticides and can then be eaten by fish, which are then eaten by humans and other animals. The UK government is to introduce a ban of microbeads from all cosmetics by the end of 2017. They will be banned from sale in the UK from the end of 2017.

The supermarket Tesco banned their use from the start of 2017.

Although reducing the amount of plastic being washed into the seas and recycling and reusing is making a difference, it can't keep up with the amount being thrown away. About 25% of waste plastic is burned to produce energy. Burning plastics produces toxic fumes that can't be released into the atmosphere. Dumping in landfill sites is no longer feasible as we are running out of space and most plastics are not biodegradable – they won't break down for many years.

Bioplastics and biodegradable plastics: Solving the disposal problem?

With increasing public awareness of the problems of disposing of plastics, new plastics are being designed that will 'disappear' more quickly. They are mainly bioplastics and biodegradable plastics.

Bioplastics are made from natural materials such as corn starch. They quickly break down into natural materials and can be put in compost heaps and landfill sites. It is often hard to tell the difference between a bioplastic and synthetic plastics. Polylactic acid (PLA) is a bioplastic made from corn starch. It looks and behaves like polythene and polypropylene. It is used to make food containers and disposable cutlery and cups.

> **📖 Word bank**
>
> • **Microbeads**
> Tiny plastic beads that have been used in cosmetics, shower gels and toothpastes.

Figure 20.12 *Tiny plastic microbeads found in some facial scrubs and body wash.*

Figure 20.13 *Cups made from polylactic acid (PLA), a bioplastic.*

Figure 20.14 *Biodegradable plastic carrier bags.*

📖 Word bank

- **Bioplastics**

Plastics made from natural materials that break down naturally.

- **Biodegradable plastics**

Synthetic plastics that have additives which cause them to break down naturally.

Biodegradable plastics are synthetic – they are made from chemicals that come from oil, but are made in such a way that they break down more quickly than traditional synthetic plastics. They contain additives that cause them to decay more rapidly in the presence of light and the air. Natural polymers like starch and cellulose can be impregnated into plastic items like polythene bags. If the bag is dumped in a landfill site microorganisms in the soil quickly break down the natural polymers in the polythene, and the bag disintegrates.

Bioplastics and biodegradable plastics are often described as 'environmentally friendly'. There is no doubt that they have their good points, but some people say there are negatives, too. Here are some points that have been made for and against their use:

- Making bioplastics produces 30–80% fewer greenhouse gases than making synthetic plastics.

- Bioplastics used in packaging can increase the shelf life of food.

- Some biodegradable plastics produce methane, a greenhouse gas, when they decompose.

- Many bioplastics can break down quickly and form compost, which can be used as a fertiliser.

- Some biodegradable plastics take years to decompose.

- Bioplastics are made from corn and maize, which could be used as food.

- Some bioplastics are made from genetically modified (GM) corn, which many say is harmful to the environment.

- Making bioplastics reduces the need for chemicals that come from oil.

☼ Chemistry in action: Edible cling film

Scientists at the US Department of Agriculture have developed a cling film that is made from milk protein and is edible. They claim it is an environmentally friendly way to keep food fresh. Because it is made from a natural material it breaks down quickly – it is a bioplastic. Although edible plastic made from starch already exists it is not good at stopping oxygen getting through and spoiling the food. The edible cling film is said to be 500 times better than synthetic plastic at keeping the air away from food. This means the food will keep fresher for longer and so reduce food wastage. One example of how it could be used is in wrapping cheese. Individually wrapped

Figure 20.15 *Using edible cling film could mean you don't have to unwrap your sandwich before you eat it.*

cheese sticks use a lot of plastic. If an edible plastic was used you could eat the wrapper as well as the cheese! It could also stop millions of tonnes of waste plastic being sent to landfill sites. Another idea is to add flavouring and even vitamins to the cling film.

How to cut down on plastics

We can reduce the amount of plastic we use by:

- taking a reusable bag when you go shopping, e.g. a cotton (natural) bag

- buying loose fruit and vegetables rather than pre-packed

- using long-lasting plastic items, such as refillable ink cartridges, rather than disposable ones

- repairing a broken or damaged plastic item rather than replacing it

- reusing plastic items such as food tubs as storage boxes

- when buying new things, looking for items made from recycled plastics, such as refuse bags and even pencils and pens.

It may be that in the future we have plastics that break down quickly, but until then we have to be smarter about how we use plastics and how we get rid of them when we are finished with them.

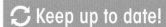

⟳ Keep up to date!

Scientists are aiming to have edible cling film made from milk on the market by 2019. You can check how they are progressing by searching for 'edible plastics' online.

GO! Activity 20.3

☺ **1.** Look at the section 'Bioplastics and biodegradable plastics'. Read the part on page 308 about the positives and negatives of using bioplastics and biodegradable plastics.

    (a) Present the information in a table.

    (b) The section could be considered as giving a balanced view on the use of bioplastics and biodegradable plastics.

        Suggest why this is the case.

☺☺ **2.** You can work with a partner or in a small group.

    Your task is to prepare a presentation about the issues surrounding the increasing amount of plastics we are using. Your presentation should include:

- examples of plastic waste
- how we get rid of plastic waste
- the problems with the way we get rid of plastics
- what bioplastics and biodegradable plastics are
- how bioplastics and biodegradable plastics are helping to solve the problem of disposing of plastic waste
- two things we can all do to reduce the amount of plastic we are using.

    You may wish to use PowerPoint or some other form of technology to do your presentation. You should think about including photographs and maybe a spider diagram to help you deliver your presentation.

    Discuss your presentation with your teacher.

(continued)

N3	L3	**N4**	L4	**309**

3. Read the Chemistry in action section on 'Edible cling film', then answer the following questions.

(a) What type of plastic is edible cling film?

(b) What is edible cling film made from?

(c) Apart from being able to eat it, give **two** other properties of the cling film that make it useful.

Ceramics

National 4

Curriculum level 4

Materials: Properties and uses of substances SCN 4-16a; Topical science SCN 4-20a

Learning intentions

In this section you will:

- learn about the bonding in traditional ceramics
- learn how modern advanced ceramics are different from traditional ceramics
- learn about the properties of composite materials and how they are used.

Ceramics have been made for thousands of years from types of clay. Clay is a very fine powdered rock that is a mixture of compounds, mainly silicates, such as aluminium silicate and silicon dioxide. It is difficult to find an area of modern life that doesn't involve ceramics in one form or another (see pages 289–291).

Their uses depend on the properties of ceramics. They are generally:

- extremely hard
- heat and electrical insulators
- heat resistant, so have high melting points and low expansion
- chemically unreactive.

These properties are a result of the bonding in ceramics and the way the atoms pack together. Ceramics usually have a combination of strong ionic and covalent bonding (see Chapter 4 pages 51–53). This is why they are hard, have high melting points and are chemically unreactive.

There are generally two types of ceramic – crystalline and amorphous (without shape). Silicon dioxide can exist in crystalline form (e.g. quartz) and an amorphous form (e.g. glass).

Ceramics have a number of properties that make them even more useful than metals and plastics. Table 20.7 shows a comparison of the general properties of ceramics, metals and plastics.

N3 | L3 | **N4** | L4

Table 20.7 *The general properties of ceramics, metals and plastics.*

Property	Ceramics	Metals	Plastics
hardness	very high	low	very low
resistance to bending	very high	high	low
resistance to chemical attack	very high	low	high
wear resistance	high	low	low
electrical conductivity	low	high	low
heat conductivity	low	high	low

Modern advanced ceramics

Ceramics have not just proved useful in everyday situations. The properties of modern advanced ceramics have made them important for some much more remarkable uses. The toughened silicon carbide ceramic used in hip replacements is not only extremely hard wearing, but it also encourages natural bone growth and tissue formation around the artificial joint.

Ceramic engine parts are used in 'lean burn' car engines that burn fuel more cleanly. Catalytic converters, which convert polluting exhaust gases into less harmful gases, are made from light but strong aluminosilicate ceramics that can withstand the high temperatures in car exhausts.

The latest lightweight, deep-sea submersibles are being built not from steel, but from ceramics originally made for military purposes.

> ★ **You need to know**
>
> Ceramics are such a versatile group of materials that it is difficult to state for sure that ceramics don't have a particular property. For example, although generally ceramics don't conduct electricity, some modern advanced ceramics can be made to conduct.

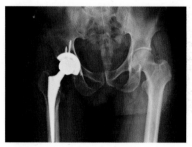

Figure 20.16 *X-ray showing an artificial hip with ceramic ball and socket.*

Figure 20.17 *Submersibles like this one inspecting an underwater oil installation are now being made from ceramic material instead of steel.*

One of the most innovative uses of ceramics is a new kind of paint made from a piezoelectric ceramic. **Piezoelectric material** produces a tiny electric current when it undergoes stresses and strains, and its Japanese inventors believe it could be used to detect metal failures or even earthquakes.

> 📖 **Word bank**
>
> • **Piezoelectric materials**
> Materials that produce a tiny electric current when put under stress.

☀ Chemistry in action: Ceramic composites

Composite materials are made from two or more different types of material. The materials for a composite are chosen because they have different properties that combine to make a more useful material. For instance, ceramic particles or fibres can be added to an existing ceramic to improve its properties. For example, ceramic composites are not brittle so don't shatter and they resist sudden extremely high temperatures. They are very hard wearing and resistant to chemical attack. This makes them useful in brake systems, which have to be extremely hard wearing and withstand sudden high temperatures.

Figure 20.18 *Advanced ceramic car disc brakes are extremely hard wearing and heat resistant.*

The NASA experimental space vehicle X-38 has a heat shield made from carbon fibres and silicon carbide (C/SiC). It is the only composite ceramic that can withstand the heat shock of temperatures above 1500 °C when it re-enters Earth's atmosphere.

📖 Word bank

- **Composite materials**
Materials that are made from two or more different types of material.

Figure 20.19 *The NASA space vehicle X-38 relies on composite ceramics for its heat shield.*

GO! Activity 20.4

☺ 1. Which **two** types of bonding are usually found in ceramics?

2. The table compares the properties of ceramics and metals.

Type of material	Name of material	Melting point (°C)	Hardness*
ceramic	aluminium oxide	2050	9
ceramic	silicon carbide	2800	9
ceramic	zirconium oxide	2660	8
ceramic	porcelain	1840	7
metal	steel	1370	5
metal	aluminium	660	3

* The higher the number the harder the material.

(a) Draw a bar chart of name of material against hardness.

(b) Make a general statement about the hardness of ceramics compared to metals.

(c) Suggest why the melting points of ceramics are generally higher than metals.

3. Knives with ceramic blades can now be bought.

   (a) State **two** advantages ceramic blades have over steel knives (you may wish to refer to Table 20.7).

   (b) Some manufacturers state that you should avoid dropping ceramic knives. Suggest why this is.

   (c) Suggest why there are very few materials that can be used to sharpen a ceramic blade.

4. Read the section 'Modern advanced ceramics'. Make up a table or spider diagram to show four uses of advanced ceramics and what property of the ceramic each uses.

Novel materials

Learning intentions

In this section you will:

- learn about the relationship between the properties of a novel material and its uses
- learn about recent applications of novel materials and their possible impact on society.

National 4

Curriculum level 4

Materials: Properties and uses of substances
SCN 4-16a; Topical science
SCN 4-20a

Some of the plastic and ceramic materials that have been developed and have unusual properties are known as 'novel' materials. Examples already covered in this chapter include edible cling film (page 308) and piezoelectric paint (page 311). In this section we will look at other novel materials. They either have properties that have been known for some time but a use only found recently, or they have been designed to do a specific job. Some applications have been found by accident.

★ **You need to know**

There are more examples of novel materials on page 292.

Hydrogels – the water-loving polymers

Hydrogels are polymers (page 297) that can absorb water. They can do this because of the structure of the hydrogel molecules. Part of the molecule is ionic (page 53). The polymer chain has negatively charged parts which attract water molecules to its surface.

Different hydrogels can be made to absorb different amounts of water. This property gives them a number of uses, including contact lenses, hair gel and as the absorbent part of a disposable nappy.

📖 **Word bank**

- **Hydrogels**
  Polymers that can absorb water.

Figure 20.20 *Soft contact lenses are made from a type of hydrogel.*

❓ Did you know ...?

Scientists are researching the possibility of hydrogels being used to carry drugs to specific parts of the body to treat diseases.

☀ Chemistry in action: Fighting fires with hydrogels

A fire-fighter called John Bartlett was attending at a fire in the USA where a whole house was destroyed. When he went into the wreckage, he discovered that the only surviving item was a nappy. He then realised that the hydrogel inside the nappy was responsible for its survival. This led to the development of a fire-fighting gel. The gel is dissolved in water then sprayed ahead of the fire, sticking to anything it touches. This way, when the fire reaches the gel, the surface is already wet, so the fire is put out. The water in the hydrogel

Figure 20.21 *A log cabin being sprayed with hydrogel.*

evaporates more slowly than water on its own, and also does not get absorbed into the ground, making it far more effective for fire-fighting than just plain water.

Seawater can be used to put out fires. However, fire-fighters found that if hydrogel was dissolved in seawater it no longer worked. This is because seawater is ionic and positive ions like sodium (Na^+) stick to the surface of the hydrogel instead of water.

Envirobond – the oil-loving polymer

Envirobond is a polymer that attracts hydrocarbons like oil (see Chapter 14) but not water. This makes Envirobond useful for cleaning up oil spills. The long-chain molecules are shaped in such a way that the oil is absorbed into the polymer structure. Once the polymer chains are 'full' of oil they stick to each other and form a solid mass that can be handled and disposed of. Envirobond also floats on the surface of the water, which makes it easier to collect.

| **Oil in water** | **Add Envirobond** | **Wait 2 hours** | **Lift out solid block** |

Figure 20.22 *Envirobond attracts oil so can be used to clean up oil spillages.*

Ooho – the water bottle you can eat

It has been estimated that over 13 billion synthetic plastic water bottles need to be disposed of every year in the UK alone. Not all of them are recycled. Many still go into landfill sites, where they will not break down for many years. A company has developed a biodegradable bottle they have called Ooho. To emphasise how natural their product is they say you can eat it! The newspaper article below gives details of how Ooho was developed and how its inventors see it being used.

Water bottle you can eat

Edible water bottles made from seaweed that are an alternative to plastic have topped the UK round of an EU competition for sustainable products.

The spherical form of packaging, called Ooho and described by its makers as "water you can eat", is biodegradable, h ygienic and costs 1p a unit to make. It is made chiefly from calcium chloride and sodium alginate, a seaweed derivative.

Ooho designer Pierre Paslier, described the product as similar to a "man-made fruit". It uses a double membrane to contain water and the capsules can be packed together - much like an orange - to carry large quantities. Global sales of packaged water are expected to reach 223bn litres this year.

"People are really enthusiastic about the fact that you can create a material for packaging

matter that is so harmless that you can eat it," he said. "So many things are wrong about plastic bottles: the time they take to decompose, the amount of energy that goes into making them and the fact we are using more and more."

You don't have to eat Ooho, Paslier told the Guardian, but the fact the packaging was edible, showed how natural it was.

"[Ooho] is a good replacement packaging that would be really widely applicable across lots of different products," said Ian Ellerington, head of innovation delivery at the UK Department of Energy and Climate Change and one of the judges of the competition. "The potential for packaging reduction is really high."

Source: *Guardian, Emma Howard, 10 September 2015*

Sugru – the mouldable glue

Sugru is a material which looks and feels like Play-Doh, and which sticks to most things and after a few hours turns into a strong flexible rubber. Its makers claim that it is ideal for DIY because it can stick to almost anything – you don't need to use different types of Sugru for different materials like you have to with many glues. Its flexibility means it can be shaped and used to fix things that are too awkward for regular glue or duct tape.

Figure 20.23 *Sugru is a mouldable glue that sets like rubber.*

| N3 | L3 | N4 | L4 |

Other advantages of Sugru include:

- it is hard wearing and waterproof – it can be used to fix items that need to be waterproof outdoors, such as hiking boots

- it can withstand high temperatures – it can be used inside a dishwasher (it's also waterproof!)

- it has some electrical resistance – it can be used to repair low-voltage cables on, say, a laptop

- it sets like rubber – it can be used on the sharp corners of tables and worktops to protect children.

Graphene – the new form of carbon

Graphene is a one-atom-thick, transparent, conductive sheet of carbon atoms arranged in hexagonal rings or a honeycomb pattern. Graphene's structure gives it unique properties. It is stronger and more flexible then steel, conducts heat 10 times faster than copper and can carry 1000 times the electrical current of copper wire. Graphene is stronger than diamonds, yet is flexible and can be stretched by a quarter of its length. It is a remarkable material with many exciting potential applications.

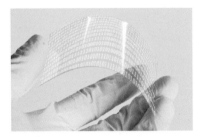

Figure 20.24 *A transparent sheet of graphene showing how flexible it is.*

There are high hopes for using graphene in the electronics industry for making the likes of flexible electronic screens. It also has potential for making advanced, lightweight batteries. Graphene is so thin that graphene paint could possibly act as a rust protector (see page 344) or an electronic ink. A small amount of graphene mixed into plastics makes them conduct electricity.

In 2010 Andre Geim and Konstantin Novoselov of the University of Manchester were awarded the Nobel Prize in Chemistry for their work on graphene. In 2016 their team announced that they had developed a graphene sieve which can separate the salts in sea water from the water they are dissolved in. The water can get through the layers of graphene but the salts can't. The water can then be used as drinking water.

Keep up to date!

You can find out how the development of graphene-based items is progressing by searching for 'graphene' online.

Figure 20.25 *Professor Andre Geim and a model of graphene.*

GO! Activity 20.5

☺ **1. Experiment**

Investigating the structure of a hydrogel

You will need: hair gel, two petri dishes, salt, sugar, a spatula.

Method

(a) Squeeze out about 2 cm length of hair gel into each petri dish.

(b) Sprinkle a spatulaful of salt onto one hair gel sample.

(c) Sprinkle a spatulaful of sugar onto the other hair gel sample.

(d) Note what you see happening to the hair gel in each dish.

Use the information about salt and sugar (below) and your knowledge of the structure of hydrogels to explain your observations.

Salt is sodium chloride (Na⁺Cl⁻), an ionic compound. Sugar is sucrose, a covalent compound.

> 🔍 **Hint**
>
> The Chemistry in action section 'Fighting fires with hydrogels' will help you.

2. Experiment

Investigating how much water a hydrogel can absorb

You can work in a group to complete this activity.

Your task is to make up a plan to accurately find the volume of water absorbed by 1 g of hydrogel.

You will need: a disposable nappy (or hydrogel beads), scissors, an electronic balance, distilled water, a burette, filter funnel, dry 250 cm³ beaker.

The following information will help you.

- The nappy contains hydrogel – cutting it open and gently shaking it will release the tiny hydrogel beads.

- 1 cm³ of water weighs 1 g. So, for example, if 1 g of hydrogel absorbs 30 cm³ of water, then the hydrogel absorbs 30 times its weight of water.

You can use any of the usual laboratory equipment. Your plan should be detailed enough so that anyone could follow it and get the same results as you.

Discuss what to do with your group and write your plan.

 Discuss your plan with your teacher and ask if you can do the experiments.

You can work in a group to do the experiments, but you must record your own observations and write your own report.

⚠ Remember! You must follow the normal safety rules.

Do not breathe in the dust from the nappy when you are shaking out the hydrogel.

Write a report that includes:

- the aim of the experiment
- what you did (method)
- how you knew when to stop the experiment
- your conclusion
- one thing you could do to improve the experiment.

(continued)

| N3 | L3 | **N4** | **L4** | **317** |

☺ **3.** Write a report about the polymers hydrogel and Envirobond.

Include: what they are and what they do; what it is about their structure that gives them unusual properties; at least one example of how we use each of them.

Do the same as above for **one** other novel material.

4. Read the newspaper report about Ooho ('Water bottle you can eat', page 315) and answer the following questions.

(a) Name the chemicals used to make Ooho.

(b) State **three** reasons given in the article to suggest Ooho could be a replacement for traditional plastic bottles.

(c) State **three** things wrong about using plastic bottles, according to the article.

Learning checklist

After reading this chapter and completing the activities, I can:

N3 L3 **N4 L4**

- state that polymers are long-chain molecules formed when small molecules join. **Activity 20.1 Q1(b)**   ○ ○ ○

- state that the small molecules that join in polymerisation are known as monomers.
  Activity 20.1 Q1(e)   ○ ○ ○

- state that polymerisation is the name given to the reaction in which monomers join to form polymers.
  Activity 20.1 Q1(c)   ○ ○ ○

- work out the name of a polymer from the name of its monomer. **Activity 20.1 Q1(d)**   ○ ○ ○

- given the structure of a monomer, draw part of the structure of the polymer formed. **Activity 20.1 Q1(a)**   ○ ○ ○

- work out the name of a monomer, from the name of the polymer. **Activity 20.1 Q2(b)**   ○ ○ ○

- explain what is meant by thermosoftening (thermoplastic). **Activity 20.1 Q2(a)**   ○ ○ ○

- explain what is meant by thermosetting.
  Activity 20.1 Q3(b)   ○ ○ ○

- state that some thermosoftening polymers can be recycled but thermosetting polymers cannot.
  Activity 20.1 Q4(b), (c)   ○ ○ ○

N3 L3 **N4 L4**

- explain that thermosetting polymers can't be recycled because the polymer chains are cross-linked. **Activity 20.1 Q4(d)**   ○ ○ ○

- state that when plastics burn they produce toxic gases such as carbon monoxide, hydrogen chloride and hydrogen cyanide. **Activity 20.2 Q1(a)**   ○ ○ ○

- state the names of the gases that will be produced when a plastic burns from the elements present in the monomer or polymer chain. **Activity 20.2 Q1(a), Q3(a), (b)**   ○ ○ ○

- state that plastics used in furniture such as a sofa have to undergo fire-resistance tests. **Activity 20.2 Q2**   ○ ○ ○

- use data about how plastics react to being held in a flame to help identify the plastic. **Activity 20.2 Q1(b)**   ○ ○ ○

- state that most plastics are not biodegradable, which makes them difficult to dispose of. **Activity 20.3 Q1**   ○ ○ ○

- state that bioplastics are made from natural materials and break down easily. **Activity 20.3 Q1**   ○ ○ ○

- state that biodegradable plastics have additives which result in them breaking down quicker than traditional plastics. **Activity 20.3 Q1**   ○ ○ ○

- state that positives about bioplastics and biodegradable plastics include: fewer greenhouse gases are produced when they are made, packaging increases food shelf life, they break down quickly, they are not made from valuable oil; and that negatives include: they are made from plants, which could be used as food, some are made from GM plants, which may be harmful to the environment, not all biodegradable plastics break down quickly and some produce greenhouse gases when they break down. **Activity 20.3 Q1**   ○ ○ ○

- state two ways we can help reduce the amount of plastic we use, from the following: reusing plastic tubs and bags; repairing broken items; buying less pre-packed food; using recycled bags and ink cartridges. **Activity 20.3 Q2**   ○ ○ ○

N3 L3 **N4 L4**

- state that ceramics usually have a combination of strong ionic and covalent bonding. **Activity 20.4 Q1** ○ ○ ○

- state that the properties of ceramics such as hardness, high melting point and lack of chemical activity are due to their bonding. **Activity 20.4 Q2(c)** ○ ○ ○

- state that modern advanced ceramics have unusual properties, which give them particular uses in society. **Activity 20.4 Q4** ○ ○ ○

- state that the unusual properties of novel materials give them particular uses in society. **Activity 20.5 Q3** ○ ○ ○

- *present information in the form of a table/diagram.* **Activity 20.1 Q4(a), Activity 20.3 Q1, Activity 20.4 Q4** ○ ○ ○

- *extract scientific information from a report.* **Activity 20.1 Q5(a), (b), Activity 20.3 Q3, Activity 20.5 Q4** ○ ○ ○

- *apply my scientific knowledge in an unfamiliar situation.* **Activity 20.1 Q5(c), Q6(a), Activity 20.2 Q1(c), Activity 20.4 Q3(b), (c)** ○ ○ ○

- *obtain information from tables of data.* **Activity 20.4 Q2(a)** ○ ○ ○

- *make predictions and generalisations.* **Activity 20.1 Q6(b); Activity 20.4 Q2(b)** ○ ○ ○

- *work with others to plan and carry out a practical activity.* **Activity 20.5 Q1, Q2** ○ ○ ○

- *present information in the form of a graph.* **Activity 20.4 Q2(a)** ○ ○ ○

21 Properties of metals 1

This chapter includes coverage of:

N3 Properties of materials • Materials SCN 3-19b

You should already know:

- that metals are elements listed on the left-hand side of the periodic table
- that metals are good conductors of heat and electricity
- that the scientific name for a battery is a cell
- that a battery is a portable power source that has a store of chemical energy
- that when two different metals are connected in a salt solution electricity is produced
- that batteries have a positive (+) end and a negative (−) end.

Metals

Learning intentions

In this section you will:

- learn about the properties of metals that make them one of our most important materials
- learn about chemical reactions of metals
- learn about corrosion of metals and how to prevent it.

National 3

Curriculum level 3

Materials: Chemical changes SCN 3-19b

Metals are one of our most widely used materials because they have properties that are useful to us. They are good conductors of heat and electricity. They are generally strong and can be pressed into different shapes, like a car body. They can be drawn out into wires which can be twisted together and used in electrical cable and cables that are strong enough to hold up large structures such as the Forth Road Bridge.

Figure 21.1 *The main cables holding up the Forth Road Bridge are each made of over 11 000 steel wires twisted together.*

Properties such as strength and conductivity are known as physical properties. The way materials react chemically, known as chemical properties, can be just as important as physical properties. This is the case for metals.

Reactions of metals

Metals react with water to form metal hydroxides and hydrogen gas. How well a metal reacts depends on the metal. For example, gold is very unreactive. Gold jewellery can be worn, even in the shower, without reacting and losing its shine. However, sodium, a group 1 metal, is so reactive that it bursts into flames when it comes in contact with water.

Word equation:

sodium + water → sodium hydroxide + hydrogen

(a)

Sodium and water (immediate reaction).

(b)

Iron nail and water after a week (some reaction).

(c)

Some copper coins in water after a week (no reaction).

Figure 21.2 *Metals react with water at different rates.*

Figure 21.3 *Calcium reacts quickly with oxygen in the air, indicating it is a reactive metal.*

Metals also react with oxygen to form metal oxides. Again, how well the metal reacts depends on the metal. Metals like gold and silver are used for jewellery because they appear not to react with oxygen in the air. Hot magnesium and calcium react with the oxygen in the air, quickly producing a bright flame, indicting they are reactive metals.

Word equation: calcium + oxygen → calcium oxide

Corrosion

The air contains oxygen and water vapour, which will react with metals when they are left exposed to the air. This is called **corrosion**. The more reactive the metal, the faster it will corrode.

📖 **Word bank**

• **Corrosion**

The reaction of a metal with water and oxygen in the air.

Corrosion weakens a metal's structure, which can be dangerous when a metal is used in buildings and bridges. In these cases it would seem the less reactive a metal used the better. However, other factors have to be taken into consideration, such as cost and strength of the metal. Although the metals gold and platinum are very unreactive, they are among the most expensive metals in the world. Even if they weren't, they lack strength and also the quantities needed for the car industry and for building projects.

Figure 21.4 *Ancient gold coins, like this one showing Alexander the Great, have survived in the ground for thousands of years because gold is very unreactive.*

Iron is our most widely used metal – it has the strength required, is relatively cheap and there is lots of it. Unfortunately, it corrodes easily. When iron corrodes it forms rust, a complicated mixture of iron compounds. Rust has no strength and crumbles easily when touched. Rust has a very recognisable brown colour. If left exposed to the air long enough an object made of iron will rust away completely.

Word equation: iron + oxygen + water → rust

Rusting is speeded up if iron is exposed to water with salt dissolved in it. Therefore, salty water is a problem for seagoing ships, and cars and iron structures near the sea. Rusting is also speeded up by acid rain.

Figure 21.5 *A bike left outside will soon rust if not protected from the air.*

Protecting iron against rusting

Iron can be protected against rusting by stopping the air and water reaching the iron. The bodywork of most cars is made from iron. It is painted to protect the steel from the air and so prevent rusting. New cars in storage are covered in a thin layer of grease, which further protects it from the air. Food cans are made of iron, which will corrode inside and spoil the food. To prevent this the insides of cans are coated with tin, which is less reactive than iron, or a thin layer of plastic material. Both stop air and water reaching the iron and so prevent rusting.

Figure 21.6 *Car bodywork is painted to stop water and air getting to the steel below, but, if damaged, rust quickly develops.*

Iron can be protected from rusting by covering it with less reactive metals like copper and silver. They stop air and water reaching the iron. In addition because they are not very reactive, the coating will last a long time. This coating process is known as **electroplating**.

Applying a protective zinc coating to steel or iron does the same job. The most common method is called hot-dip **galvanising**. The item to be protected is dipped in a bath of molten zinc.

Protecting steel structures against rusting is expensive. However, the cost has to be balanced against the need to protect steel structures. A rusting car could break up when being driven at speed, and a bridge might collapse if the rusting gets too bad. Both could lead to serious injury or death.

> ### 📖 Word bank
>
> - **Electroplating**
>   Coating a metal with a less reactive metal.
> - **Galvanising**
>   Coating a metal by dipping it in molten zinc.

? Did you know ...?

The Forth Bridge had to be painted all year round, every year, to protect it from rusting. In 2002 the bridge was repainted fully for the first time, at a cost of £130 million. Special paint based on the type used to coat North Sea oil rigs was used. It is estimated that the bridge will not need repainting until at least 2022 and even then will only need one coat. Engineers have said the bridge itself could have a lifespan of another 100 years.

Figure 21.7 *The Forth Bridge, famous around the world for its design and red oxide paint colour, was last painted in 2002.*

Where do metals come from?

Metals are natural materials that exist in the earth. Only unreactive metals like gold and platinum are found in their pure acompounds are found in rocks. When there is enough of the metal compound in the rock to make it worth separating it out from the rock then the compound is known as an ore. Figure 21.8 shows some common ores.

Lead ore   Iron ore   Copper ore

Lead metal (Pb)   Iron metal (Fe)   Copper metal (Cu)

Figure 21.8 *Some common metal ores and their metals.*

To get iron from its compound in the ore, the ore has to be heated with carbon. In industry this is done on a large scale in a blast furnace. A form of coal called coke is the source of carbon, and the limestone removes impurities (slag waste).

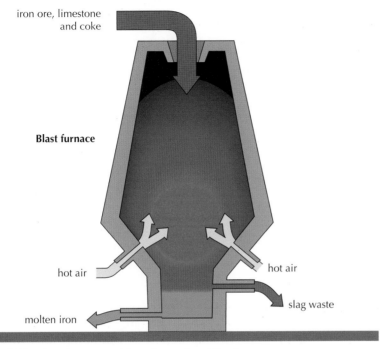

iron ore, limestone and coke

Blast furnace

hot air

hot air

slag waste

molten iron

Figure 21.9 *Iron is extracted from its ore in a blast furnace.*

Reactive metals like sodium and aluminium are obtained by passing electricity through melted compounds containing the metals.

GO! Activity 21.1

☻ 1. Metals are among our most useful materials.
   (a) State **two** useful properties of metals.
   (b) Give an example of a metal that reacts quickly with water.
   (c) Name the compound formed when magnesium reacts with oxygen.

2. Most metals corrode when left outside.
   (a) Give the meaning of the word corrosion.
   (b) State **two** ways in which corrosion is speeded up.
   (c) What special name is given to the corrosion of iron?
   (d) State three ways in which a piece of iron can be prevented from corroding.

3. Metals are natural materials found in the earth.
   (a) Why does gold exist in its pure form in the earth?
   (b) In what form are most metals found in the earth?
   (c) Describe how iron is obtained from its ore.
   (d) State how a reactive metal like sodium is obtained from its ore.

(continued)

| N3 | L3 | N4 | L4 | **325** |

4. The bar chart shows the amount of our most common elements, as a percentage, found in the Earth.

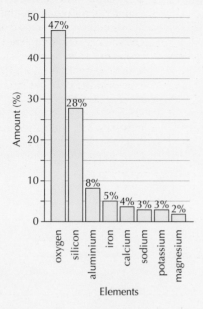

Elements

(a) Use the periodic table on pages 34–35 to identify the metals in the bar chart.

(b) Which is the most common metal?

(c) Draw up a table and list the metals you identified and their percentages.

National 3

Metals and batteries

Learning intention

In this section you will:

- learn how metals are used in batteries.

Figure 21.10 *Different shapes and sizes of batteries.*

Batteries produce electricity from chemical reactions. They are vital to our everyday life. They come in all shapes and sizes. Miniature 'button' batteries are used to power hearing aids, heart pacemakers and wristwatches. Thin batteries are used in smartphones and computers, and large heavy batteries are used to start our cars. Most batteries are portable – they can be taken anywhere, into outer space and to the bottom of the oceans. Others are the size of football pitches, designed to supply electricity to a whole town in the event of a blackout.

The one thing they have in common is that the chemical reactions that take place usually involve metals. These reactions don't happen until the battery is connected into the device you are using.

Disposable batteries

Some batteries are disposable – once the chemicals are used up no more electricity can be produced so they are 'thrown away' (see 'You need to know' opposite). They are designed for use in the likes of remote controls and watches because they don't use much electricity. Examples of common disposable batteries are zinc–carbon and alkaline, which also contain zinc.

Rechargeable batteries

Some batteries can be recharged. They are put into a charger, which is then plugged into the mains to be charged up for a few hours. During this time electrical energy is stored in the chemicals. When the batteries are put into, say, a torch, chemical reactions take place and electricity is produced straight away. A typical rechargeable battery can be charged up hundreds of times and can last for 3 or 4 years. They save you money and are better for the environment than disposable batteries. Most rechargeable batteries contain the likes of lithium and nickel metals.

The batteries in items like smartphones and laptops are built into the appliance and don't need to be removed to be charged.

Wireless chargers containing copper wire coils are becoming more common for smartphones and have been used for many years for charging electric toothbrushes. Some coffee shops have introduced wireless table top chargers into some of their shops, and furniture manufacturers like IKEA are already building wireless chargers into some furniture.

★ You need to know

Disposable batteries should not be thrown away into landfill. They contain chemicals that could cause environmental problems, such as chemicals leaking out into the soil. The metals in the batteries can also be recycled. Shops and supermarkets that sell batteries now have to have collection points for used batteries. Many local authorities collect used batteries with other recycled items.

Figure 21.11 *This used battery bin allows you to test the battery to see if there is any charge left before putting it in the bin.*

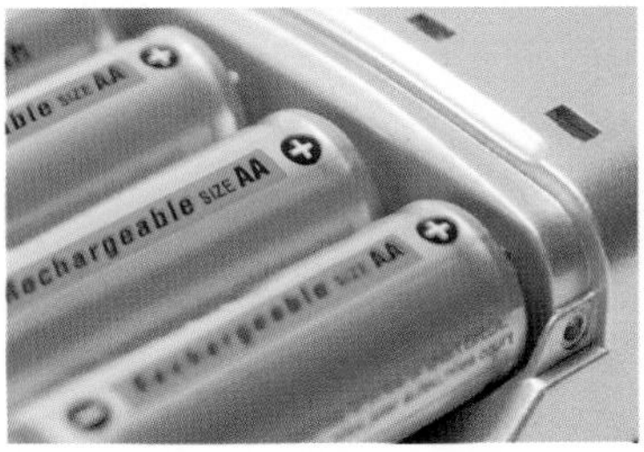

Figure 21.12 *Rechargeable batteries can be charged and used many times.*

Figure 21.13 *The batteries inside smartphones are recharged by plugging the phone directly into the mains supply.*

Figure 21.14 *Wireless battery chargers contain coils of copper wire inside them.*

Cars and other vehicles use rechargeable lead-acid batteries to start the vehicle. Car manufacturers are developing electric-powered cars. They use electricity instead of petrol or diesel. Electric cars can use lead-acid batteries, but rechargeable lithium batteries are lighter and better at producing the energy needed to power a car. Advances in rechargeable technology and concerns about the environment have made electric cars more attractive.

↻ Keep up to date!

You can keep up to date with the development of the BMW *i8* electric sports car and electric cars in general by searching online.

BMW has announced that it's developing a wireless charging system for its *i8* electric sports car. It can charge the battery in as little as 2 hours and provides enough charge to allow the car to travel 300–350 miles.

Figure 21.15 *Car batteries are rechargeable. They contain lead, which makes them very heavy.*

Figure 21.16 *The battery in an electric-powered car can be recharged at recharging points in town and supermarket car parks.*

? Did you know ...?

The Daniell cell was invented in 1836 by British chemist John Frederic Daniell. It was the first battery that produced a reliable source of electricity over a long period of time. It was widely used as a power source for electrical telegraph networks. It was made from a copper pot filled with a copper sulfate solution. Inside that was a ceramic container filled with sulfuric acid and a piece of zinc. Modern batteries are still based on the Daniell cell.

Figure 21.17 *Daniell cells, first made in 1836, were early types of battery.*

GO! Activity 21.2

☺ 1. Batteries are a portable source of electricity.

    (a) State what takes place inside a battery to produce electricity.

    (b) Why do disposable batteries stop producing electricity?

    (c) Give an example of one type of disposable battery.

    (d) Give **two** reasons why disposable batteries should not be put into landfill.

    (e) State how, when a rechargeable battery runs out, it can be made active again.

    (f) Give **two** examples of how rechargeable batteries are used in everyday life.

2. Experiment

Making a rechargeable battery

You can work with others to do this experiment, but you must record your own observations and write your own report.

Remember! You must follow the normal safety rules.

You should also wear disposable gloves and safety glasses to do this experiment. It should be carried out in a well-ventilated room or in a fume cupboard. Your teacher may demonstrate the experiment – ask your teacher.

You will need: two lead plates, a 500 cm³ beaker, four crocodile clips, two leads, a bottle of dilute sulfuric acid, a power pack, stop clock, a small bulb.

Method

Part 1: Charging up the battery.

(a) Pour the acid into the beaker until it reaches the 250 cm³ mark.

(b) Attach the leads to the lead plates using the crocodile clips.

(c) Connect the other end of the leads to the power pack.

(d) Put the plates into the acid as in the diagram.

STOP! Ask your teacher to check your arrangement and voltage on the power pack.

(e) Switch on the power pack and time 2 minutes, using the stop clock.

(f) Note what you see happening at the electrodes.

(g) After 2 minutes, switch off the power pack and stop the clock.

Part 2: Using the battery.

(h) Take the leads out of the power pack and attach them to the bulb, as in the diagram.

(i) Note what you see happening to the bulb as soon as it is connected.

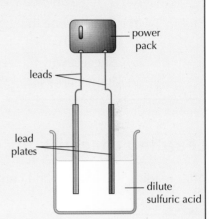

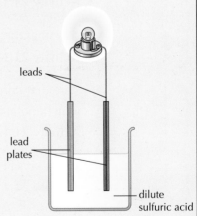

(continued)

Part 3: Recharging the battery.

(j) Reconnect the power pack to the plates and repeat steps (e)–(i).

Now do the following.

(a) Write a report about your experiment. Include the aim of your experiment, what you did and what you saw.

(b) Answer the following questions.

(i) What evidence was there to suggest that a chemical reaction had taken place?

(ii) What evidence was there to suggest that electricity was produced *after* the power pack was disconnected?

(iii) What evidence was there that the battery had been recharged?

(iv) Suggest **one** thing you could have tried to see if you could produce more electricity.

(v) Suggest why the experiment has to be done in a well-ventilated room or a fume cupboard.

Learning checklist

After reading this chapter and completing the activities I can:

N3 L3 N4 L4

- state two useful properties of metals. **Activity 21.1 Q1(a)**   ◯ ◯ ◯

- state that group 1 metals react quickly with water. **Activity 21.1 Q1(b)**   ◯ ◯ ◯

- state that metals react with oxygen to form metal oxides. **Activity 21.1 Q1(c)**   ◯ ◯ ◯

- state that corrosion is the name given to the process in which metals react with oxygen and water in the air. **Activity 21.1 Q2(a)**   ◯ ◯ ◯

- state that salt water and acid rain speed up corrosion. **Activity 21.1 Q2(b)**   ◯ ◯ ◯

- state that rusting is the name given to the corrosion of iron. **Activity 21.1 Q2(c)**   ◯ ◯ ◯

- state that rusting of iron can be prevented by doing any one of the following: oiling/greasing; painting; coating in plastic or another metal. **Activity 21.1 Q2(d)**   ◯ ◯ ◯

N3 L3 N4 L4

- state that gold is so unreactive that it exists in its pure form in the earth. **Activity 21.1 Q3(a)**

- state that most metals exist in the earth as compounds. **Activity 21.1 Q3(b)**

- state that iron is obtained from its ore by heating with carbon. **Activity 21.1 Q3(c)**

- state that reactive metals are obtained from their compounds by passing electricity through them. **Activity 21.1 Q3(d)**

- state that electricity is produced in a battery when a chemical reaction takes place. **Activity 21.2 Q1(a)**

- state that in a disposable battery when the chemicals are used up no more electricity is produced. **Activity 21.2 Q1(b)**

- state that zinc–carbon and alkali are examples of disposable batteries. **Activity 21.2 Q1(c)**

- explain why disposable batteries should not be thrown away. **Activity 21.2 Q1(d)**

- state that when a rechargeable battery runs out it can be recharged by connecting it to mains electricity. **Activity 21.2 Q1(e)**

- give two examples of how a rechargeable battery is used in everyday life. **Activity 21.2 Q1(f)**

- *present information in the form of a table.* **Activity 21.1 Q4(c)**

- *interpret information presented in a graph.* **Activity 21.1 Q4(a), (b)**

- *work with others to plan and carry out a practical activity.* **Activity 21.2 Q2**

22 Properties of metals 2

This chapter includes coverage of:

N4 Metals and alloys • Forces, electricity and waves SCN 4-10a • Materials SCN 4-16a, SCN 4-19b • Topical science SCN 4-20b

You should already know:

- that metals are grouped together on the left of the periodic table
- that metals are good conductors of electricity and heat
- that group 1 metals react quickly with water
- that metals react with oxygen to form metal oxides
- that gold is so unreactive it exists in the earth in its pure form
- that most metals exist in the earth as compounds, found in ores
- that iron is obtained from its ore by heating with carbon
- that reactive metals are obtained by passing electricity through their compound
- how to work out the formulae for elements and compounds
- that corrosion is the name given to the process in which metals react with oxygen and water in the air
- that salt water and acid rain speed up corrosion
- that rusting is the name given to the corrosion of iron
- that rusting of iron can be prevented by doing any one of the following: oiling/greasing; painting; coating with plastic or another metal
- that electricity is produced in a battery when a chemical reaction takes place
- that when the chemicals in a disposable battery are used up no more electricity is produced
- that zinc–carbon and alkaline are examples of disposable batteries

| N3 | L3 | **N4** | **L4** |

Reactivity of metals

National 4

Curriculum level 4

Materials: Chemical changes SCN 4-19b; Topical science SCN 4-20b

Learning intentions

In this section you will:

- learn how metals react with oxygen, water and acid
- learn about the reactivity series of metals
- learn how the method used to extract metals from their ores is related to the reactivity series
- learn about the environmental and social impact of mining for ores.

Metals are one of our most useful materials. They conduct heat and electricity and many are strong. They are **malleable** (can be shaped) and **ductile** (can be made into wires). Some, like aluminium, have low **density** (lightweight), and others, like lead, have high density (heavy). These are physical properties. The ways in which metals react with chemicals such as acids, water and oxygen are also important when deciding how we are going to use them. These are chemical properties. The way metals react chemically allows us to put them in order of chemical activity. This is known as the **reactivity series**. In the following sections metals are reacted with oxygen, water and acid and how well they reacted used to put them in order of reactivity.

Reaction of metals with oxygen

When a metal reacts with oxygen heat energy is produced – the reaction is exothermic (see page 69). How fierce the reaction is can be used to put the metals in order of reactivity. Figure 22.1 shows what happens when magnesium, iron and copper are heated and then put into a gas jar full of oxygen.

📖 Word bank

- **Malleable**
Can be shaped.
- **Ductile**
Can be made into wires.
- **Density**
The relationship between the mass of a substance and the amount of space it takes up (volume).
- **Reactivity series**
The list of metals arranged in order of their chemical activity.

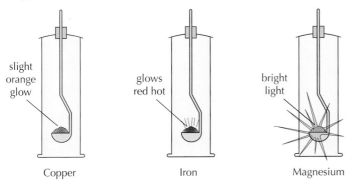

slight orange glow

Copper

glows red hot

Iron

bright light

Magnesium

Figure 22.1 *Metals reacting with oxygen.*

The order of reactivity from this experiment is:

magnesium;   iron;   copper.

(most)   (least)

? Did you know ...?

Metals like lithium, sodium and potassium (group 1) are so reactive that they are stored in the laboratory under oil. This stops oxygen and water in the air reacting with the metal.

Figure 22.2 *Potassium stored under oil.*

It is too dangerous to react metals like potassium and sodium with oxygen, which gives an indication of how very reactive they are. At the other extreme, gold coins and ornaments exposed to the air for thousands of years show no sign of chemical reaction, which indicates how unreactive they are.

When a metal reacts with oxygen a metal oxide is produced.

$$metal \ + \ oxygen \ \rightarrow \ metal \ oxide$$

Using magnesium as an example, word and chemical equations can be written.

Word equation: magnesium + oxygen → magnesium oxide

Formula equation: $Mg \ + \ O_2 \ \rightarrow \ MgO$

(Remember! The rules for working out formulae (pages 58–61) must be followed.)

Reaction of metals with water

Pieces of magnesium, iron and copper were added to water in a test tube. No reaction appeared to be taking place. After being left for a week:

- some bubbles of gas were seen in the magnesium test tube and the surface of the metal had become dull

- in the test tube with the iron there were no bubbles of gas, but the iron had started to rust

- in the copper test tube there were no bubbles and no sign of reaction.

This order of reactivity fits in with the results from the experiments of metals with oxygen.

When lithium, a group 1 metal, was added to water it reacted violently and bubbles of gas were given off. When universal indicator was added to the water it turned purple (Figure 22.3), showing that the water had become alkaline (see pages 77–78). This is because during the reaction lithium hydroxide (an alkali) was formed. Hydrogen is the gas produced.

Word equation:

lithium + water → lithium hydroxide + hydrogen

Formula equation:

$Li \ + \ H_2O \ \rightarrow \ LiOH \ + \ H_2$

If you have a gold or silver ring you will know that washing your hands or doing dishes has no effect on them. Gold and silver are very unreactive metals.

Figure 22.3 *Lithium reacts violently with water. The purple colour is from universal indicator showing an alkali is formed.*

Reaction of metals with acid

Pieces of magnesium, iron and copper were added to hydrochloric acid in a measuring cylinder. The detergent was only there to create a foam, which made it easier to see how much gas was produced.

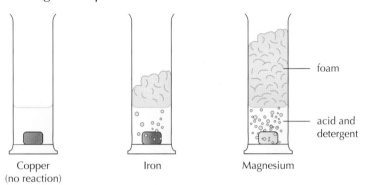

Copper (no reaction)   Iron   Magnesium

Figure 22.4   *Magnesium and iron react with acid but copper does not.*

Figure 22.5   *Magnesium produces lots of bubbles of gas when it reacts with acid.*

The results are shown in Table 22.1.

Table 22.1   *The results when magnesium, iron and copper are added to hydrochloric acid.*

Metal	Observations when acid added
magnesium	lots of bubbles of gas: very reactive
iron	some bubbles of gas: reactive
copper	no bubbles: unreactive

This order of reactivity fits in with the results from the experiments of metals with oxygen and with water.

When the gas produced was collected and a flame brought close to the test tube it burned with a 'pop'. This is the test for hydrogen gas. A salt was also produced.

Using magnesium as an example, equations can be written:

Word equation:

magnesium + hydrochloric acid $\rightarrow$ magnesium chloride + hydrogen

Formula equation:

$$Mg + HCl \rightarrow MgCl_2 + H_2$$

The reactivity series

Using the results from the above experiments and other experiments, an order of reactivity for common metals can be worked out. This is known as the reactivity series (Figure 22.6). It matches the order in which metals were thought to have first been extracted from their ores – the least reactive first (Figure 22.7).

★ You need to know

You need to know that reactive metals are near the top of the reactivity series and unreactive metals are near the bottom. You do not need to memorise the exact order of the metals.

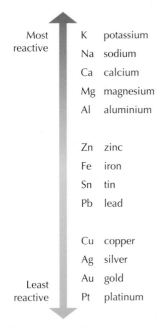

Figure 22.6   *The reactivity series.*

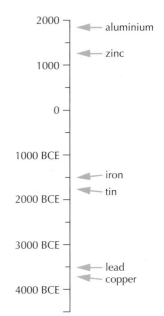

Figure 22.7   *The dates that common metals were first thought to have been extracted from their ores (BCE means Before the Common Era).*

Extraction of metals from their ores

The methods used to **extract metals** from their **ores** are dependent on the position of the metal in the reactivity series.

The more reactive the metal is the harder it is to separate it from its compound. This is because the more reactive metals have stronger bonds with the other element(s) in the compound than the less reactive metals.

The dates when metals are first thought to have been extracted from their ores (Figure 22.7) matches their position in the reactivity series.

Extracting unreactive metals

Metals like platinum and gold are found at the bottom of the reactivity series. They are so unreactive that they exist in the earth in their pure form (not in a compound).

Other unreactive metals like silver are not only found in the pure form, but also exist in compounds. Silver is so unreactive that it does not bond strongly in a compound. Silver oxide, for example, only needs to be heated to extract it from its ore.

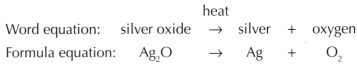

★ You need to know

The section on extracting metals from their ores relates to the position of the metal in the reactivity series. The industrial extraction of metals is very complicated and doesn't necessarily follow the method used in the laboratory.

📖 Word bank

- **Extracting metal**

Breaking down a metal compound to obtain the metal.

- **Ore**

Rock containing enough of a metal compound to make it worth extracting the metal.

Extracting more reactive metals

Metals like copper and iron are higher in the reactivity series than silver. They cannot be extracted from their ore simply by heating. This is because copper and iron have stronger bonding within their compounds. However, they can be extracted when heated with carbon. The carbon is more reactive than these metals so it displaces (takes the place of) the metal from its compound.

Word equation:

copper oxide + carbon → copper + carbon dioxide

Formula equation:

$$CuO + C → Cu + CO_2$$

Word equation:

iron oxide + carbon → iron + carbon dioxide

Formula equation:

$$Fe_2O_3 + C → Fe + CO_2$$

The extraction of iron from its ore (haematite) is carried out industrially in a blast furnace (see page 325).

Extracting very reactive metals

Metals like sodium and aluminium are found near the top of the reactivity series. They cannot be extracted from their ore by heating or reacting with carbon. This is because metals like sodium are very strongly bonded within a compound and a lot of energy is needed to break the compound down. This energy comes from electricity. Passing electricity through the molten compounds break them down. This is known as **electrolysis**.

> 📖 **Word bank**
>
> - **Electrolysis**
>   Breaking down a compound by passing electricity through it in the molten state.

electricity

Word equation: aluminium oxide (ℓ) → aluminium + oxygen

GO! ## Activity 22.1

☻ **1.** Select your answers to the following questions from these metals:

   calcium   platinum   tin   zinc

   (a) Which metal would react quickly with water?

   (b) Which metal would **not** react with acid?

   (c) Which metal is most likely to react with oxygen?

> 🔍 **Hint**
>
> You may wish to refer to the reactivity series, Figure 22.6 on page 336.

   (d) The compounds of which **two** metals would have to be heated with carbon in order to extract the metal?

   (e) The oxide of which metal would have to be melted and then electrolysed in order to extract the metal?

   (f) Which metal can be extracted from its ore by heating the ore?

2. A technician dropped her ring into an acid solution. She told her colleague there was no rush to get the ring out of the acid because it wouldn't react.

   Suggest which metal the ring could have been made of.

3. Group 1 metals should not be added to acid.

   Using your scientific knowledge suggest why this is the case.

4. A number of experiments were carried out on four metals, P, Q, R and S. Some of the results are as follows.

   • Metals Q and S react with hydrochloric acid. Metal P does not react.

   • Metals Q and R give off a bright flame when reacted with oxygen. Metal S only glows in oxygen.

   • Metal R reacts quickly in water. Metal Q reacts only slowly with water.

   (a) Place the metals in order of their reactivity (most reactive first).

   (b) Suggest names for each of the metals (you may wish to refer to the reactivity series, Figure 22.6).

5. Suggest why sodium and potassium are stored under oil but magnesium and zinc can be stored in air.

6. The table shows the percentage of copper in a number of copper ores.

Copper ore	Percentage of copper (%)
azurite	55
bornite	63
chalcocite	80
malachite	57
tetrahedrite	39

   Draw a bar chart of copper ore against percentage copper. (Your teacher may have a blank graph prepared for you.)

The impact of mining for metal ores

Some metals, like gold, exist in their pure form in the earth. Most metals exist in the earth as compounds trapped within rocks, known as ores. Regardless of how the metals exist, they have to be dug out of the earth. This is called **mining**. There are over 2500 large-scale mines around the world, and the environmental and social impact is huge.

📖 Word bank

• **Mining**

Digging into the earth.

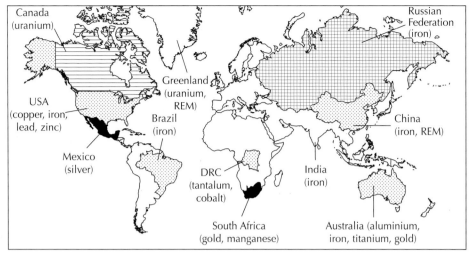

DRC = Democratic Republic of the Congo

REM = Rare Earth Metals

Figure 22.8   *Major metal deposits in the world.*

Mining involves removing thousands of tonnes of waste rock, which has to be put somewhere. Much of it can be crushed and used for road filler or building material. The hole that is created by surface mining is unsightly and can be dangerous. Mine operators have to **reclaim** the land when they stop mining an area. This sometimes involves filling in some areas or flooding it to create an artificial lake. The area has to be planted with trees and other plants to restore the landscape.

Sometimes the waste rock contains small amounts of toxic materials. These can dissolve out of the rock over time. Toxic materials enter the soil, rivers and lakes and harm fish and other animals. Waste liquids from the mining process have to be stored in specially lined pools to prevent them from getting into the ground or water system.

Mines are also noisy places. Blasting, heavy machinery and trucks transporting to and from the mine all contribute to noise pollution. Large amounts of energy are used to get the ores out of the ground and transported to where the metal is to be extracted.

Greenland is one of the latest areas of the world where huge reserves of important metals have been found. Geologists claim that Greenland has 50% of the world's rare earth metals (REMs). REMs are used in everything from mobile phones to solar panels and wind turbines. China currently supplies most of the REMs, but there is concern that they will be using more themselves instead of exporting them to the rest of the world. Greenland also has gold and uranium deposits. Uranium is essential to the nuclear energy industry.

> 📖 **Word bank**
>
> • **Reclaim**
> Make usable again.

Figure 22.9   *An open-pit copper mine in Utah, USA.*

Figure 22.10   *Mining for metal ores in the mountains in Greenland has raised environmental concerns.*

Make the link

For more about rare earth metals, look at Chapter 3, page 40.

Mining metal ores generally brings great wealth to the country. However, many local people are concerned that mining will destroy the environment and/or their traditional way of life.

Tantalum is a heat-resistant metal used in mobile phones and computer chips. It is one of the rare metals mined in the Democratic Republic of the Congo (DRC). The DRC is rich in many minerals, including metal ores. This is one of the reasons behind a war that has claimed millions of lives. The war has also forced farmers from their land, caused children to leave school to work in the mining industry and destroyed the environment, including endangering species such as the lowland gorilla. With its mineral wealth the DRC should be a wealthy country but its people are some of the poorest in the world. This is because the money made from mining is not shared with everyone.

Activity 22.2

1. Look at Figure 22.8, which shows the major metal deposits in the world.

   Draw a table to show each metal highlighted on the map and the country that has major deposits of each metal.

2. Discuss with a partner some of the environmental and social issues resulting from mining for metal ores.

   Record the results of your discussion by doing one of these:

   • writing a report summarising the issues

   • drawing a 'mind map' linking the issues

   • making a presentation to deliver to the rest of the class.

   Ask your teacher to check your final report/mind map/presentation.

National 4

Curriculum level 4

Materials: Properties and uses of substances SCN 4-16a

Word bank

• **Alloy**
A mixture of metals or of a metal mixed with non-metals.

• **Bronze**
An alloy of copper and tin.

Alloys

Learning intentions

In this section you will:

• learn about alloys and how they are used

• learn about smart alloys.

Alloys are mixtures of metals or of a metal mixed with non-metals. Alloys tend to have more useful properties than pure metals. **Bronze** is an alloy of copper with tin and small amounts of other metals. It has been known for thousands of years. It is harder than pure copper, so was used to make weapons and tools like axe heads.

N3 L3 **N4** L4

Common alloys

Steel – our most important alloy

Steel is an alloy of iron with carbon. Pure iron breaks relatively easily, whereas steel is much stronger. There are many different types of steel, depending on the other elements in the mixture and the quantity added. Most steel is used to make girders for the building trade, and in car production and ship building. **Stainless steel** has chromium and other metals like manganese added. Cutlery is usually made of stainless steel as it resists rusting (see page 344). If you see cutlery with stains or marks on it then it is likely to be a poor-quality alloy. The Kelpies in Falkirk are statues made of structural steel covered with stainless steel. The structural steel is cheaper and it rusts, whereas the stainless steel covering resists rusting and so protects the structural steel underneath. Stainless steel also reflects the light (because it is uncorroded), which makes the statues shine in the sunlight.

Figure 22.11 *The Kelpies in Falkirk are covered in stainless steel.*

📖 Word bank

- **Steel**
An alloy of iron and carbon.
- **Stainless steel**
An alloy of iron, carbon and other metals like chromium and manganese.

Gold alloys

Pure gold is a relatively soft metal so it is not usually used to make jewellery. In order to make it stronger it is mixed with other metals such as silver and copper. Alloying also changes the colour of gold. Although alloying gold reduces its value, it is still valuable. The percentage of gold in an alloy is indicated by the number of carats (ct) it has. Pure gold is said to be 24 ct. The percentage of gold and other metals in some different alloys of 'yellow' gold is shown in Table 22.2.

Figure 22.12 *Gold rings tend to be made of 9 ct gold because it is harder wearing than pure gold.*

Table 22.2 *The amount of gold and other metals in different alloys of yellow gold.*

Carats (ct)	Percentage gold (%)	Percentage other metals (%)
22	91.7	Ag: 5.0; Cu: 2.0; Zn: 1.3
18	75.0	Ag: 15.0; Cu: 10.0
14	58.3	Ag: 30.0; Cu: 11.7
9	37.5	Ag: 42.5; Cu: 20.0

Coins

All British coins are alloys. 1 p and 2 p coins, which look like copper, are actually stainless steel coated with copper. Copper is so expensive nowadays and 1 p and 2 p coins of such low value that if they were made from pure copper the metal itself would be worth more than the value of the coin.

Figure 22.13 *All British coins are made from alloys.*

Table 22.3 lists the coins currently in circulation and the percentage of each metal in the alloy.

Table 22.3 *The percentages of different metals used in British coins.*

Coin	Alloy
1p, 2p	Steel (covered in copper)
5p, 10p, 50p	Cupro–nickel (75.0% Cu : 25.0% Ni)
20p	Cupro–nickel (84.0% Cu : 16.0% Ni)
£1 (round)*	Nickel–brass (70.0% Cu : 5.5% Ni : 24.5% Zn)
£1 (12-sided) and £2	Outer: Nickel–brass (76.0% Cu : 4.0% Ni : 20.0% Zn) Inner: Cupro–nickel (75.0% Cu : 25.0% Ni)

* Round £1 coin removed from circulation in October 2017.

Alloy wheels

Alloy wheels on cars are made from an alloy of aluminium and magnesium. This alloy is lighter than steel so gives improved performance but lacks the strength of steel.

Figure 22.14 *Aluminium alloy wheels are lighter than steel wheels.*

Smart alloys

Smart materials have properties that can change depending on their surroundings. **Smart alloys** (shape-memory alloys) can remember their original shapes. When a smart alloy is deformed (bent or twisted) it keeps its new shape until it is heated. When the temperature reaches a certain level the alloy returns to its original shape. Nitinol, a smart alloy of nickel and titanium, is not only the most successful smart alloy, but also the most expensive. It is used to make dental braces, frames for glasses and medical stents used to widen blood vessels. Small nitinol stents are put into a damaged blood vessel cold and expand inside the blood vessel to the correct size as the stent warms up to body temperature. One disadvantage of smart alloys, apart from their high cost, is that continual bending and twisting can cause them to break.

Some of the reasons for the continuing advances in alloy technology are:

- the availability of materials

- new manufacturing techniques

- the ability to 'test' alloys before they are ever produced; most modern alloys are preplanned using computer simulations, which help predict what properties the alloy will have.

> 📖 **Word bank**
>
> - **Smart alloys**
>   Alloys that can remember their original shapes.

Figure 22.15 *A tiny metal stent; such stents are often made of the smart alloy nitinol.*

GO! Activity 22.3

☻ **1.** Bronze is an alloy that has been used for thousands of years.

   (a) State what is meant by an alloy.

   (b) Suggest why bronze was known thousands of years ago. (You may wish to look at the reactivity series (Figure 22.6) to help you answer.)

   (c) Give **one** reason for using an alloy instead of a pure metal.

2. The table shows the composition of a 1 p coin over the last 300 years.

Period	Composition
1707–1796	Silver
1797–1859	Copper
1860–1970	Bronze
1971–present	Copper-plated steel

   (a) Identify which metals are pure and which are alloys.

   (b) State which elements are in the metals you identified as alloys.

   (c) Suggest why 1 p coins are no longer made of silver.

3. The pie chart shows the proportion of the metals in one of the coins in Table 22.3.

   Look at the table and work out:

   (a) which coin is represented by the pie chart

   (b) which metals are represented by A, B and C

   (c) the percentage represented by A, B and C.

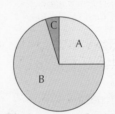

4. Use the information in Table 22.2 to help you answer these questions.

   (a) Estimate the percentage of gold in 10 ct yellow gold.

   (b) A necklace made of which caratage of gold would be the most expensive? Explain your answer.

   (c) 18 ct white gold contains the same percentage of gold as yellow gold. The other metal is palladium.

   Calculate the percentage of palladium in the white gold.

   (d) Rose gold gets its name from the colour of gold when copper is added.

   (i) Suggest which would have more copper – 22 carat or 9 carat.

   (ii) Give a reason for the answer you gave to part **(i)**.

   (iii) Calculate the percentage of copper in 14 ct rose gold.

5. Smart alloys have a number of uses.

   (a) What is meant by a 'smart alloy'?

   (b) Name a smart alloy and state **one** use for it.

Corrosion

Learning intentions

In this section you will:
- learn what happens to iron atoms when iron rusts
- learn which experiments can be set up to investigate the rusting of iron
- learn ways to protect iron (steel) from rusting.

📖 **Word bank**

- **Rusting**
Special name given to the corrosion of iron and steel.

Rusting

When metals corrode a chemical reaction takes place, which results in the metal atoms changing into metal ions. The corrosion of iron is called **rusting**. Each iron atom loses two electrons and forms an iron ion with a 2⁺ charge.

Word equation: iron atoms   →   iron ions   +   electrons

Formula equation: Fe → Fe^{2+} + $2e^-$

In order for this to happen both water and oxygen (from the air) must be present. The electrons from the iron atoms go to the oxygen and water molecules, and a complicated mixture of iron oxides and hydroxides is formed. This is rust.

An experiment can be set up to prove that both oxygen and water are needed. Three test tubes are set up with iron nails in different conditions. After 1 week the nails were as shown below.

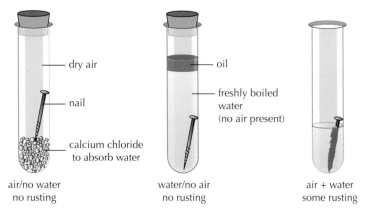

dry air

nail

calcium chloride
to absorb water

oil

freshly boiled
water
(no air present)

air/no water
no rusting

water/no air
no rusting

air + water
some rusting

Figure 22.16   *An iron nail rusts only when both oxygen and water are present.*

Physical methods of preventing rusting

If water and oxygen are physically prevented from reaching the iron then rusting will not take place. This can be done by coating the metal in plastic or another metal (electroplating), or oiling or greasing the surface.

Painting is a widely used method of protecting iron. The Forth Bridge famously needed to be constantly painted (see page 324). However, polymer-based paints are now available that contain ingredients that react chemically with rust to form a tough polymer layer that bonds well to the metal surface. These have the following features:

- only the loose flakes of rust need be removed before treatment

- they are more easily applied

- some kinds can be applied under water (for small areas)

- very long lifetimes are claimed for these paints.

Chemical methods of preventing rusting

Rusting is caused by the iron atoms losing electrons, so if electrons are supplied back to the iron then rusting will be prevented. This is a chemical way of preventing rusting.

When metals form ions, they give away one or more electrons. Some metals, such as lithium and sodium, lose their electrons very easily – they corrode quickly. Other metals, such as silver and gold, do not give away electrons easily – they corrode very slowly. The **electrochemical series** is a list of metals arranged in order of how easily the metal atoms lose electrons (page 348).

An experiment can be set up to test if connecting an iron nail to different metals in the electrochemical series has any effect on the rate of rusting. A chemical indicator, called ferroxyl indicator, can be used to indicate rust because it changes from yellow/green to blue in the presence of iron ions.

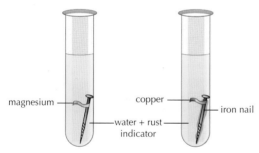

Figure 22.17 *Magnesium protects iron from rusting, but copper speeds up rusting.*

Figure 22.17 shows there is no blue colour around the iron nail connected to the magnesium. This means the iron is not rusting. Connecting the iron nail to magnesium prevents the iron rusting. This is because magnesium is higher in the electrochemical series than iron and is able to give electrons to the iron. There is a blue colour around the nail attached to copper. This shows the nail is rusting – the copper is not protecting the iron. In fact, the copper actually speeds up the rusting of iron because iron is higher than copper in the electrochemical series and so gives electrons to the copper.

Protecting a metal in this way is often called **sacrificial protection**. Magnesium can be used to protect underground

> ### 📖 Word bank
> - **Electrochemical series**
> A list of metals arranged in order of how easily the metal atoms lose electrons.

> ### 📖 Word bank
> - **Sacrificial protection**
> Protecting a metal by connecting it to a metal higher in the electrochemical series.

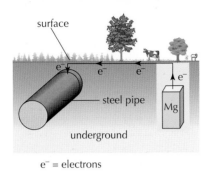

Figure 22.18 *Magnesium protects underground steel pipes from rusting.*

📖 Word bank

- **Galvanising**

Coating a metal in zinc to stop the metal from corroding.

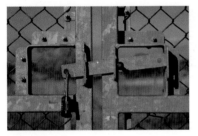

Figure 22.19 *These old galvanised steel gates show some rust spots, but the steel is still being protected by the zinc coating.*

pipelines. The magnesium is connected to the steel pipeline at intervals along its length. The magnesium itself corrodes when it gives electrons to the steel (iron) and has to be replaced regularly.

Zinc is commonly used in the sacrificial protection of steel, especially if the metal is likely to be scratched. Steel railings are often dipped in molten zinc. This is known as **galvanising**. Zinc is higher than iron in the electrochemical series. If the zinc and steel (iron) are both exposed to air and water, then the zinc will give electrons to the steel and stop it rusting. The zinc is being sacrificed to protect the steel.

At one time chromium was used to cover the steel bumpers on cars. However, if the bumper was scratched the chromium speeded up the rusting of iron. This is because iron is higher than chromium in the electrochemical series and gives electrons to the chromium.

Although large structures like oil rigs and ships can be protected from rusting by bolting bocks of zinc to their surface, it is not enough to give the protection they need. Connecting large structures to the negative terminal of a direct current (DC) supply gives the steel electrons and so protects it against rusting.

☀: Chemistry in action: The Forth Road Bridge

The Forth Road Bridge is a steel suspension bridge that links Fife and the Lothians. It was opened in 1964 and predicted to last for 120 years. However, in recent years there have been structural issues due to rusting, despite over 500 tonnes of zinc being used to coat the steel at the time it was built. A study completed in 2005 found that the main cables holding up the carriageway had lost 8–10% of their strength as a result of rusting. In 2009 a system designed to reduce the moisture in the air around the main cables was

Figure 22.20 *The Queensferry Crossing uses modern rust detection and prevention techniques.*

installed. Fears over the state of the bridge led to its closure for 3 weeks in December 2015. An additional bridge, known as the Queensferry Crossing, was completed in 2017. It is a cable-stayed bridge – this means the steel cables supporting the carriageway run straight from the supporting towers, which results in a distinctive fan-like pattern. This design means damaged cables can be replaced more easily. The rust detection and prevention measures now in place on the Forth Road Bridge will be used from the start on the Queensferry Crossing.

🔵 Activity 22.4

☺ **1. (a)** Write word and formula equations showing what happens to iron atoms when iron rusts.

   (b) Name the indicator used to show iron is rusting.

   (c) State the colour change you would see happening to rust indicator when iron rusts.

   (d) Look at Figure 22.16. Suggest why oil is poured onto the surface of the water in test tube 2.

2. Magnesium can protect iron from rusting by sacrificial protection.

   (a) State what is meant by sacrificial protection.

   (b) Give an example of a situation where sacrificial protection is used to protect iron.

3. Explain why connecting a steel structure to the negative terminal of a DC power supply helps protect it against rusting.

4. Read the Chemistry in action section 'The Forth Road Bridge' and answer the following questions.

   (a) Suggest why the steel used in the construction of the Forth Road Bridge was covered in zinc.

   (b) What effect does rusting have on steel?

   (c) Suggest why it is important to reduce the amount of moisture around the bridge cables.

Chemical cells and the electrochemical series

Learning intentions

In this section you will:
- learn how to make a chemical cell
- learn what the electrochemical series is
- learn how to use the position of metals in the electrochemical series.

National 4
Curriculum level 4
Forces electricity and waves: Electricity SCN 4-10a

When metals form ions electrons are also produced. If two metals are connected by a wire and placed in an **electrolyte** (ionic solution), the electrons move through the wire from one metal to another. The electrolyte is needed to complete the circuit. This is a **chemical cell** and is shown in Figure 22.21. A battery contains one or more chemical cells (see pages 52–53).

📖 Word bank

- **Electrolyte**
  An ionic liquid or solution.
- **Chemical cell**
  Produces electricity from a chemical reaction.

Table 22.4 *The voltages of some metals when connected to copper.*

Metal connected to copper	Voltage (V)
magnesium	2.3
zinc	1.4
iron	0.9
lead	0.5

Table 22.5 *The position of some metals in the electrochemical series.*

Metal
lithium
potassium
calcium
sodium
magnesium
aluminium
zinc
iron
nickel
tin
lead
copper
silver
gold

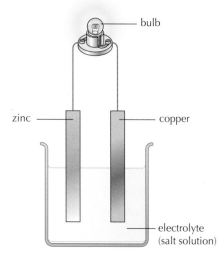

Figure 22.21 *A simple chemical cell.*

A voltmeter can be placed in the cell instead of a bulb to measure the voltage produced when different metals are used. Table 22.4 shows the voltages produced when different metals are connected to copper.

This can be used to put metals in order of how easily they give away electrons (form ions). This order of metals is known as the electrochemical series.

The further apart two metals are in the electrochemical series the higher the voltage produced in the cell will be.

Table 22.5 shows common metals and their position in the electrochemical series.

The direction of electron flow in a cell is from the metal higher in the electrochemical series to the lower metal, through the wire, as shown in Figure 22.22.

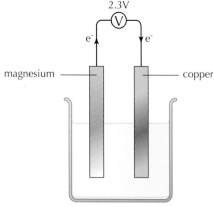

Figure 22.22 *Chemical cell showing direction of electron (e⁻) flow from magnesium to copper, through the wire.*

★ You need to know

The order of metals arranged in the reactivity series and electrochemical series are similar but not exactly the same. You do not need to memorise the exact order of the metals in either series, but you do need to be able to make predictions using both series of metals.

GO! Activity 22.5

☻ **1.** You are given a piece of zinc, a piece of copper, a voltmeter and any other chemicals and equipment you wish.

    (a) Draw a labelled diagram to show how you would make a chemical cell.

    (b) Mark on your diagram the direction of electron flow.

    (c) Suggest what the voltage might be.

    (d) Select a metal from the electrochemical series (Table 22.5) that could replace zinc to produce a higher voltage in the cell.

    (e) Suggest what will happen if one of the metals is removed from the solution.

       (You can check if you are right if you do the experiment in Activity 22.5 Question 3.)

2. Four metals (W, X, Y and Z) are connected in turn to make a chemical cell with copper. The voltages produced by each pair are shown in the table.

Metal connected to copper	Voltage (V)
W	0.6
X	1.9
Y	1.1
Z	0.3

Place the metals in the order they would appear in the electrochemical series (highest first).

3. Experiment

Your task is to plan an experiment to compare the voltage obtained when different metals are connected to copper to form an electrochemical cell.

You will need: strips of clean copper, magnesium, iron, zinc and lead, salt solution, a 250 cm³ beaker, voltmeter, two leads, two crocodile clips.

Your plan should be detailed enough so that anyone could follow it and get the same results as you.

Discuss what to do with your group and write your plan.

Discuss your plan with your teacher and ask if you can do the experiments.

You can work in a group to do the experiments, but you must record your own observations and write your own report.

⚠ Remember! You must follow the normal safety rules.

Write a report that includes:

- the aim of the experiment
- what you did (method)
- your results, in a table
- your conclusion
- one thing you could do to improve the experiment.

Question

How do your results compare with those in Table 22.4? Suggest why they are different.

Learning checklist

After reading this chapter and completing the activities, I can:

N3 L3 **N4 L4**

- state that the higher a metal is in the reactivity series the faster it will react with oxygen. **Activity 22.1 Q1(c)**   ○ ○ ○

- state that very reactive metals like potassium, sodium and calcium react quickly with water. **Activity 22.1 Q1(a)**   ○ ○ ○

- state that unreactive metals like copper, silver, gold and platinum do not react with acid. **Activity 22.1 Q1(b)**   ○ ○ ○

- use the results of experiments involving metals and oxygen, water and acid to arrange metals in order of reactivity. **Activity 22.1 Q4(a)**   ○ ○ ○

- use the reactivity series to predict how a metal will react with oxygen, water and acid. **Activity 22.1 Q4(b)**   ○ ○ ○

- state that unreactive metals like silver can be extracted from their ore by heating them. **Activity 22.1 Q1(f)**   ○ ○ ○

- state that the ores of metals high in the reactivity series, like aluminium, magnesium and sodium, have to be melted then electrolysed in order to extract the metal. **Activity 22.1 Q1(e)**   ○ ○ ○

- state that ores of metals in the middle of the reactivity series like copper and iron have to be heated with carbon in order to extract the metal. **Activity 22.1 Q1(d)**   ○ ○ ○

- state that mining for ores causes environmental and social issues. **Activity 22.2 Q2**   ○ ○ ○

- state that alloys are mixtures of metals or metals with non-metals. **Activity 22.3 Q1(a), Q2**   ○ ○ ○

- state that alloying metals changes their properties. **Activity 22.3 Q1(c)**   ○ ○ ○

- state that smart alloys can change their shape when the temperature changes. **Activity 22.3 Q5(a)**   ○ ○ ○

- give an example of a smart alloy and how it is used. **Activity 22.3 Q5(b)**   ○ ○ ○

- state that when a metal corrodes the atoms lose electrons and form ions. **Activity 22.4 Q1(a)**   ○ ○ ○

N3 L3 **N4 L4**

- state that ferroxyl indicator can be used to show rusting occurring. **Activity 22.4 Q1(b), (c)**   ◯ ◯ ◯

- state that sacrificial protection prevents iron from rusting. **Activity 22.4 Q2(a)**   ◯ ◯ ◯

- give an example of where sacrificial protection is used. **Activity 22.4 Q2(b)**   ◯ ◯ ◯

- state that attaching iron to the negative terminal of a DC supply prevents it from rusting. **Activity 22.4 Q3**   ◯ ◯ ◯

- state that a chemical cell can be made from pairs of metals connected together and placed in an electrolyte. **Activity 22.5 Q1(a)**   ◯ ◯ ◯

- state that the voltage produced between metals in a chemical cell can be used to arrange the metals in the electrochemical series. **Activity 22.5 Q2**   ◯ ◯ ◯

- state that the further apart metals are in the electrochemical series the higher the voltage produced when they are in a chemical cell. **Activity 22.5 Q1(c)**   ◯ ◯ ◯

- state that the position of a metal in the electrochemical series can be used to predict the direction of electron flow in a chemical cell. **Activity 22.5 Q1(b)**   ◯ ◯ ◯

- *present information in the form of a table.* **Activity 22.2 Q1**   ◯ ◯ ◯

- *present information in the form of a graph.* **Activity 22.1 Q6**   ◯ ◯ ◯

- *interpret information presented in a graph.* **Activity 22.3 Q3**   ◯ ◯ ◯

- *extract scientific information from a report.* **Activity 22.4 Q4**   ◯ ◯ ◯

- *make a prediction or generalisation.* **Activity 22.3 Q4, Activity 22.5 Q1(c), (e)**   ◯ ◯ ◯

- *apply my scientific knowledge in an unfamiliar situation.* **Activity 22.1 Q2, Q3, Q5, Activity 22.3 Q1(b), Q2(c), Activity 22.4 Q1(d)**   ◯ ◯ ◯

- *work with others to design and carry out an experiment.* **Activity 22.5 Q3**   ◯ ◯ ◯

23 Properties of solutions

You should already know:

- that when a solute dissolves in a solvent a solution is formed
- the factors that can affect how well and how much of a substance dissolves: temperature, particle size, stirring, amount of solute and solvent
- everyday examples of dissolving
- that when no more solute will dissolve the solution is said to be saturated
- that water is our most widely used solvent, but not our only solvent.

National 3

Curriculum level 3

Materials: Properties and uses of substances SCN 3-16b

Solubility of substances

Learning intentions

In this section you will:
- learn about the solubility of everyday chemicals
- learn about factors that affect solubility
- learn about using the property of solubility in everyday life.

Figure 23.1 *Soft drinks are solutions where water is the solvent.*

Solubility is an important property that affects our everyday life. Most people will drink a solution every day. Tea, coffee and soft drinks are all solutions where water is the solvent. Water is our most important solvent. It is known as the 'universal solvent' because it dissolves so many different substances. However, imagine going out in the rain and the jacket you are wearing starts to dissolve. This is a case of the material of your jacket being insoluble in water, which, in this case, is a useful property of the material.

Different solutes have different solubilities and the temperature of the solvent will have an effect on how much will dissolve. Table 23.1 shows the solubility of some substances.

Solubility tests can be carried out on substances (see Activity 23.1 Question 8). The results can then be put into solubility tables to make it easy for us to find them. Table 23.1 gives the solubility of some compounds. The solubility of sodium chloride is highlighted in blue in the table as an example (very soluble).

Table 23.1 *The solubility of some compounds.*

	carbonate	chloride	nitrate	sulfate
calcium	insoluble	very soluble	very soluble	soluble
copper	insoluble	very soluble	very soluble	very soluble
iron	insoluble	very soluble	very soluble	very soluble
silver	insoluble	insoluble	very soluble	soluble
sodium	very soluble	very soluble	very soluble	very soluble

Sparkling Low Calorie Soft Drink with Vegetable Extracts with Sweeteners
Ingredients: Carbonated Water, Colour (Caramel E150d), Sweeteners (Aspartame,Acesulfame K), Natural Flavourings Including Caffeine, Phosphoric Acid, Citric Acid. Contains a Source of Phenylalanine. Best before end: See side of cap or bottle neck for date. Store cool and dry. Please open by hand.
best served

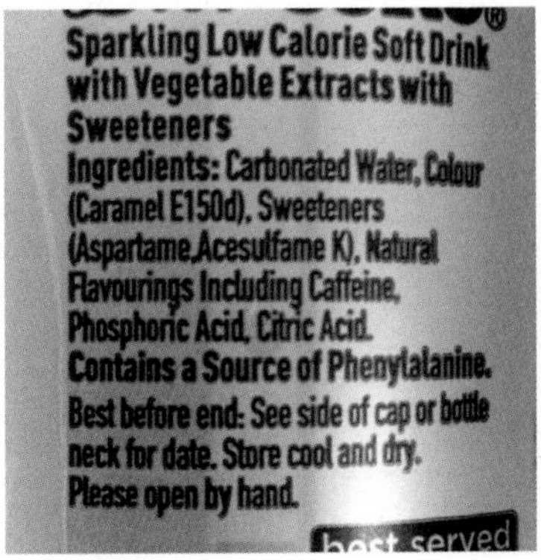

Figure 23.2 *All of the ingredients listed on this drinks label are soluble in water.*

Table 23.2 gives the actual solubility of some common substances.

Table 23.2 *The amount of some substances that will dissolve in 100 g of water.*

Substance	Mass dissolving in 100 g of water at 20 °C (in grams)
sodium chloride	36
calcium sulfate	0.2
copper sulfate	32
sugar (sucrose)	204

Factors affecting solubility

Changing the concentration

The concentration of a solution is a measure of the amount of substance (solute) dissolved in the solvent. For example, if someone likes the strong taste of coffee, they will add more coffee to the cup of water. This is increasing the concentration of the coffee solution. Concentrated orange juice being diluted by adding water is an example of decreasing the concentration of a solution (diluting).

★ You need to know

To make sure all of a substance dissolves a mixture is usually stirred. This does not increase the amount dissolving but does speed up the dissolving process.

Changing the temperature

Increasing the temperature of the solvent generally increases the amount of a substance which dissolves. Table 23.3 shows how the solubility of copper sulfate changes with temperature.

Table 23.3 *The solubility of copper sulfate increases as the temperature increases from 20 °C to 40 °C.*

Temperature of water (°C)	20	30	40
Mass of copper sulfate dissolving in 100 g of water (g)	32	38	45

Figure 23.3 *Alcohol is used as a solvent in the perfume industry.*

Types of solvent

Water is not the only important solvent. In industry, alcohol and acetone are important solvents for the likes of paints and plastics, healthcare products and cosmetics. Alcohols can dissolve chemicals that have colour, flavour and odour (smell) to make scented products like perfumes. When the perfume is sprayed on the skin the alcohol and the pleasant smelling odours **evaporate** easily from the skin into the air around the body.

Vanilla, used to flavour the likes of ice cream, does not dissolve in water. However, it does dissolve in a type of alcohol to make vanilla extract. Some vitamins, like vitamin C, dissolve best in water, while others, like vitamin E, dissolve best in oil.

The chemicals that give the vegetable asparagus its flavour are very soluble in water. However, they are not very soluble in vegetable oil. Therefore, it is recommended that asparagus be cooked in oil instead of water so that more of the flavour stays in the vegetable. Broccoli is the opposite, so cooking it in water keeps most of the flavour in the broccoli and not the water.

> ### 📖 Word bank
>
> - **Evaporate**
>
> Change state from liquid to gas.

Figure 23.4 *Asparagus loses a lot of its flavour when cooked in water.*

Finding the right solvent

Although water is our most common solvent, not all substances dissolve in water. Other liquids are widely used in industry as solvents. Table 23.4 gives some examples of common solvents and their uses.

Figure 23.5 *Some paints use solvents that are not water.*

Table 23.4 *Some common solvents and their uses.*

Solvent	What it dissolves (solute)
water	sugar, food colour, emulsion paint
alcohol (ethanol)	ballpoint pen ink, perfume, herbs and spices
acetone (propanone)	nail polish, polystyrene
white spirit	grease, oil-based paint

Some clothes have labels that say they are 'dry clean only'. Dry cleaning involves using a solvent that is not water to dissolve grease and oils, which trap the dirt. A solvent known as perc is widely used for dry cleaning in the UK. It has excellent cleaning power, is non-flammable and is gentle on most materials. However, perc is harmful to humans and the environment. The clothes are washed in the solvent at around 30 °C before being tumbled in warm air at around 60 °C to remove it. All the vapours produced are collected and cooled so that they turn back into a liquid. Dry cleaners can recycle nearly 100% of the solvent. This is important because of perc's high toxicity to humans and the harm it does to the environment.

On 14 July 2016, there was an article in the *Edinburgh Evening News* about a chemical leak at an Edinburgh hotel, thought to be caused by the solvent in a cleaning product.

Figure 23.6 *Dry-cleaning machines collect and recycle the dry-cleaning solvent.*

Top hotel evacuates guests in chemical leak scare

Hundreds of guests were forced to flee one of the Capital's top hotel after staff fell ill amid a chemical scare.

Emergency services rushed to the Sheraton Grand Hotel and Spa at around 1pm yesterday, following reports of an incident on the third floor.

A small number of staff were treated at the scene after breathing in fumes from a substance that had been used by cleaners to decontaminate one of the bedrooms.

According to hotel officials, the cleaning products had been hired from a third-party supplier for a one-off deep clean.

She added "When we got outside, one of the paramedics told us it was something to do with the chemicals."

Donna Carlson, 68, from Canada, was in her room getting ready to head out when she realised something was wrong.

She said: "My friends and I were just in our rooms when all the alarms started going off."

A spokeswoman for the Sheraton Grand Hotel and Spa insisted that the safety of guests was "paramount" and hotel officials would work with each individual affected on a case-by-case basis.

She said: "Guests were evacuated yesterday afternoon. It was due to an incident with cleaning products, which caused non-toxic fumes."

A spokeswoman for the Scottish Fire and Rescue Service said: "We attended the Sheraton Hotel yesterday afternoon following reports of a chemical incident in one of the rooms.

"On arrival, the ambulance service were there and were dealing with a number of casualties who had become unwell after inhaling an unknown chemical."

Source: *Courtney Cameron, Edinburgh Evening News, 14 July 2016*

N3 | L3 | N4 | L4 | **355**

✺ Chemistry in action: Using differences in solubility to obtain salt

Salt (sodium chloride) is commonly used to give food flavour and to melt ice on the roads in winter. It is also an important industrial chemical for making chlorine, hydrogen and sodium hydroxide. Salt is not only obtained from seawater, it also exists as underground deposits. Most of the salt produced in the UK comes from salt deposits in Cheshire in the north west of England.

One form of the salt is rock salt, a mixture of salt and materials like clay. It is obtained by mining – digging down into the earth. This involves workers going into the mines, which can be hundreds of metres below the surface. It is a difficult and dangerous job. Most of the rock salt produced is used to spread on roads to melt ice. To obtain pure salt from rock salt the soluble salt is separated from the insoluble clay by dissolving the salt in water, and filtering off the clay.

Figure 23.7 *Rock salt is a mixture of soluble salt and insoluble materials like clay.*

Most of the UK's table (white) salt used to flavour food is extracted from underground salt deposits by a process called solution mining. As the name suggests it uses the salt's high solubility to separate the salt from rock impurities. Water is pumped down a pipe and dissolves the salt but not the insoluble rocks. The salty water (brine) is forced up a central pipe and collected on the surface. The water is evaporated leaving the salt behind. This is shown in Figure 23.8. Solution mining is a continuous process and is safer than sending miners underground.

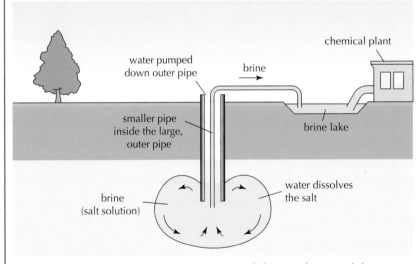

Figure 23.8 *Using solution mining to extract salt from underground deposits.*

Environmental issues

It is claimed that extracting underground salt produces little or no waste products. However, the results of mining for salt can cause the ground above the mine to collapse. This is called subsidence. This happens because the mine can't support the weight of the ground above it. Huge sink holes can suddenly appear. If there are buildings and roads above the abandoned mines then they risk collapsing, too.

Figure 23.9 *Sinkholes above old mineworks can appear without warning.*

? Did you know ...?

Indigo dye gives denim jeans their blue colour. Indigo is insoluble in water so keeps its colour. However, indigo does not stick well to the cotton fibres and so comes off the denim as it gets worn. This is one of the reasons there are often warnings on jeans telling you to avoid sitting on light-coloured chairs. Your skin can also turn blue, especially when the jeans are new.

Figure 23.10 *Indigo dye used to give some jeans their blue colour is insoluble in water.*

? Did you know ...?

The Wieliczka salt mines in Poland produced table salt from the thirteenth century until it stopped production in 2007. The mine is over 300 metres deep and over 178 miles long! It is now a tourist attraction with over one million visitors a year. The mine's attractions include dozens of statues and four chapels carved out of the rock salt by the miners.

Figure 23.11 *One of the underground chapels carved out of rock salt in the Wieliczka salt mine in Poland.*

GO! Activity 23.1

☻ **1.** Water is the basis for making all soft drinks. Some of the ingredients in a soft drink are: carbon dioxide, sugar, citric acid and artificial colouring.

    (a) Suggest **one** property that all of the ingredients have in common.

    (b) Complete the sentence by choosing the correct word highlighted in bold:

        In the manufacture of soft drinks water is the **solute / solution / solvent**.

2. An industrial chemist found that the new paint he was developing didn't dissolve in water. Suggest two other liquids he could have tried.

3. Angela made a pot of coffee for her friends. She used powdered instant coffee.

    (a) Sarah said she preferred the coffee to taste stronger (more concentrated).

        Suggest what Sarah could do to increase the concentration of her coffee.

    (b) John said he preferred coffee that didn't taste so strong (more dilute).

        Suggest what John could do to dilute his coffee.

    (c) Suggest what effect it would have on the concentration of the coffee if Angela used the same mass of coffee granules instead of powder to make the pot of coffee.

4. (a) A student predicted that exactly the same masses of copper sulfate and sodium chloride would dissolve in 100 cm³ of water.

        Is he correct?

    (b) Use the information in Table 23.2 to find the mass of each chemical in part **(a)** that would dissolve at 20 °C.

    (c) What could you do to the water to make it dissolve more of the chemicals?

(continued)

5. The graph shows how soluble sodium nitrate and potassium nitrate are in 100 cm³ of water at different temperatures.

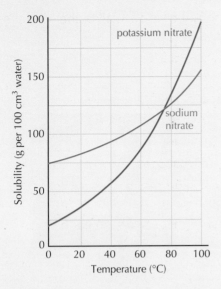

(a) Which of the chemicals is more soluble at 40 °C?

(b) Write a general statement describing the effect of temperature on the solubility of potassium nitrate.

(c) (i) If 50 g of potassium nitrate was added to 100 cm³ of water at 20 °C, would it all dissolve?

   (ii) Explain your answer to part **(c)(i)**.

(d) (i) State what is meant by a saturated solution.

   (ii) From the graph, how many grams of sodium nitrate would you need to add to 100 cm³ of water at 50 °C to make a saturated solution?

6. The table shows how the solubility of aluminium nitrate changes with temperature.

Temperature (°C)	Solubility (g/100 cm³ water)
20	74
40	89
60	106
80	132
100	

Predict the solubility of aluminium nitrate at 100 °C.

7. (a) Describe how table salt is obtained from underground deposits by solution mining.

(b) Explain why it is possible to obtain pure salt this way even though underground the salt is mixed with clay and other materials.

8. Experiment

Testing solubility

Your task is to plan how to find out if one solvent could dissolve all of the substances tested.

Substances: salt, sugar, bicarbonate of soda, plastic, wax, sand.

Solvents: water, alcohol, acetone, vegetable oil.

You can use any of the usual laboratory equipment. Your plan should be detailed enough so that anyone could follow it and get the same results as you.

Discuss what to do with your group and write your plan.

 Discuss your plan with your teacher and ask if you can do the experiments.

You can work in a group to do the experiments, but you must record your own observations and write your own report.

 Remember! You must follow the normal safety rules. In addition, do not have any flames near the solvents.

Write a report that includes:

- the aim of the experiment
- what you did (method)
- how you knew if the substance had dissolved
- your results, in a table
- your conclusion
- one thing you could do to improve the experiment.

9. Using your results from the experiment in Question 8, answer the following questions.

(a) Which of the solvents would you not keep in a plastic bottle?

(b) Which solvent could you use to help remove candle wax from a table?

(c) Suggest why some outdoor jackets are waxed.

(d) Suggest why there should be no flames near the solvents.

10. Copy and complete the table to show which solvent dissolves the substance shown.

Substance	Solvent
sugar	
perfume	
nail polish	
oil-based paints	

(continued)

11. (a) Suggest why some graffiti can't be removed from buildings by scrubbing it with water.

   (b) A painter was going to clean his paint brushes with water, but his co-worker said he should use white spirit for paint.

   Explain why both painters might be correct.

   (c) A car cleaner was going to try to remove some tar spots from the bodywork of a car using white spirit.

   Suggest why this might not be a good idea.

   > **🔍 Hint**
   >
   > You may wish to use the information in Table 23.4 to help you.

12. (a) Dry-cleaning solvent is toxic to humans and harmful to the environment, so it is important that it is not released into the atmosphere.

   Suggest another reason for recycling dry-cleaning solvent.

   (b) Suggest why it is important that dry-cleaning solvent is non-flammable.

   (c) Read the article about a chemical leak at an Edinburgh hotel on page 355.

   (i) How were the fumes described by the spokeswoman for the hotel?

   (ii) What did her description mean?

Learning checklist

After reading this chapter and completing the activities, I can:

N3 **L3** N4 L4

* state that solubility is an important chemical property for ingredients in drinks to have. **Activity 23.1 Q1(a)**   ○ ○ ○

* state that water is the solvent used to make drinks. **Activity 23.1 Q1(b)**   ○ ○ ○

* state that alcohol and acetone are two solvents used in industry. **Activity 23.1 Q2**   ○ ○ ○

* state that equal volumes of the same solvent will not dissolve the same mass of different solutes. **Activity 23.1 Q4(a)**   ○ ○ ○

* state that the concentration of a solution can be increased by dissolving more solute in the solvent. **Activity 23.1 Q3(a)**   ○ ○ ○

* state that the concentration of a solution can be decreased by adding more solvent (dilution). **Activity 23.1 Q3(b)**   ○ ○ ○

N3 L3 N4 L4

- state that increasing the temperature of a solvent generally increases the amount of solute it will dissolve. **Activity 23.1 Q4(c)** ○ ○ ○

- state that a solution in which no more solute can dissolve is described as a saturated solution. **Activity 23.1 Q5(d)** ○ ○ ○

- describe how table (white) salt is obtained from underground salt deposits. **Activity 23.1 Q7(a)** ○ ○ ○

- explain why it is possible to obtain pure salt by solution mining. **Activity 23.1 Q7(b)** ○ ○ ○

- give examples of solvents used in everyday situations: sugar dissolves in water; perfumes are dissolved in alcohol; nail polish can be removed with acetone; white spirit can dissolve oil-based paint. **Activity 23.1 Q10** ○ ○ ○

- state that not all substances dissolve in water. **Activity 23.1 Q11(a)** ○ ○ ○

- *extract scientific information from a report.* **Activity 23.1 Q12(c)** ○ ○ ○

- *apply my scientific knowledge in an unfamiliar situation.* **Activity 23.1 Q3(a), (b), Q9, Q11, Q12(a), (b)** ○ ○ ○

- *obtain information from tables of data.* **Activity 23.1 Q6** ○ ○ ○

- *interpret information presented in a graph.* **Activity 23.1 Q5** ○ ○ ○

- *make predictions and generalisations.* **Activity 23.1 Q3(c), Q5(b), Q6** ○ ○ ○

- *work with others to plan and carry out a practical activity.* **Activity 23.1 Q8** ○ ○ ○

24 Fertilisers

This chapter includes coverage of:
N4 Fertilisers

You should already know:
• that fertilisers help plants grow • about the importance of fertilisers in world food production.

National 4

Growing healthy plants

Learning intention
In this section you will: • learn about the essential elements needed by plants.

Figure 24.1 *Plants are essential food for animals, including humans.*

📖 Word bank

• **Nutrients**

Elements needed by plants for healthy growth.

By the end of 2016 the world population, according to the United Nations (UN), was around 7.5 billion. By 2066 – in your lifetime – it is estimated to be over 10 billion. One of the consequences of this is the increased demand for food. All of our food comes from plants, either directly, by eating the plant, or indirectly, by eating animals that have eaten plants.

Plants need 'food' for healthy growth. This comes in the form of certain elements, known as the plant **nutrients**. Different plants need different nutrients in different proportions. Table 24.1 lists the 16 elements needed by plants for healthy growth.

Table 24.1 *The elements that plants need for healthy growth.*

Main elements	Trace elements
nitrogen	iron
phosphorus	manganese
potassium	copper
carbon	zinc
hydrogen	molybdenum
oxygen	boron
sulfur	chlorine
calcium	
magnesium	

Carbon, hydrogen and oxygen are used by the plant to make carbohydrate by photosynthesis (see Chapter 10). Three other elements vital for healthy plant growth are nitrogen (N), phosphorus (P) and potassium (K). Nutrients dissolve in water and are taken in through the roots of the plant.

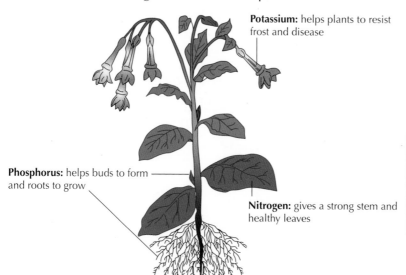

Potassium: helps plants to resist frost and disease

Phosphorus: helps buds to form and roots to grow

Nitrogen: gives a strong stem and healthy leaves

Figure 24.2 *The essential elements N, P and K are needed for healthy plant growth.*

Plants use up such large amounts of these elements that they need to be replaced on a regular basis. Anything that replaces nutrients taken out of the soil is known as a **fertiliser**.

📖 Word bank

- **Fertiliser**
  A substance that replaces nutrients removed from the soil.

⚗️ Make the link

For more about fertilisers and their role in growing healthy plants, see *S1–N4 Biology Student Book*, Chapter 29.

🟢 Activity 24.1

☺ **1. (a)** State the **three** essential elements needed for healthy plant growth.

  (b) State why each of the elements in part **(a)** is needed.

2. The table shows how the world population has changed since 1804. Use the information in the table to answer the questions that follow.

World population (billions)	1	2	3	4	5	6	7	8	9	10
Year	1804	1927	1960	1974	1987	1999	2011	(x)	2038	2056

  (a) How does the rate of increase in the world population in the second half of the twentieth century compare with the first half?

  (b) Predict the year (x) in which the world population will reach 8 billion, assuming the rate of increase is the same as it has been between 1999 and 2011.

  (c) State the effect of the change in world population on food production requirements.

  (d) What does the estimated number of years it is predicted to take for the world population to rise from 9 billion to 10 billion suggest about the rate of change of the world population in the future?

Natural and synthetic fertilisers

Learning intentions

In this section you will:

- learn how fertilisers supply essential elements to plants
- learn about the role of chemists in ensuring our food supply
- learn about advantages and disadvantages of using different types of fertilisers
- learn how to calculate the percentage of an element in a fertiliser.

📖 Word bank

- **Compost**

Plant material that has rotted, leaving nutrients behind.

- **Manure**

Waste from cattle and horses that is rich in nutrients.

- **Natural fertilisers**

Compost and manure that replace nutrients to the soil.

⭐ You need to know

Some farmers still use compost and manure as fertiliser. Many households have compost bins. Waste fruit and vegetables are put in a bin where they rot over many months and are then used as a fertiliser in the garden.

Natural fertilisers

In nature, when plants and animals die the elements they contain go into the soil. These elements are taken up by new plants as they grow. Nature provides its own fertiliser for plants to grow the following year.

If plants are taken out of the ground before they die, then the nutrients are not being replaced. This is the case when plants are produced and removed from the ground for us to eat. This is what happens in farming.

A hundred years ago farmers replaced the essential nutrients in the soil by adding compost and manure. **Compost** is plant material that has rotted, leaving nutrients behind. **Manure** is waste from cattle and horses, and is rich in nutrients. Compost and manure are examples of **natural fertilisers**. Natural fertiliser also adds structure to the soil, helping it to soak up water and also to keep the nutrients in the soil. It is a renewable source of nutrients and environmentally friendly.

Figure 24.3 *Leaves that fall from trees in autumn rot and act as fertiliser for plants to grow again in spring.*

Figure 24.4 *Animal manure – a natural fertiliser – ready to spread on a field.*

Figure 24.5 *Many people compost waste food for use as a fertiliser in the garden.*

However, natural fertilisers release the nutrients into the soil very slowly, and because they need warmth and moisture in order to break down, they are not always available all year round. The constantly increasing world population means there is not enough natural fertiliser for the amount of plant material needed to feed the people and animals on the planet.

Synthetic fertilisers

Synthetic fertilisers are so called because various chemical treatments are needed for their manufacture. Chemists have had a key role in the development of synthetic fertilisers. The manufacture of synthetic fertilisers was a result of the large-scale production of ammonia early in the twentieth century. The use of commercial fertilisers has increased steadily in the last 50 years, in line with the increase in population. It has been estimated that almost half the people on the planet are currently fed as a result of the use of synthetic nitrogen fertiliser.

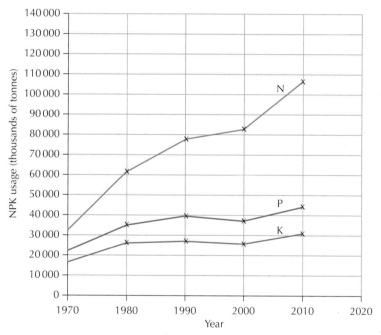

Figure 24.6 *Graph of world nitrogen, phosphorus and potassium (NPK) usage.*

> ★ **You need to know**
>
> The main nutrients supplied by synthetic fertilisers are nitrogen (N), phosphorus (P) and potassium (K).

Synthetic fertilisers give a supply of nutrients almost immediately and can be made in the quantities needed throughout the world. Farmers can buy fertilisers that have the mixture of nutrients needed for the particular crop they are growing. Some disadvantages are that most are very soluble and are easily washed out of the soil (see 'Environmental impact of fertilisers', page 369) and don't improve the structure of the soil. They are also made from non-renewable resources.

Figure 24.7 *Soybean roots with nodules containing bacteria that can 'fix' nitrogen.*

📖 Word bank

- **Nitrogen fixation**

The process by which some plants, which have tiny nodules on their roots containing a bacterium, can convert nitrogen in the air into a form plants can use.

- **Ammonia (NH_3)**

A compound high in nitrogen which can be used as a fertiliser.

- **Ammonium nitrate (NH_4NO_3)**

The most common nitrogen fertiliser.

★ You need to know

A solution of ammonia is sometimes called ammonium hydroxide (NH_4OH). Using this name and formula makes it easier to recognise that ammonia is a base.

❓ Did you know ...?

Without synthetic fertilisers it is estimated that about one-third of the food produced now could not be produced. This could result in billions of people not having enough food.

Nitrogen

The air is almost 80% nitrogen. Unfortunately, most plants can't take nitrogen directly from the air. Some, like beans, peas and clover, can. They have tiny nodules on their roots which contain a bacterium that can convert nitrogen in the air into a form plants can use. This is known as **nitrogen fixation**.

At the start of the twentieth century chemists came up with a way of reacting the unlimited supply of nitrogen in the air with hydrogen to make **ammonia** (NH_3). Ammonia itself can not only be used as a fertiliser, but it can also be used to make other fertilisers.

The most common nitrogen fertiliser used in the UK is **ammonium nitrate** (NH_4NO_3). It has a high percentage of nitrogen and is very soluble. This means it can easily be taken up by a plant through its roots. High solubility can be a disadvantage, depending on weather conditions. Heavy rain can cause nitrate to be washed out of the soil and so be lost to the plants. This can also cause environmental problems (see page 369).

❓ Did you know ...?

Ammonium nitrate can be mixed with other chemicals to make explosives. For this reason ammonium nitrate has been banned for use as a fertiliser in some parts of the world.

Ammonium nitrate is produced by reacting ammonia (NH_3) with nitric acid (HNO_3).

Ammonia is a base, so when it reacts with nitric acid a neutralisation reaction (see page 83) takes place. Ammonium nitrate is the salt formed in the reaction. Ammonium nitrate is sold under the name Nitram®.

ammonia	+	nitric acid	→	ammonium nitrate
NH_3	+	HNO_3	→	NH_4NO_3
base	+	acid	→	salt

Phosphates

Phosphate fertilisers are made by neutralising phosphoric acid. If ammonia is used, a mixture of phosphates is formed. Ammonium phosphate is one of them.

ammonia	+	phosphoric acid	→	ammonium phosphate
base	+	acid	→	salt

Ammonium phosphate is used when both nitrogen and phosphorus are needed.

Potassium

Potash is the most widely used potassium fertiliser. It is mainly highly soluble potassium chloride (KCl). It has been formed naturally deep underground and has to be mined. Potash is mined in the UK in Boulby, Yorkshire. However, in 2012 the owners of the company started mining a mineral known as polyhalite. It is a naturally occurring mixture of potassium, calcium and magnesium sulfates. It is sold as polysulfate. It has the advantage over potash in that it contains more of the elements needed by plants than potash does. It is being marketed as the multi-nutrient fertiliser of the future.

NPK fertilisers

The most common type of synthetic fertilisers used contain two or more nutrients. Mixtures of nitrogen, phosphorus and potassium compounds are known as **NPK fertilisers**. The bag usually has three numbers, which show the percentage of each nutrient in the mixture. A 20-14-14 NPK fertiliser, for example, contains 20% nitrogen, 14% phosphorus and 14% potassium by weight.

Calculating percentage composition

The percentage composition of an element in a compound, based on the mass of the element, can be calculated from relative atomic mass (RAM) of the element and formula mass of the compound (see page 63). The relationship used is:

Percentage composition of element $= \dfrac{\text{RAM} \times \text{number of atoms in formula}}{\text{formula mass}} \times 100$

The percentage of nitrogen in ammonium nitrate (NH_4NO_3) can be calculated using the relationship.

RAM = 14; no. of atoms of N = 2;

formula mass = 14 + 4 + 14 + 48 = 80

Percentage composition of element $= \dfrac{\text{RAM} \times \text{number of atoms in formula}}{\text{formula mass}} \times 100$

$$= \frac{14 \times 2}{80} \times 100$$

Percentage composition of nitrogen = 35%

> **📖 Word bank**
>
> • **Potash**
>
> A mixture of soluble potassium compounds, mainly potassium chloride.

Figure 24.8 *NPK fertilisers contain a mixture of the essential nutrients.*

> **📖 Word bank**
>
> • **NPK fertilisers**
>
> Fertilisers that contain mixtures of nitrogen, phosphorus and potassium compounds in different proportions.

GO! Activity 24.2

☻ 1. (a) Explain why soil has to be fertilised.

 (b) Plants in the wild can be said to be self-fertilising.
 Describe how this happens.

2. (a) Explain why farmers have to add fertilisers to the soil.

 (b) Give **two** examples of natural fertilisers used by farmers.

 (c) State why there is a need for synthetic fertilisers.

3. Many synthetic fertilisers are made by reacting an acid and a base.

 (a) Name this kind of reaction.

 (b) Potassium nitrate (KNO_3) is a compound that can be used as a fertiliser.

 (i) Name the acid used to make potassium nitrate.

 (ii) State **two** properties a fertiliser should have to be considered a good fertiliser.

 (c) (i) The formula mass of potassium nitrate (KNO_3) is 101.
 Calculate the percentage of nitrogen in potassium nitrate, to the
 nearest whole number.
 (Relative atomic masses can be found on page 63.)

 (ii) The percentage of nitrogen present in KNO_3 is less than in ammonium nitrate.
 Suggest why potassium nitrate is still used as a fertiliser instead of ammonium
 nitrate.

 (iii) Some fertilisers are described as NPK.
 Explain what is meant by NPK.

4. Look at the graph of world fertiliser use (Figure 24.6).

 (a) State the general trend for nitrogen, phosphorus and potassium use worldwide,
 between 1970 and 2010.

 (b) Assuming the trend between 2000 and 2010 continues at the same rate, estimate
 how many thousands of tonnes of each nutrient will be used in 2020.

5. Nearly 80% of the air is nitrogen.
 Explain why only certain plants can use nitrogen directly from the air.

6. Summarise the benefits and problems associated with using natural and synthetic
 fertilisers by doing one of the following:

 • write a report

 • make up a mind map

 • prepare a presentation.

Environmental impact of fertilisers

Learning intentions

In this section you will:

- learn about the problems caused by overuse of fertilisers
- learn about feeding the world in the future.

Using too much fertiliser not only costs farmers money, but can also damage plants and pollute rivers and lochs. Nitrate fertilisers and some phosphates are very soluble. If too much is spread on the surface of the soil or at a time when there is heavy rain, much of the fertiliser can dissolve in rain water and be washed out of the soil into rivers and lochs. This is known as **leaching**. This contributes to the formation of **algal blooms**. They take oxygen out of the water, which kills off marine life. The presence of algal blooms can be recognised because the water turns green or blue-green.

📖 Word bank

- **Leaching**
In agriculture, the loss of water-soluble plant nutrients from the soil.
- **Algal bloom**
A rapid increase in the growth of algae in water.

Figure 24.9 *Algal bloom caused by fertiliser being washed into a lake.*

Algal blooms can be so bad that they create 'dead zones', where sea life no longer exists. Once formed, an algal bloom is very difficult to get rid of. One exception is the Black Sea, which has recovered since the early 1990s. Less fertiliser is being used, so less is being run off into the sea, allowing oxygen levels to rise.

Algal blooms have become common in some parts of the world.

- In 2015 in Rio de Janeiro, Brazil, an estimated 50 tonnes of dead fish were removed from the lagoon where water events at the 2016 Olympics were planned to take place.

- In 2016, 23 million salmon that were being farmed in Chile died from a toxic algal bloom. To get rid of the dead fish, the ones fit for consumption were made into fishmeal and the rest were dumped 60 miles offshore, to avoid risks to human health.

Chemistry in action: Feeding the world in the future – The Sahara Forest Project

The Sahara Forest Project is a scheme that aims to make use of desert land, seawater and the sun to produce fresh water, food and renewable energy, as well as re-vegetating areas of uninhabited desert. A pilot plant was set up in Qatar in 2012. Qatar, in the Middle East, only has about 8 cm of rainfall a year and has huge desert areas. It imports over 90% of its food. The project involved growing vegetables in greenhouses, which were cooled using seawater. Solar power was used as a source of renewable energy. Fresh drinking water was obtained from seawater and used to water the plants. The project is now complete and was so successful that it has been set up in Aqaba in Jordan, and in Tunisia, North Africa. It is hoped that in the future large areas of desert land in many countries will be using this method to grow food.

Figure 24.10 *The greenhouses used in the Sahara Forest Project use seawater to cool the air and to produce fresh water.*

↻ Keep up to date!

You can keep up to date with the Sahara Forest Project in Jordan and Tunisia by searching online.

GO! Activity 24.3

☺ **1.** Copy and complete the summary below. Use the following words to help you.

algal blooms   dissolve   fertilisers   leaching   nitrates   oxygen   plants   pollute

The overuse of **(a)** _____ not only costs farmers money, it can damage the **(b)** _____ and **(c)** _____ rivers and lochs. **(d)** _____ in fertilisers and some phosphates are very soluble. If too much is spread on the surface of the soil or at a time when there is heavy rain, much of the fertiliser can **(e)** _____ in rain water and end up in rivers and lochs. This is known as **(f)** _____. This contributes to the formation of **(g)** _____ _____. They take **(h)** _____ out of the water, which kills off marine life.

2. Write a short report (50–70 words) about what is being done to try to grow food in desert areas.

You may wish to use the internet to get information in addition to what's covered in the Chemistry in action section 'Feeding the world in the future – The Sahara Forest Project'.

Learning checklist

After reading this chapter and completing the activities, I can:

N3 | L3 | **N4** | L4

- state that nitrogen (N), phosphorus (P) and potassium (K) are nutrients essential for healthy plant growth. **Activity 24.1 Q1(a)** ○ ○ ○

- state what each of the elements N, P and K do in a plant. **Activity 24.1 Q1(b)** ○ ○ ○

- state that plants take nutrients out of the soil. **Activity 24.2 Q1(a)** ○ ○ ○

- state that in nature nutrients are replaced when a plant dies and rots. **Activity 24.2 Q1(b)** ○ ○ ○

- state that natural fertilisers include compost and manure. **Activity 24.2 Q2(b)** ○ ○ ○

- state that farmers have to replace nutrients using fertilisers. **Activity 24.2 Q2(a)** ○ ○ ○

- state that synthetic fertilisers are needed to meet the increasing world demand for food. **Activity 24.1 Q2(c)** ○ ○ ○

- state that many synthetic fertilisers can be made by neutralisation reactions. **Activity 24.2 Q3(a)** ○ ○ ○

- state that a good fertiliser has to contain the essential elements and be soluble in water. **Activity 24.2 Q3(b)(ii)** ○ ○ ○

- state that NPK fertilisers are mixtures with different proportions of nitrogen, phosphorus and potassium. **Activity 24.2 Q3(c)(iii)** ○ ○ ○

- explain why only some plants can fix nitrogen from the air. **Activity 24.2 Q5** ○ ○ ○

- give a balanced view of reasons for using natural and synthetic fertilisers. **Activity 24.2 Q6** ○ ○ ○

- calculate the percentage composition of an element in a fertiliser. **Activity 24.2 Q3(c)(i)** ○ ○ ○

N3 L3 **N4 L4**

- state that using too much fertiliser can cause environmental problems such as algal bloom, as a result of leaching of nitrates and phosphates into rivers and lochs. **Activity 24.3 Q1** ○ ○ ○

- give an example of what's being done to grow food in desert areas. **Activity 24.3 Q2** ○ ○ ○

- *interpret information presented in a graph.* **Activity 24.2 Q4(a)** ○ ○ ○

- *apply my scientific knowledge in an unfamiliar situation.* **Activity 24.2 Q3(b)(i), Q3(c)(ii)** ○ ○ ○

- *obtain information from tables of data.* **Activity 24.1 Q2(a)** ○ ○ ○

- *make predictions and generalisations.* **Activity 24.1 Q2(b), (d), Activity 24.2 Q4** ○ ○ ○

25 Nuclear chemistry

This chapter includes coverage of:

N4 Nuclear chemistry

You should already know:

- that elements are arranged in the periodic table in order of atomic number
- that the atomic number is the same as the number of protons in an atom.

Formation of elements

National 4

Learning intention

In this section you will:

- learn how elements were formed.

Our universe is thought to have formed nearly 14 billion years ago as a result of a violent explosion of a very small amount of matter at an extremely high temperature. This idea is the basis of the Big Bang theory. The Big Bang also produced all of the hydrogen and most of the helium we have in the universe. These are our lightest elements. As the clouds of cosmic dust and gases from the Big Bang cooled, stars formed and these then grouped together to form galaxies.

For most of their life, stars **fuse** (join) hydrogen atoms to form helium. This is where most of a star's energy comes from. This is what is happening in our Sun, a very young star.

📖 Word bank

- **Nuclear fusion**

A reaction where two or more atoms join to form one or more new atoms.

	nuclear fusion		
two hydrogen atoms	$\rightarrow$	one helium atom
H + H	$\rightarrow$	He
Atomic number: 1 + 1		2

Our Sun is by far the brightest object in the sky. It was formed over 4 billion years ago and it has been estimated it will exist for another 5 billion years. The heat and light that come from the Sun are produced 93 million miles away. It takes over 8 minutes for the light from the Sun to reach Earth.

Over 98% of the Sun is made up of hydrogen and helium, with the other elements being mainly oxygen, carbon, neon and iron.

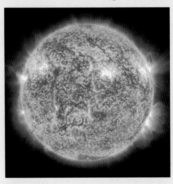

Figure 25.1 *Most of the energy from the Sun is produced by the fusion of hydrogen atoms to form helium.*

When a star's core runs out of hydrogen, the star begins to die. At this point helium atoms fuse to form carbon.

	three helium atoms			→	one carbon atom
	He +	He +	He	→	C
Atomic number:	2 +	2 +	2		6

In big stars, heavier elements, from oxygen up to and including iron, are formed.

	one carbon atom	+	one helium atom	→	one oxygen atom
	C	+	He	→	O
Atomic number:	6	+	2	→	8

During a supernova – the explosion of most of the material in a star – the heaviest elements such as uranium and gold are formed.

Earth was formed nearly 4.5 billion years ago, when clouds of dust and gas containing elements formed in stars came together.

Figure 25.2 *The Milky Way — the galaxy that contains our solar system. A supernova can be seen near the middle of the picture.*

GO! Activity 25.1

☻ 1. **(a)** State how hydrogen and most of the helium in the universe was formed.

   (b) State where most of a star's energy comes from.

2. Elements up to and including iron are formed in stars, where huge amounts of heat cause smaller atoms to join to form larger atoms. Use the periodic table on pages 34–35 to help you complete the examples in the table below (atomic number in brackets). The first one has been done for you.

Atoms joining	Atoms formed
helium (2) and lithium (3)	boron (5)
nitrogen (7) and beryllium (4)	**(a)** _____ (_)
hydrogen (1) and aluminium (13)	**(b)** _____ (_)
boron (5) and magnesium (12)	**(c)** _____ (_)
carbon (6) and **(d)** _____ (_)	potassium (19)

Background radiation

National 4

Learning intentions

In this section you will:
- learn what radiation is
- learn the sources of background radiation
- learn how background radiation is detected.

The nuclei of very big atoms are unstable. They try to become more stable by emitting (throwing out) particles and energy. This is known as **radiation**. Radiation can damage animal cells and make them cancerous.

Radiation surrounds us all the time. This is known as **background radiation**. Most of the background radiation comes from natural sources, some from outer space and some from Earth. Rocks, soil, plants and the air all contain radioactive material. Most of the background radiation we produce is artificial and comes from medical procedures, such as taking X-rays.

📖 Word bank

- **Radiation**
  Particles or energy emitted (thrown out) by unstable atoms.
- **Background radiation**
  Radiation that surrounds us all the time.

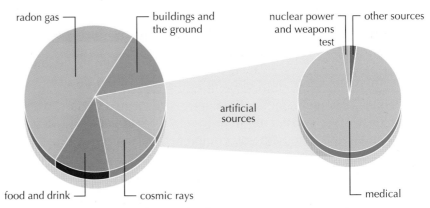

Figure 25.3 *Sources of background radiation.*

Natural sources of background radiation

For most people, natural sources contribute the most to their background radiation dose.

Natural sources of background radiation include:

* rocks and soil: some rocks are radioactive and also produce radioactive radon gas

* living things: plants absorb radioactive materials from the soil and these pass up the food chain when they are eaten by animals, including humans

* cosmic rays: radiation that reaches Earth from space.

Rocks and soil

Elements like uranium, which is found in rocks and soil in the Earth's crust, emit particles and energy which contribute to background radiation. In addition uranium and thorium produce radioactive **radon** gas which we breathe in. Radon is often the single largest contributor to an individual's background radiation dose, and is the most variable from location to location.

Aberdeen is built on radioactive rock, and is often called a 'radioactive city'. However, the radioactive radon gas is trapped in the rock and so is harmless. The most radioactive city in the world is thought to be Ramsar in Iran. This is mainly due to using local naturally radioactive limestone as a building material. On average, the residents receive six times the average annual dose worldwide. In one house the residents were receiving over 80 times the average annual dose of radiation worldwide because of radioactive radon trapped in the house.

Word bank

* **Radon**
A radioactive naturally occurring noble (group 0) gas.

Figure 25.4 *Some of the buildings in Aberdeen are built of granite, which contributes to background radiation.*

Living things

The mountains in Greenland have lots of radioactive uranium, which companies want to mine. Uranium is needed for the nuclear power industry. Many residents are concerned about the possible effects of radioactive dust released into the atmosphere. Local farmers are worried about their animals eating grass contaminated with the dust and the possibility of radioactivity being passed on when the animal meat and milk are consumed.

In April 1986 one of the reactors at the Chernobyl nuclear power station in Ukraine exploded. It was the worst nuclear accident in history. Radioactive material released into the atmosphere was blown over Northern Europe and brought down to the ground by rain. In Scotland, one of the most affected areas was Dumfries and Galloway, in the south-west of the country. Fruit, vegetables and water supplies in the area were monitored to make sure they were not affected by the radioactive fallout. Restrictions were brought in on the movement and slaughter of sheep. The monitoring of radioactivity in foods lasted until June 2010.

Figure 25.5 *A giant air-tight steel cover was placed over the Chernobyl nuclear plant in September 2016. The cover is designed to stop radioactive material getting into the atmosphere.*

Cosmic rays

Earth and all living things on it are constantly bombarded by radiation from outer space. Earth's atmosphere reduces the amount of cosmic radiation reaching its surface. As a result, the further you are away from the surface of Earth, the greater the exposure. For example, people who live in the US city of Denver live at 1650 metres above sea level. They are estimated to get twice as much daily exposure to radioactivity than someone living at sea level. Airline crew and frequent flyers have even greater exposure. Similarly, astronauts on the International Space Station have greater exposure to cosmic radiation than humans on the surface of Earth.

Figure 25.6 *Frequent flying increases our exposure to background radiation.*

Artificial sources of background radiation

A small percentage of the radiation in the environment is caused by human activities in science and industry. Most of the artificial background radiation comes from radioactive medical devices, especially X-rays. The use and testing of nuclear weapons has resulted in an increase in global levels of background radiation. Nuclear power plants also contribute to environmental radiation, as does the recycling of nuclear fuels. Small sources, such as radioluminescent paints, cause very little background radiation. Radioluminescent paints glow in the dark because they contain a chemical that gives off light when in contact with radiation. It was used to coat the hands

Figure 25.7 *Torness nuclear power station outside Edinburgh.*

⟳ Keep up to date!

You can keep up to date with testing for radiation in the area around Torness nuclear power station by checking the FSA or other websites.

Figure 25.8 *An environmental scientist using a Geiger counter to test the radiation emitted by an apple.*

📖 Word bank

- **Geiger–Muller tube**
A machine used to detect radiation.

of watches and aircraft instruments so they could be seen in the dark. Because of health concerns linked to radioactivity radioluminescent paints are seldom used in watches or clocks nowadays.

People who work with radioactive materials are more at risk of the effects of exposure to radioactivity than other members of the public. Radioactive emissions from Torness nuclear power station near Dunbar, East Lothian, are carefully monitored. The Food Standards Authority (FSA) regularly tests the radioactivity levels in the local environment.

In 2014 the results of testing a local person exposed to the air and water near the power station, and who regularly ate locally grown food and fish, showed very low radiation levels. The results were similar when the local environment was tested.

Detecting radiation

Human senses cannot detect radiation, so we need equipment to do this. The **Geiger–Muller tube** – often called a Geiger counter – detects radiation. Each time it absorbs radiation, it transmits an electrical pulse to a counting machine. This makes a clicking sound or displays the count rate. The greater the frequency of clicks, or the higher the count rate, the more radiation the Geiger–Muller tube is absorbing.

Photographic film goes darker when it absorbs radiation, just like it does when it absorbs visible light. The more radiation the film absorbs, the darker it is when it is developed. People who work with radiation wear film badges, which are checked regularly to monitor the levels of radiation absorbed. A more accurate device for monitoring exposure to radiation, called a thermoluminescent dosimeter (TLD), uses a crystal that gives out light when exposed to radiation.

❓ Did you know ...?

Bananas contain naturally occurring radioactive potassium. The units of radioactivity measurement are quite complicated. An informal unit, called banana equivalent dose (BED), is sometimes used to give consumers an idea of how much radioactivity is in the food we eat. One BED is approximately 1% of our daily exposure to radiation. Radioactivity from food doesn't build up in the body as it is excreted.

Figure 25.9 *Bananas contain naturally occurring radioactive potassium.*

☀: Chemistry in action: Using radiation in the home!

Smoke detectors are now seen as an essential safety device that should be in all homes. One type uses radioactive americium-241 to detect smoke. Radioactive particles emitted by the americium-241 atoms pass between two charged metal plates. This causes the gas molecules in the air to split into positive and negative ions. The ions are attracted to the oppositely charged metal plate causing an electrical current to flow.

Figure 25.10 *A smoke detector being fitted to the ceiling. The alarm part can be clearly seen.*

When smoke enters between the plates, some of the radioactive particles are absorbed, causing less ionisation of the molecules in the air to take place. This means a smaller-than-normal current flows, so the alarm sounds.

Although this type of smoke alarm contributes to background radiation it is a very small amount and is not considered a health hazard. This is because small amounts of the radioactive substance are used and the radioactive particles are enclosed in the smoke detector, which is usually fixed to the ceiling.

☺ Activity 25.2

☺ **1.** The atoms of many heavy elements are radioactive.

(a) State how radioactivity is produced in an atom.

(b) Background radiation can come from both natural and artificial sources. One of the natural sources of background radiation is cosmic rays.

(i) State **two** other natural sources of background radiation.

(ii) State the main natural source of background radiation.

(iii) State the main artificial source of background radiation.

2. (a) Explain why people who live in La Paz, Bolivia, which is at a height of over 3500 m above sea level, are exposed to more cosmic radiation than someone who lives in Glasgow, which is at sea level.

(b) Suggest why an astronaut on the Moon is exposed to more cosmic radiation than when they are on Earth.

3. We cannot see background radiation, but we can use machines to detect radiation.

Give the name of a machine used to detect radioactivity.

4. Ask your teacher if you can use a radioactivity detector to detect if there is radioactivity in some foods.

You can work in a group to carry out this activity

You could test the following: banana; potato; sunflower seeds; Brazil nuts; kidney beans; reduced sodium salt.

Present your findings in a table.

You could use the internet to find out which is the most radioactive.

Learning checklist

After reading this chapter and completing the activities, I can:

N3 | L3 | **N4** | **L4**

- state that hydrogen and helium were formed in stars.
  Activity 25.1 Q1(a)
  ○ ○ ○

- state that most of a star's energy comes from the fusion of hydrogen atoms to form helium. **Activity 25.1 Q1(b)**
  ○ ○ ○

- describe how elements in the periodic table up to iron are formed by the fusion of lighter atoms. **Activity 25.1 Q2**
  ○ ○ ○

- state that radioactivity is the emission of particles and energy by unstable heavy atoms. **Activity 25.2 Q1(a)**
  ○ ○ ○

- state that natural sources of background radiation include cosmic rays, radon gas and living things.
  Activity 25.2 Q1(b)(i)
  ○ ○ ○

- state that radon gas is the main natural source of background radiation. **Activity 25.2 Q1(b)(ii)**
  ○ ○ ○

- state that the main artificial background radiation is due to medical uses. **Activity 25.2 Q1(b)(iii)**
  ○ ○ ○

- state that a Geiger–Muller tube (Geiger counter) can be used to detect radiation. **Activity 25.2 Q3**
  ○ ○ ○

- *present information in the form of a table.* **Activity 25.2 Q4**
  ○ ○ ○

- *apply my scientific knowledge in an unfamiliar situation.* **Activity 25.2 Q2**
  ○ ○ ○

- *work with others to plan and carry out a practical activity.* **Activity 25.2 Q4**
  ○ ○ ○

26 Chemical analysis 1

You should already know:

- that special safety measures have to be followed when working in the laboratory
- how to deal with acid/alkali spillages.

Chemical hazards

National 3

Learning intentions

In this section you will:

- learn about the importance of Scotland's chemical industry
- learn how to recognise chemical hazards
- learn how to test for some polluting chemicals.

> ### 📖 Word bank
>
> - **Fungicide**
>   A chemical that kills fungi such as mould.
> - **Fine chemicals**
>   Chemicals that are made in small amounts for special use such as making medicines.

The chemical and related industries in Scotland employ over 70,000 people and make billions of pounds for the country. It produces fuels for our transport system and chemicals that are used to make the likes of plastics, solvents and vital medicines. Figure 26.1 shows the location of some of our most important chemical industries and what they make.

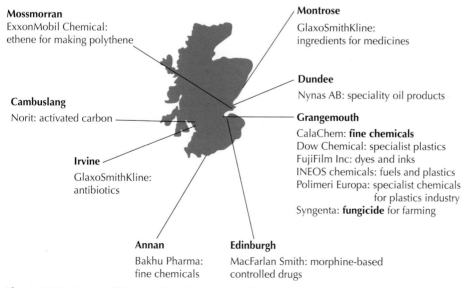

Mossmorran
ExxonMobil Chemical: ethene for making polythene

Montrose
GlaxoSmithKline: ingredients for medicines

Cambuslang
Norit: activated carbon

Dundee
Nynas AB: speciality oil products

Grangemouth
CalaChem: **fine chemicals**
Dow Chemical: specialist plastics
FujiFilm Inc: dyes and inks
INEOS chemicals: fuels and plastics
Polimeri Europa: specialist chemicals for plastics industry
Syngenta: **fungicide** for farming

Irvine
GlaxoSmithKline: antibiotics

Annan
Bakhu Pharma: fine chemicals

Edinburgh
MacFarlan Smith: morphine-based controlled drugs

Figure 26.1 *Some of Scotland's major chemical industries.*

Although chemicals are essential to us and we make useful materials from them, some of the chemicals can be harmful to us and the environment. A lot of waste chemicals are produced and solvents used in chemical processes are left over. They, too, can harm us and cause environmental problems. They cannot be allowed into the sewage system as they could contaminate rivers and kill fish and other animals. There are very strict laws to protect workers, the public and the environment.

Protecting workers

By law, employers have to put risk-management measures in place to make sure chemicals are used safely in the workplace. All workers have to be provided with information about hazards involving the chemicals used and produced in a factory. They have to be trained about the danger of the chemicals they will be in contact with. Workers also have to wear protective clothing such as safety glasses, gloves and overalls or laboratory coats. However, it is not enough for workers to know the names of chemicals and the dangers. There must be labels on the containers of **hazardous** chemicals which highlight the possible dangers when handling the chemical or if the chemical escapes. Labels explain what the hazards are and how to avoid them. Packaging is also important to ensure that chemicals are stored and disposed of safely.

There are Chemical Labelling and Packaging (CLP) regulations. They make sure that the hazards presented by chemicals are clearly communicated to workers and consumers in the European Union through classification and labelling of chemicals. A symbol is usually given with a short description of the hazard, as shown in Table 26.1.

> ### 📖 Word bank
>
> - **Hazardous**
>
> Dangerous; hazardous chemicals are dangerous in some way.

Table 26.1 *Chemical hazards and their symbols.*

Symbol	Exploding bomb	Flame	Flame over symbol for oxygen
Meaning	Explosive	Flammable	Oxidising
Explanation	Explosives are materials like gunpowder that produce violent chemical reactions	The vapours of flammable chemicals like petrol catch fire easily	Oxidising agents like potassium permanganate produce oxygen, which is a fire risk

(continued)

Symbol	Corrosion	Skull and crossbones	Dead tree and fish
Meaning	Corrosive	Acute toxicity	Hazardous to the environment
Explanation	Corrosive chemicals like acids and alkalis can cause burns to the skin and eyes	Toxic chemicals like hydrogen cyanide cause damage to our vital organs, such as liver and kidneys	Substances like white spirit used to clean paint brushes can harm animal and plant life
Symbol	Exclamation mark	Health hazard	Gas cylinder
Meaning	Health hazard/hazardous to the ozone layer	Serious health hazard	Gas under pressure
Explanation	Harmful substances can be in paints and cleaning agents, which can cause irritation, but if used properly can be safely used	Lead compounds can cause serious health problems, especially in young children	Portable gas cylinders have gas under pressure, which adds to the hazard caused by the gas's flammability

Transporting chemicals

Most chemicals are produced in factories or industrial plants which are often far away from where they are needed. Petrol and diesel, for example, are transported in huge tankers from where they are processed to petrol stations all over the country.

Petrol and diesel are highly flammable. In case of accidents and spillages it's not only the driver who needs to know what he/she is delivering. Emergency services who deal with chemical spillage need to know what chemical they are dealing with and what hazards they can cause. The easiest way to have the information is on the tanker itself. Tankers have hazard warning signs and other information on them.

Figure 26.2 *A tanker displaying hazard symbols.*

Figure 26.3 *Fire-fighters attending a chemical spill need to know what the chemical is and what hazards there are.*

Analysing waste products

Chemical companies and outside agencies test the waste products coming from chemical plants. The companies themselves have laboratories with chemists testing their products for quality and the waste products for hazardous materials. The Scottish Environmental Protection Agency (SEPA) and Scottish Water also test waste water in the chemical plant and in the canals, rivers and lochs nearby. There are a variety of tests, including testing for dangerous metals and pH.

Testing for polluting metals

There are many metals that could end up in waste water, some of which are very harmful to us. There are simple chemical tests for some metals.

Lead

Lead can get into the water supply from industries that extract or reclaim lead. Lead is toxic and attacks our nervous system. It affects children's brain development. Its effects cannot be reversed.

Test: Add a few drops of yellow potassium chromate solution to some water. A yellow/orange colour is formed if even very small amounts of lead are in the water.

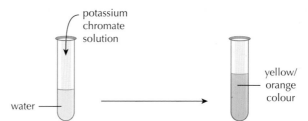

Figure 26.4 *Testing water for the presence of lead.*

Chromium 6+

Chromium compounds are used in the leather industry. Chromium compounds cause itching of the skin and can damage the kidneys and the liver if swallowed.

Test: Add a few drops of barium chloride solution to some water. A light yellow colour is formed if chromium 6+ is present.

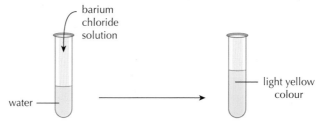

Figure 26.5 *Testing water for the presence of chromium 6+.*

Testing for acidity and alkalinity

Pure water has a pH of 7. Tap water is usually slightly acidic. If waste water from industrial plants is too acidic or alkaline it can kill fish and other animals in the water. Universal indictor or pH paper will turn red if the water is too acidic and dark blue if it's too alkaline.

Acids and alkalis, like fuels, are transported around the country in tankers that have hazard signs indicating there are hazardous chemicals on board.

Chemical hazards in the laboratory and at home

In a school laboratory you will never be exposed to toxic or very harmful chemicals. However, you are likely to handle corrosive chemicals like acids and alkalis and flammable chemicals like alcohol, which can be harmful if you don't take simple precautions. These include always washing off any chemical that splashes on your skin or clothes, and wearing safety goggles when doing experiments. Like workers in a chemical factory or research laboratory you have to be able to recognise what the hazard symbols mean and how to handle the chemicals.

Chemical hazards are not just something that can occur in industry or in the laboratory. A number of household items can be hazardous if not used properly. Oven cleaners contain corrosive alkalis to dissolve grease. They also give off fumes which can affect your breathing. Even washing-up liquid can cause serious irritation if it gets in your eye.

Make the link

For more about universal indicator, acids and alkalis, look at Chapter 6.

Figure 26.6 *Washing-up liquid has a health hazard symbol as it irritates the eyes.*

Activity 26.1

1. (a) State **two** common hazards faced by us when working with household chemicals in everyday life.

   (b) State some precautions that workers in a chemical factory should take when working with chemicals.

2. White spirit is a solvent commonly found in homes. It is used to clean paint brushes. Its label has a number of hazard symbols.

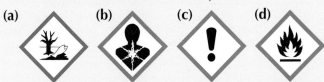

   (a)   (b)   (c)   (d)

   Write down the hazards indicated by each symbol.

3. Acids and petrol are chemicals that are transported around the country in huge tankers. Each presents a different hazard.

   Suggest which symbol should go on each tanker (Table 26.1 will help you choose.)

(continued)

4. Suggest which symbols should appear on the bottles of the following household chemicals (Table 26.1 will help you choose.)

   (a) Bleach, which contains alkali and chemicals that are harmful to fish and other animals.

   (b) Furniture polish, which contains chemicals that could go on fire if sprayed near a flame.

   (c) Bathroom cleaner, which could irritate the skin.

5. In March 2016, a chemical company was found guilty of releasing caustic soda (alkali) into the Manchester Ship Canal.

   Describe how you would test the canal water to show that it was alkaline.

6. You are given three samples of water, labelled A, B and C. Each test tube contains water polluted with one of the following: acid; chromium 6+; lead.

   Describe how you would test each sample to see which pollutant it contained.

   (You should read 'Analysing waste products' on pages 384–385 before attempting this activity.)

 Ask your teacher if you can test some polluted water samples.

★ You need to know

You may see the old 'harmful/irritant' symbol on some household chemicals, as shown in Figure 26.7.

Figure 26.7 *This old tin of teak oil has the old health hazard symbol.*

This has been replaced by the exclamation mark symbol, as shown in Figure 26.8.

Figure 26.8 *The new symbol to indicate health hazard/ harmful to the ozone layer.*

✺ Chemistry in action: INEOS

INEOS is a chemicals company that operates all over the world. Its site in Grangemouth produces the raw materials used to make products like plastics and medicines. The company makes polythene on site, and also produces fuels from crude oil in its refinery (see Chapter 14).

Figure 26.9 *Tankers carrying fuel must have the flammable hazard warning clearly visible.*

Crude oil from the North Sea is piped to the site. It is separated into more useful parts (fractions). Those used as fuels are transported by road and rail all over Scotland and Northern Ireland. INEOS supplies 70% of all the fuel used in these areas. Fuels are highly flammable, so tankers carry hazard symbols to warn us of possible dangers when handling the chemicals or as a result of spillages.

Some of the gases found in natural gas from the North Sea are used to make plastics. However, they are running out and expensive.

N3 | L3 | N4 | L4

In September 2016 INEOS started importing ethane (a hydrocarbon) obtained from shale gas from the USA. The ethane is processed to make the raw material needed to make polythene. It is brought in by specially designed tankers called Dragon ships. The ships are able to sail up the Forth estuary and dock at the INEOS site at Grangemouth.

Figure 26.10 *An INEOS Dragon ship brings fracked shale gas from the USA to Grangemouth.*

The engines on the ship are dual fuel – they can run on the ethane gas it is carrying or diesel. When using ethane as fuel the amount of polluting gases is reduced. The nitrogen oxides emissions are 85% lower than the regulation maximum and the carbon dioxide emissions are 25% less than if using diesel. The sulfur dioxide and solid particle emissions are practically zero.

Scotland has its own deposits of shale gas, much of it in the Grangemouth area. It would be much cheaper and more convenient to extract shale gas from the area. It is thought that there is a lot of it so would last a long time. However, there is concern that the method of obtaining the gas, known as 'fracking', will cause environmental damage. As a result, the Scottish Government has put a hold on any further moves to extract shale gas.

Grangemouth has the good transport links by road, rail and sea that are essential for getting raw materials in and end products out to where they are needed. It also has a skilled workforce, built up over many years of the chemical industry being in the area.

All companies have to follow strict guidelines regarding safety and possible environmental effects.

↻ Keep up to date!

You can keep up to date with the fracking debate by searching for 'fracking in Scotland' online.

⁙ Make the link

For more about fracking, see Chapter 10.

GO! Activity 26.2

☻ 1. The chemicals obtained by processing crude oil at the Petroineos oil refinery are shown below, with the percentage of each:

- petrol and diesel: 46%
- fuel and gas oil: 23%
- chemicals for making plastics: 12%
- fuel gas and jet fuel: X%

(a) Match A, B and C in the pie chart with the chemicals listed above.

(b) Work out the percentage of fuel gas and jet fuel produced.

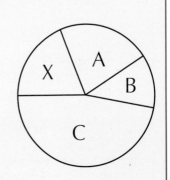

(continued)

| N3 | L3 | N4 | L4 | **387** |

2. The Dragon ships used by INEOS to transport ethane from the USA use ethane as a fuel for its engines.

   State some of the environmental benefits of burning ethane instead of diesel.

3. Discuss with a partner why INEOS want to extract shale gas in the local area but the Scottish Government has put a hold on any further moves to extract shale gas.

   Write a report of about 30–50 words long summarising your findings.

Learning checklist

After reading this chapter and completing the activities, I can:

N3 L3 N4 L4

- state that the hazards associated with working with chemicals include damage to the skin, eyes and airways. **Activity 26.1 Q1(a)**   ○ ○ ○

- state the precautions that should be taken when working with chemicals include wearing: eye protection, gloves, laboratory coat. **Activity 26.1 Q1(b)**   ○ ○ ○

- recognise the warnings on hazard pictograms to include: hazardous to the environment, health hazard, severe health hazard, flammable. **Activity 26.1 Q2**   ○ ○ ○

- select a pictogram to go with a particular hazard to include: corrosive, hazardous to the environment, health hazard, flammable. **Activity 26.1 Q3, Q4**   ○ ○ ○

- outline how to test the pH of waste water using pH indicator. **Activity 26.1 Q5, Q6**   ○ ○ ○

- outline how to test for the presence of lead and chromium 6+ in waste water. **Activity 26.1 Q6**   ○ ○ ○

- appreciate the views of those who wish to develop chemical production and those with environmental concerns. **Activity 26.2 Q3**   ○ ○ ○

- *interpret information presented in a graph.* **Activity 26.2 Q1**   ○ ○ ○

- *extract scientific information from a report.* **Activity 26.2 Q2**   ○ ○ ○

27 Chemical analysis 2

You should already know:

- about the importance of Scotland's chemical industry
- how filtration, evaporation and distillation are used to separate mixtures
- some of the hazards associated with working with chemicals and the precautions that should be taken when working with them
- how to recognise the warnings on hazard pictograms
- how to test the pH of waste water
- how to test for the presence of lead and chromium 6+ in waste water.

Analytical chemists

National 4

Learning intention

In this section you will:

- learn the reasons for carrying out chemical analysis.

Chemicals are an important part of our everyday life. They are in our food and drink, and in the household cleaning products and cosmetics we use. We need to be sure that these products are safe for us to use. We rely on chemists to carry out chemical tests to find out what is in the products and how much of each chemical there is. This is called **chemical analysis**.

📖 Word bank

- **Chemical analysis**

Tests carried out to find chemicals present in a substance and how much of a chemical there is.

Figure 27.1 *Chemists use various techniques to carry out chemical analysis.*

Figure 27.2 *Forensic scientist collecting evidence from a crime scene for analysis.*

Chemical analysis is done for many different reasons, such as:

- research chemists need to accurately analyse new products to make sure they have made the right chemical
- geologists need to find out if rocks and ores contain enough of a metal to make it worth mining
- oil has to be analysed to see what fractions it contains
- soil samples have to be analysed so that farmers know the pH of the soil and what kind of fertilisers need to be added
- the police have forensic scientists who collect and analyse evidence from crime scenes.

Although manufacturers have their own chemists to carry out analysis, there are also government and local authority agencies that carry out chemical analysis to make sure that manufacturers' claims about their products are correct. They also test waste chemicals to ensure that they are not going to harm the environment. These agencies include:

- the Food Standards Agency (FSA): the FSA is an independent government department whose job it is to make sure the food we eat is safe and what it says it is
- the Scottish Environmental Protection Agency (SEPA): SEPA assesses the quality of our environment by monitoring our air, land and water.

Chemistry in action: SEPA

In 2012 and 2013, SEPA received complaints from members of the public about the quality of water in the Bothlin Burn and Luggie Water. The water was brown in colour and contained a lot of sediment.

SEPA officers visited a nearby construction site several times between 26 June 2012 and 4 July 2013 to speak to the company about stopping the discharge.

In order to stop more water contamination, SEPA issued warning letters, carried out regular visits to the site, and offered advice and guidance.

However, the company continued to discharge contaminated water into the Bothlin Burn. As a result SEPA submitted a report to the Procurator Fiscal recommending prosecution.

At the court case on 23 November 2016, the company pled guilty to a number of offences, including discharging discoloured and contaminated surface water run-off into the Bothlin Burn, and failing to provide adequate and effective surface water run-off drainage/control.

The construction company was fined £4000 for discharging contaminated surface water run-off.

GO! ## Activity 27.1

😊 **1.** Write a report about the role of an analytical chemist. Your report should be 50–70 words long.

You should include: what they do; why we need them; public protection agencies and what they do.

Simple analytical techniques

National 4

Learning intention

In this section you will:
- learn how to carry out some of the techniques used in chemical analysis.

Distillation, filtration and evaporation

Analytical chemists use a variety of techniques to find out what is in a substance and how much there is. A major part of chemical analysis is separating mixtures. This involves some techniques you should already know:

- distillation: separating a mixture of liquids

- filtration: separating solids from liquids

- evaporation: separating a dissolved substance (solute) from the solvent.

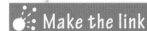

Make the link

For more about these techniques to separate mixtures, see Chapter 3.

Problem solving using filtration and evaporation

Rock salt is a mixture of salt and other material like clay. Rock salt is spread on the roads in winter to stop them freezing over. The amount of salt in rock salt varies. Generally, the more salt there is, the more effective the rock salt is at preventing freezing.

The problem

A local authority wants a rock salt with a high percentage of salt in it. Their analytical chemist decides to find out the percentage of salt in a particular rock salt. Here's what she did.

Figure 27.3 *Rock salt.*

Analysing rock salt

1. Weigh an empty evaporating basin on an electronic balance.
2. Add 50 g of rock salt to the evaporating basin (the exact amount doesn't matter so long as the actual mass is recorded).
3. Transfer the rock salt to a mortar and grind the rock salt into a fine powder using a pestle.

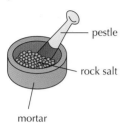

pestle

rock salt

mortar

4. Add the powdered rock salt to a 250 cm³ beaker and add water up to the 100 cm³ mark; stir with a glass rod for 1 minute.
5. Filter the mixture into a conical flask. The 'rock' is trapped in the paper.
6. Keep the salt solution which drips through into the conical flask.

Getting solid salt

1. Pour the salt solution into the evaporating basin weighed earlier.
2. Gently heat the basin until most of the water has boiled off.
3. Leave the basin to cool then transfer to an oven overnight.
4. Weigh the basin containing pure salt.

Calculating the percentage of salt

1. Weight of evaporating basin + pure salt = 241.5 g
2. Weight of evaporating basin = 200.0 g
3. Weight of pure salt = 41.5 g
4. Weight of rock salt = 50.0 g
5. Percentage of salt in rock salt

$$= \frac{\text{weight of pure salt}}{\text{weight of rock salt}} \times 100$$

$$= \frac{41.5}{50.0} \times 100$$

Percentage of salt in rock salt = 83%

GO! Activity 27.2

☺ 1. A chemist wants to find the percentage of pure salt in a rock salt sample.

(a) Describe how he could separate the salt from the rock salt.

(b) The chemist started off with 40 g of rock salt. The mass of pure salt obtained was 37.6 g. Calculate the percentage of salt in the rock salt.

(You may wish to look at 'Analysing rock salt' above.)

| N3 | L3 | **N4** | L4 |

Paper chromatography

A widely used separation technique is called **chromatography**. There are many different types of chromatography that can detect and separate very small amounts of substance. Simple paper chromatography can be used to separate the likes of food colouring used in making sweets. There are a number of sweets that have coloured dyes in their coating. Some dyes are natural and others are synthetic. These dyes have to be identified to make sure they are safe.

> **Word bank**
>
> • **Chromatography**
> Method of separating out substances in a mixture.

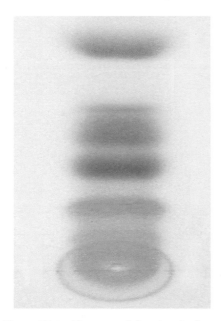

Figure 27.4   *The natural dyes in a leaf separated by paper chromatography. The yellow dye is carotene, found in carrots and used as a food colouring.*

Problem solving using paper chromatography

The problem
A company has bought green food colouring from abroad. The label says it is a mixture of two permitted dyes. The chemists at the food company use paper chromatography to see if the colouring they have bought contains permitted dyes. If the mixture of dyes in the imported colouring is the same as the permitted mixture, then the number of spots for each dye on the chromatography paper will be the same and they will travel the same distance up the paper.

Using paper chromatography

1. A pencil line is drawn 2 cm from the bottom of a piece of filter paper cut to fit into a 250 cm³ beaker.

2. The colouring to be tested (T) is placed on the pencil line on the filter paper.

3. A spot of the permitted colouring (P) is placed next to the test dye.

4. Water is added to the beaker to a depth of 1 cm.

5. The paper is placed in the water and left until the water has moved to near the top of the paper.

6. The paper is removed and dried.

Word bank

- **Chromatogram**

The pattern formed when substances are separated by chromatography.

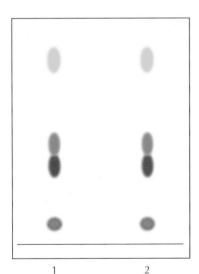

1 Crime scene paint
2 Suspect car paint

Figure 27.6 *Paper chromatogram showing matching spread of dyes in paint.*

Results

Figure 27.5 shows that there are two chemicals in the dye the company bought (T), but only one of these chemicals matches the chemicals in the permitted dye (P).

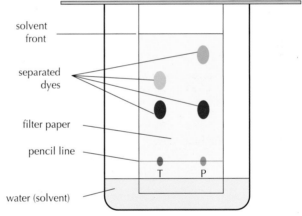

Figure 27.5 *Separating dyes used in a sweet using paper chromatography.*

Conclusion

The bought colouring contains one permitted dye and one unknown chemical so should not be used. Further tests should be done to identify the unknown chemical.

Forensic scientists can also use paper chromatography to help solve crimes. Figure 27.6 shows a **chromatogram** of paint found at a crime scene (1) and paint from a car (2) suspected of being used in a robbery. The spots match exactly so it can be concluded that the paint from the crime scene matches the paint from the car.

The dyes used in ink can be separated in a similar way.

GO! Activity 27.3

1. A threatening letter was sent to a member of the public. The police have found two water-soluble ink pens at the home of a suspect.

Your task is to use paper chromatography to identify if the ink from either of the pens matches the ink used to write the note.

Discuss with a partner how you would carry out the experiment. Make up a plan that is detailed enough for someone else to be able to follow it and get the same results.

Show your plan to your teacher and ask if you can do the experiment. Your teacher will supply the pens you are to use.

2. Food dyes are given E numbers in order to identify them. A paper chromatogram was run to analyse three colours used in foods and the E numbers of the dyes in the colour.

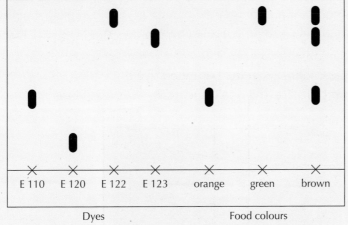

Dyes   Food colours

(a) Which food colour is a mixture of dyes?

(b) Which food colour contains only dye E110?

Testing pH

Analytical chemists test waste water to check how acidic or alkaline it is (see pages 77–78). Universal indicator or pH paper is used to give a rough idea of the pH of a solution. In order to get a more accurate measurement a pH meter is often used.

(a)   **(b)**

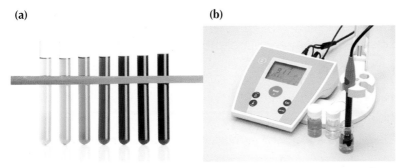

Figure 27.7 **(a)** *The changing colour of universal indicators give a rough measure of the pH of a solution.* **(b)** *A more accurate measurement is given by a pH meter.*

Figure 27.8 *Environmental scientists often use portable battery-operated meters which can measure pH and other factors such as salt content.*

An added benefit of using a pH meter is it can be used to measure the pH of coloured solutions. Also, a pH meter does not contaminate the sample with chemicals so the pH probe can be put directly into rivers or lochs. A battery-operated pH meter is portable – it can be taken out to where the water is to be tested. This means samples don't have to be collected and taken back to the laboratory for analysis.

Problem solving using pH

The problem

A gardener was having little success growing potatoes. Potatoes grow best in soil that is quite acidic. A friend suggested his soil might be the wrong pH, resulting in important nutrients (see Chapter 24) not being taken in by the plant. He bought a home analysis kit from the garden centre, with the following instructions.

Testing the pH of soil

1. Collect some soil from the area where you wish to grow vegetables.
2. Add about 2 cm depth of soil to the test tube provided.
3. Add water to the test tube up to the halfway mark.
4. Stopper the test tube and shake it for 30 seconds.
5. Allow the soil to settle to the bottom of the test tube.
6. Dip the pH paper into the water.
7. Compare the colour of the pH paper with the pH chart.

If the result shows the soil is too acidic add lime.

If the result shows the soil is too alkaline add sulfur.

Result

The colour of the pH paper indicated that the pH of the soil was slightly alkaline.

This was the reason for the poor growth of the potatoes.

Activity 27.4

🙂 1. (a) State why an environmental scientist would use a portable battery-operated pH meter.

   (b) A farmer wants to grow a crop which grows best over a very narrow pH range. Suggest why the farmer should use a pH meter rather than universal indicator to test the soil.

🙂🙂 2. The pH of garden soil has an effect on how well certain vegetables will grow.

   You will need: pH paper, pH meter (optional).

   Collect samples of soil from two different areas in the school grounds and test their pH. You should follow the instructions from the garden centre on page 396.

   Ask your teacher if you can use a pH meter.

🙂 3. Using your scientific knowledge, answer the following questions related to the garden centre instruction card on page 396.

   (a) Suggest what the pH might be if the soil sample was slightly alkaline.

   (b) The soil was too alkaline for growing potatoes.

   What should the gardener add to the soil to make it more suitable for growing potatoes?

   (c) Suggest why the instructions said to let the soil settle before testing the pH of the water.

   (d) Suggest why it might be a good idea to test the pH of the water before it was added to the soil sample.

Flame tests

Some metals produce characteristic flame colours when their compounds are heated in a Bunsen flame. Table 27.1 lists some metals and their flame colour. Chemists can use this to identify the metals that could be present in a rock sample. If the test is positive, further analysis is needed to confirm whether the metal is present in quantities that make it worth mining.

Table 27.1   *Some metals and their flame colour.*

Metal	Colour
barium	green
calcium	orange-red
copper	blue-green
lithium	red
nickel	light green
potassium	lilac
sodium	yellow
strontium	red

Figure 27.9   *Copper compounds produce a blue-green flame when heated.*

| N3 | L3 | **N4** | L4 | **397** |

Problem solving using flame tests

The problem

A technician had samples of copper carbonate and nickel carbonate, which look very similar, and she couldn't remember which was which. She decided to carry out flame tests in order to tell them apart. The method she used is outlined below.

Metal flame test

1. Collect a clean wire loop.
2. Dip the loop in the compound to be tested.
3. Place the loop in a hot Bunsen flame and note the colour of the flame.
4. Clean the loop by dipping it in hydrochloric acid and placing the loop in the hot flame.
5. Repeat steps 2 and 3 for another compound.

Result and conclusion

One sample produced a blue-green flame, and so it was the copper carbonate (see Figure 27.9). The other sample produced a light green flame, so it was the nickel carbonate.

GO! Activity 27.5

☻ 1. Lithium chloride and sodium chloride are both white solids. A student got samples of each mixed up. She decided to carry out flame tests to tell the compounds apart.

   (a) State the colours she would see when each compound was tested.
       (You may wish to look at Table 27.1.)

   (b) Describe how she should carry out a flame test on each compound.

? Did you know ...?

Special machines are commonly used to analyse the content of a substance. X-rays, for example, can be used to find out which metals are present in a sample and the percentage of each. This has been particularly useful in detecting fake coins. It is estimated that there are millions of pounds worth of fake coins circulating in the UK. £2 coins recently discovered were such good fakes that they could only be detected by using X-ray analysis. £2 coins have an inner metal and an outer metal designed to make them harder to forge. Both metals are alloys, but the types of alloy in the fake are different to the genuine coin.

Figure 27.10 *Fake coins can be detected using X-rays.*

Learning checklist

After reading this chapter and completing the activities, I can:

N3 | L3 | **N4** | **L4**

- give examples of when chemical analysis is used, such as research chemistry, geology, the oil industry, agriculture and forensic science. **Activity 27.1 Q1**   ○ ○ ○

- describe a situation where filtering and evaporation are used to separate substances in a mixture. **Activity 27.2 Q1(a)**   ○ ○ ○

- calculate the percentage of salt in rock salt. **Activity 27.2 Q1(b)**   ○ ○ ○

- describe how paper chromatography can be used to separate chemicals in a mixture. **Activity 27.3 Q1**   ○ ○ ○

- interpret the information on a paper chromatogram. **Activity 27.3 Q2**   ○ ○ ○

- say that reasons why an analytical chemist would use a pH meter instead of pH paper include: it is more accurate, it is portable and it doesn't contaminate the water. **Activity 27.4 Q1**   ○ ○ ○

- test the pH of soil using universal indicator/pH paper or a pH meter. **Activity 27.4 Q2**   ○ ○ ○

- describe how a flame test can be carried out to help identify a metal in a compound. **Activity 27.5 Q1(b)**   ○ ○ ○

- *apply my scientific knowledge in an unfamiliar situation.* **Activity 27.4 Q3**   ○ ○ ○

- *obtain information from tables of data.* **Activity 27.5 Q1(a)**   ○ ○ ○

Unit 3 practice assessment

Section 1 National 3 Outcomes

Total: 16 marks

N3 Properties of materials: Materials

1. A student heated a metal rod and held it against a piece of plastic. The plastic melted.

   Is this a thermosoftening plastic (thermoplastic) or thermosetting? **1**

2. Circle the correct words to complete the sentence.

   The plastic used to make a safety helmet needs to be
   light / heavy and **hard / soft**. **2**

3. Choose a novel material you have investigated.

   (a) Describe a useful property of the material. **1**

   (b) State a use for the material. **1**

N3 Properties of materials: Metals

4. **(a)** A student added different metals to water.

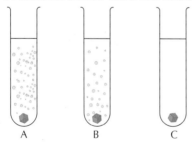

   Which test tube shows what happens when copper is added to water? **1**

   (b) Items made of iron will rust if they are left outside.

   State **one** way the iron can be protected from rusting. **1**

5. A group of students tested the conductivity of some materials.

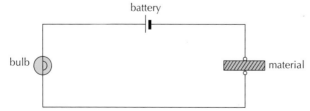

The results are shown in the table.

Material	Does it conduct?
ceramic	(X)
metal	yes
plastic	(Y)

Complete the table by giving answers for (X) and (Y).   2

6. Complete the following sentence:

In a battery, electricity comes from a _____ reaction.   1

N3 Properties of materials: Solutions

7. The table shows how much of a chemical dissolves in 100 cm³ of water at different temperatures.

Temperature of water (°C)	Amount of chemical dissolved (g)
20	5
30	10
40	15
50	20

(a) Choose the correct word(s) to complete the sentence.

As the temperature of the water increases the amount of chemical which dissolves in the water **decreases / increases / stays the same**.   1

(b) How many grams of the chemical dissolved at 40 °C?   1

N3 Chemical analysis

8. (a) Four hazard symbols are shown.

A   B   C   D

(i) Which symbol would be attached to a bottle of acid in the laboratory?   1

(ii) Which symbol should be attached to a lorry carrying petrol?   1

(b) Give **one** example of a piece of safety equipment that a worker in a chemical factory should wear.   1

9. A chemical company is suspected of letting alkaline waste water escape into a canal.

Describe how the waste water could be tested to find out if it was alkaline.   1

Total 16

A score of 8 or more is a pass!

Section 2 National 4 Outcomes

Total: 27 marks

N4 Properties of materials

1. Ethene is an example of a monomer that can be used to make a polymer.

    (a) Name the process by which polymers are made.   **1**

    (b) Name the polymer formed from ethene.   **1**

    (c) Ethene has the formula C_2H_4. Suggest the name of a harmful gas that could be produced when the polymer made from ethene burns.   **1**

2. The graph shows the amount of plastic going to landfill in Europe.

    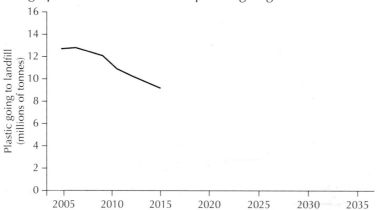

    (a) Describe the **general** trend in the amount of plastic going to landfill.   **1**

    (b) If the trend remains the same, predict, from the graph, when the amount of plastic going to landfill would be zero.   **1**

3. Some of the properties of ceramics are shown below.

    A Electrical resistance

    B Hard wearing

    C Heat resistance

    D Scratch resistance

    Select the most important property for the ceramic blade on a kitchen knife.   **1**

N4 Properties of metals and alloys

4. Metals are one of our most important materials.

    (a) Potassium is only found in compounds and electricity is needed to extract it from its compound.

    What does this indicate about the reactivity of potassium?   **1**

(b) A student added a piece of magnesium, iron and silver to test tubes of acid.

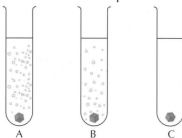

Which test tube (A, B or C) shows what would happen when magnesium was added to acid?   **1**

(c) Copper can be made into pipes used to carry water and gas in the home.

Which property of copper does this show?

A Conductor of heat

B Conductor of electricity

C Ductility (easily made into wires)

D Malleability (easily shaped)   **1**

5. Rusting is the special name given to the corrosion of iron.

(a) Name the substances which cause the rusting of iron.   **1**

(b) Ferroxyl indicator can be used as an indicator to detect iron ions produced when iron rusts.

State the colour change seen in ferroxyl indicator when iron rusts.   **1**

(c) Steel is an alloy of iron.

Explain what an alloy is.   **1**

6. Part of the electrochemical series is shown.

> magnesium
>
> aluminium
>
> zinc
>
> iron
>
> tin
>
> lead
>
> copper
>
> silver

(a) Select a metal from the electrochemical series which, when connected to iron, would protect the iron from rusting.   **1**

(continued)

(b) A chemical cell uses metals to produce electricity.

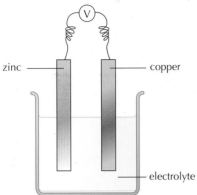

(i) Select the correct statement about the flow of electricity in the cell:

A Electricity flows from zinc to copper through the electrolyte.

B Electricity flows from zinc to copper through the wire.

C Electricity flows from copper to zinc through the electrolyte.

D Electricity flows from copper to zinc through the wire.   **1**

(ii) Select the **two** metals from the electrochemical series which would give the highest voltage when used in a chemical cell.   **1**

N4 Fertilisers

7. Fertilisers are added to the soil to provide the essential elements needed by plants.

(a) Nitrogen is one essential element needed for plant growth.

Name one other element essential for healthy plant growth.   **1**

(b) Ammonium nitrate is a synthetic fertiliser made by reacting ammonia (a base) with nitric acid.

Name the type of reaction taking place.   **1**

8. The solubility of a fertiliser in 100 g of water is shown in the table.

Temperature (°C)	Solubility (g)
20	190
30	240
40	290
50	340
60	

(a) State the effect of increasing the temperature on the solubility of the fertiliser.   **1**

(b) Predict the solubility, in grams, of the fertiliser at 60 °C.   **1**

(c) Suggest why the high solubility of the fertiliser can be a problem when it is spread onto fields.   **1**